Mining and Australia

Mining and Australia

edited by
W.H. RICHMOND
and
P.C. SHARMA

University of Queensland Press
St Lucia • London • New York

Typeset by University of Queensland Press
Printed and bound by Silex Enterprise and Printing Co, Hong Kong

Distributed in the United Kingdom, Europe, the Middle East, Africa, and the Caribbean by Prentice-Hall International, International Book Distributors Ltd, 66 Wood Lane End, Hemel Hempstead, Herts., England.

National Library of Australia
Cataloguing-in-Publication data

Mining and Australia.

Bibliography.
ISBN 0 7022 1742 5.
ISBN 0 7022 1752 2 (pbk.).

1. Mineral industries — Australia. I. Richmond, W.H. (William Henry), 1945- . II. Sharma, Pramod Chandra, 1943-

338.2'0994

Library of Congress Cataloguing in Publication Data

Main entry under title:

Mining and Australia.

Bibliography: p.
1. Mineral industries — Australia. I. Richmond, W.H., 1945- . II. Sharma, P.C. (Pramod C.), 1943- .
HD9506.A72M57 1983 338.2'0994 82-13357
ISBN 0-7022-1742-5
ISBN 0-7022-1752-2 (pbk.)

Contents

Tables

Figures

Illustrations

Acknowledgements

The editors are indebted to two organizations. The Australian Inland Mission Frontier Services (now the Uniting Church National Mission Frontier Services) was responsible for the first initiatives which led to the development by the Division of External Studies of the multi-disciplinary course "Perspectives on Mining" and ultimately to this book. The Utah Foundation provided a grant which enabled the original course to be prepared and offered on a pilot basis. Neither of these organizations is in any way responsible for (nor in any way influenced) the substance of the course or of this book. Gratitude is also extended to these publishers who granted permission to include Poems by Judith Wright, Angus & Robertson Publishers; "Time is Running Out" and "No More Boomerang" by Kath Walker, Jacaranda Wiley Ltd; and "Radiation Victim" by Colin Thiele, *Meanjin*.

Editors and contributors

W.H. Richmond, Lecturer in Economics, Division of External Studies, University of Queensland
P.C. Sharma, Lecturer in Geography, Division of External Studies, University of Queensland

Patricia Dale, Teaching Fellow, School of Australian Environmental Studies, Griffith University
Noela C. Deutscher, Senior Lecturer in History, Division of External Studies, University of Queensland
Reba Gostand, Lecturer in English, Division of External Studies, University of Queensland
J.B. Kelly, Lecturer in Government, Division of External Studies, University of Queensland
J.R. Laverty, Professor of History, Division of External Studies, University of Queensland
Hugh Saddler, Research Fellow, Centre for Resource and Environmental Studies, Australian National University
Errol Stock, Teaching Fellow, School of Australian Environmental Studies, Griffith University

Introduction

Mining forms an integral part of Australia's economic, social and political history. The discovery of gold in the middle of the nineteenth century transformed Australian economic development and Australian society; in the ensuing decades the discovery and mining of other minerals also made a distinctive impact on the history of the nation. The names of Ballarat, Bendigo, Kalgoorlie, Broken Hill, Mount Isa and the sites of other major mineral discoveries came to be as familiar to Australians as Sydney, Melbourne and Canberra. After a period of decline during the first half of the twentieth century there has been a resurgence in mining activity in recent decades. The names of Mount Tom Price, Paraburdoo, Goonyella, Weipa, Gove and Jabiluka may not evoke quite the same response, but most Australians who read newspapers and listen to their politicians on radio or television have a sense that their future will increasingly be associated with the exploitation of their country's considerable supply of mineral resources. Accordingly anyone seeking to understand the development of Australia as a social, economic and political entity must, at some point, turn to the subject of mining.

Despite its importance to Australia, however, many aspects of mining and its various implications are not widely understood. This is all the more surprising in view of the fact that in recent years mining has become a curiously controversial activity and has been the source of much dissension in the community. The importance of mining to Australia as a nation has thus been taken as the general theme of the essays in this volume, or, to be more precise, of those which constitute Part B of the book. The first three chapters, in Part A, are intended to provide some geographic, historical and economic background for the remaining chapters.

While the focus is on mining in the strict sense, the discussion in several chapters is extended where appropriate to encompass the production of petroleum and also the processing of mineral ores, both of which activities are included in what is generally known as the mineral industry.

It is widely known that Australia is a "resource-rich" country, possessing an extraordinary range of minerals. In some cases these constitute a substantial proportion of total known world reserves. Chapter 1 provides an elementary summary of the geology of the Australian continent and of the origin, location, nature and scale of Australia's major metallogenetic provinces and mineral wealth. An understanding of the principles underlying the formation of minerals allows us to comprehend why they occur in the locations they do and why mining activities have been so diverse and extensive in Australia. This in turn provides the basis for understanding many aspects of the history of mining and many of its social and economic implications, as well as giving an indication of the possibilities of future discoveries.

Chapter 2 traces the history of the development of mining over nearly one and a half centuries. Until about 1880, prospecting for minerals was often a process of trial and error and mining activity was predominantly a small-scale and technologically unsophisticated activity. After the 1880s prospecting became rather more organized; the technology of mining and mineral processing became more complex and the organization and management of mining activities more professional; improved transport and communications facilities allowed the development of mining activities in locations previously difficult of access; and substantial capital expenditure led to larger-scale mining and processing operations. However, after the first world war the industry declined and for several decades there was little progress and little optimism that mining would regain its former vitality and place in the Australian economy. But since the 1950s, and particularly since the mid 1960s, there has been a major resurgence in mining and mineral processing activities. This modern period has been characterized by highly sophisticated prospecting techniques and large scale investment in development projects, oriented principally to the export market.

The principal economic characteristics of the major industries in the mineral sector as it now exists are surveyed in chapter 3. The circumstances of production and price determination of the major minerals vary considerably, though there are some characteristics common to most industries. Production tends to be concentrated in the hands of a small number of firms, many of whom produce several minerals. There is a high level of participation by overseas companies, mainly mining companies diversifying their operations, though to an increasing extent overseas buyers of Australian minerals are seeking equity in mineral development projects. All industries (except petroleum) are highly dependent on export markets, though the nature of these markets varies considerably. In the case of bauxite/alumina/aluminium, for example, production and export is carried on in the context of a world market dominated by six major producers, whereas in other cases the world market is a relatively free one. The chapter contains a detailed analysis of the economic structure of the Australian coal, iron ore, aluminium, copper, lead-zinc-silver, petroleum and uranium industries which together now account for about 85 per cent of total mineral production.

Against this background, the impact of mining activities on various aspects of Australian life is examined in the essays in Part B. The effect of mining on the pattern of urban development in Australia is often neglected. During the pioneering and consolidation phases of Australian mining, up until the first world war, mining had a very direct and obvious impact on the settlement pattern in Australia. Many mining towns and mining fraternities with distinctive characteristics and mores became part of the Australian scene. Mining also influenced the development of colonial capitals, particularly Melbourne, Perth and Adelaide. In the inter-war period, when mining activity stagnated, few new mining towns were established and many of those which owed their existence primarily to mining underwent the debilitating process of decline and decay. It was in this period, however, that the interesting urban phenomenon of the "company town" developed noticeably. The diversification by some large mining companies into metal processing and associated industrial activities also promoted substantial, and permanent, urban settlement. With renewed mining activity after the second world war, and especially from the 1960s, a sprinkle of new mining communities has grown up, many in remote or sparsely settled regions. Many of these reflect the growing direct involvement of mining companies in urban development and in the provision of social assets and services for mining communities. They are, in different ways, new versions of the company town. Their development also represents an interesting exception to the general trend of population towards the coast and of urban concentration in or near the capital cities. Because the new mining projects have been capital rather than labour intensive, and processing operations have usually been located away from the mine site, the new mining towns have been relatively small. But the developments, as in the earlier period, have given a powerful impetus to the expansion of ports and service towns and to the capital cities, especially of those states in which the major projects have been undertaken. Chapter 4 analyzes these urban implications of mining from the mid-nineteenth century to the present, explaining the growth of mining towns themselves and the linkage effect of mining on the growth of other urban centres in terms of the extent and nature of mineral deposits, mining methods and management and the development of metallurgy and mining and transport technology.

Chapter 5 focuses more closely on the new mining towns associated with the boom of the 1960s and 1970s, particularly those developed in conjunction with iron ore mining in the Pilbara region of Western Australia and with black coal mining in the Bowen Basin region of Queensland. These towns are not large; compared to the "new" towns associated with mining (particularly gold mining) in earlier phases of development they are much less significant in relation to the overall urban pattern. But they are of interest as the only exception to the "great retreat" of population to the coast and to the capital cities in particular, and because they are contrived "outback suburbias", planned urban environments in isolated regions. Their location, their physical fabric — the design and lay-out and the nature of the buildings — and the extent to which the mining company they serve controls accommodation and commercial life, have been the result of considered planning decisions. Some of these decisions have produced results that are less than ideal and less satisfactory than they might have been. The towns also have some marked socio-demographic characteristics: male dominance, atypical age structure, a large non-Australian-born population, the dominance of a single employer and a high population turnover. All of this adds up to a distinctive physical and social environment quite at variance with the rip-roaring image of earlier mining towns: it is more planned, controlled and ordered but nonetheless not one which meets with the approbation of many of the inhabitants. The results of several studies of these new mining towns have been drawn on to analyze these characteristics and their social implications. The impact of the growth of the new towns on their surrounding regions is also examined.

Mining has played an important part in the process of economic development in Australia. In the mid-nineteenth century, mining was the most important sector of the economy; the gold rushes fundamentally altered the scale and structure of the Australian economy and transformed the process of Australian economic development. Despite the discovery and exploitation of new minerals, mining gradually became of less significance in this process over the next hundred years. Even the diversification and expansion of the industry in the post second world war period made little impact in aggregate economic terms until the 1960s, when the new mining boom — based mainly on the production for export of iron ore, coal and bauxite — considerably increased the direct contribution of the mining industry to gross domestic product. By the end of the 1970s, mining still accounted directly for only 4.4 per cent of total national production — hardly, one would think, a rate of growth to get excited about — but the consequences of the increase have been greater than this figure might suggest. The economic linkage effects of mining developments have been substantial: mineral processing industries have been encouraged and there has been a considerable demand created for the capital input requirements of both mining and mineral processing industries. Very large capital expenditures associated with the establishment of mineral projects have constituted a significant proportion of aggregate investment expenditure in the economy, and much additional production and income has been generated. At the same time, the competition for limited resources of capital and labour required by these developments has had consequences for other sectors of the economy also seeking these resources. More importantly, a major consequence of the mining boom, which has been geared very largely to the export market and has stimulated a high level of capital inflow, has been to create a tendency to surpluses in the balance of payments. While there are clearly potential advantages to be gained from such a situation, it has in turn produced a tendency for the Australian dollar to appreciate. This has had implications for the profitability of existing Australian industries producing commodities which are exported or which compete with imported goods, that is, the traditional rural export industries and those which form a large part of the manufacturing sector. Through this mechanism the growth of the mining sector has served as a force for structural change within the Australian economy. Chapter 6 outlines the nature of these economic effects, and indicates their consequences and likely magnitude, both in the sixties and seventies and in the near future.

These economic effects, together with other particular characteristics of the mineral sector, have created problems for governments and given rise to issues on which public policies have had to be formulated. These relate to — and materially influence — the size and distribution of the benefits and costs associated with the rapid growth of the mineral sector. For example, the response of the government to the pressure on the exchange rate of the Australian dollar — and in particular whether the dollar is allowed to appreciate or alternate policies (such as a general lowering of levels of protection) are pursued — has important implications for producers and employees in different sectors of the economy and for consumers generally. The government's policy with respect to foreign investment influences both the rates at which developments in the mineral sector occur and the way in which the benefits from such developments accrue. Taxation policies (particularly as they are applied to mineral rents) have similar implications, and requirements imposed upon companies concerning the provision of infrastructure associated specifically with mineral projects also determine the distribution of benefits and costs arising from them. These and other issues are discussed in chapter 7; past policies and the policy options for the 1980s are analyzed.

The mineral developments of recent decades have had considerable effect on the natural environment. The environmental consequences of mining have attracted a good deal of public attention, and chapter 8 examines them by way of several case studies of industries selected for their economic and/or environmental significance. In each case the consequences of development and production activities are outlined, together with details of reconstruction work undertaken by the mining companies involved. The issue of government control over the environmental effects of mining is then discussed, and a brief account is given of policies pursued by the commonwealth and state governments.

The economic and environmental consequences of mining — though they are in many respects no greater than those of other activities – have caused mining to be the centre of considerable political controversy in the last two decades. Governments have found themselves faced with many difficult public policy decisions associated with the mineral boom; the most controversial problem has undoubtedly been uranium. The mining and export of uranium raises a number of environmental and economic issues of the sort discussed in chapters 7 and 8 but its consequences also extend beyond these to encompass issues such as nuclear arms proliferation and the environmental implications of nuclear power generation and disposal of nuclear fuel residues. Partly because of the location of Australian uranium deposits, Aboriginal land rights has also been a major issue in the mining of uranium. In 1973 the Australian Government commissioned the Ranger Uranium Environmental Inquiry, which took a critical role in the decision-making process. Chapter 9 traces the formulation and implementation of government policy concerning uranium mining, examining the role of the Ranger Inquiry and of groups (including trade unions) which have attempted to influence that policy, and evaluates the policy-making process. Not all issues related to mining are of such magnitude and complexity as the uranium issue, but the case has some interesting implications for the formulation of mineral policy in a more general sense.

Given the pervasiveness of mining in Australian economic, social and political life, it is not surprising that the mining theme occurs frequently in Australian literature. The final chapter of the book explores the way in which this has happened. The actual processes of mining have been an important source of metaphor for Australian writers; but far more important for the shaping of our literature and the development of aspects of our national self-concept has been the "image" of mining in the broader sense. Certain values and attitudes, relating to wealth, power, adventure, danger, independence, personal relationships, hardship, disappointment, that cluster particularly (if not uniquely) around the idea and the practice of mining, have provided thematic material for much of our literature and have been used by many Australian writers as a vehicle for their expression. In one sense this literature has presented a distorted picture of reality: dramatic and "negative" aspects of a theme are often emphasized while, as the author of this study observes, "happy humans leading a blameless or uneventful life do not make for sustained literary interest". Literature with a mining theme has become increasingly polemical, and provides one of several contexts in which the impact – particularly the social and environmental impact – of mining has been debated. It is an interesting reflection of the greater impact of mining on Australian society, or at least of an increased awareness of its importance and implications for the welfare of the Australian people.

The ten essays presented here have, with one exception, been developed from material prepared for a course offered by the University of Queensland's Division of External Studies under the title of "Perspectives on Mining". The course has dual aims: to introduce students to a range of academic disciplines and fields of study, and to consider the place of mining in Australia and its impact on Australia as a nation. In terms of the latter goal there are some further aspects of mining which deserve detailed treatment. The broad-ranging impact of the discovery of minerals, especially gold, in the nineteenth century has not been analyzed in detail (apart from the implications for urban development); however this impact is such an integral part of Australian history that the subject is well covered in many general works. In the contemporary period there are also political implications of mining in addition to those discussed in our case study of uranium. The possession of extensive mineral and energy resources has become an important determinant of Australia's place and role in world affairs which are themselves being increasingly influenced by the way such resources are distributed among the countries of the world and the manner in which they are exploited; "resources diplomacy" is becoming a more important element in Australian foreign policy. On the domestic front, the new mining boom has some important implications for Australian federalism; the boom is centred to a large extent on Queensland and Western Australia but may well leave some other states relatively unaffected, or possibly even adversely affected. Nevertheless the collection does present a reasonably comprehensive review of the place of mining in Australia and its impact, both past and present; moreover it emphasizes the value of analyzing a subject from the perspective of different disciplines.

W H R
P C S

Part A

1 The Resource Base

P.C. Sharma

While this volume may quite appropriately take a national view, the popular image of mining in Australia is a regional one. Thus one hears of iron ore in the Pilbara, brown coal in Victoria, black coal in the Bowen Basin or bauxite at Weipa. This popular image can be confirmed by an examination of a mineral deposits map of Australia, which shows that there are mineral rich and mineral scarce regions. Geologists refer to mineral rich regions as metallogenetic or metalliferous provinces.

This essay examines the origin, location, nature and scale of Australia's major metallogenetic provinces, where and why particular mineral discoveries have been made and where future discoveries are likely. It then examines the origin, location and extent of the minerals which are economically most important. This discussion of the resource base establishes the framework within which the issues discussed in the remainder of the book have arisen.

BASIC CONCEPTS[1]

Minerals

In purely technical terms a mineral can be defined as a naturally occurring crystalline substance with a definite chemical composition and a characteristic crystal structure.[2] However, it is usually defined in more general terms, and in legal terms (in the United States, for example) a mineral is regarded as any substance occurring in the earth having sufficient value when separated from its *situs* to be mined, quarried or dug for its own sake or for its specific use.[3]

The above two definitions indicate some of the problems in defining minerals. It can be seen that one is based on economic value while the other is based on the chemical and physical properties of the substance itself. There have been attempts to broaden the definition to include both organic and inorganic substances as well as to include synthetically produced substances.[4] Even these broader definitions have problems in including substances such as coal (which is a solid but not a crystalline substance) or the naturally occurring fluids such as oil.

Mineral Deposits

A mineral deposit can be simply defined as a naturally occurring accumulation of minerals. Mineral deposits include deposits formed in the past as well as those still being formed. Of the deposits formed in the past a further distinction is often made between those that are being maintained in a steady state without additions or losses and those which are now being destroyed. While "deposit" is a useful term the related concept of "reserves" is used more widely.

Reserves are quantities of minerals measured in tonnes, cubic metres, barrels or litres that can reasonably be assumed to exist and which are producible with existing technology and under present economic conditions. Various systems have been used to express the *degree of reliability* of these quantities. Perhaps the most useful of these classifications is one which distinguishes between *measured, indicated* and *inferred reserves.*[5]

Whatever definitions are used, the important distinction is between known and workable deposits, and deposits which are unknown or are to some degree marginal and hence unworkable. It is also important to appreciate the degree of judgement that goes into calculation of reserves. Measured or proven reserves are calculated on sound engineering principles and geological data and should be subject only to small errors. Indicated and inferred reserves are progressively less reliable, as the judgement of the geologist is applied to the probabilities of continuation of mineralization beyond the measured limits. The speculation and the subsequent crash in fortunes of several mining companies in the Australian mining boom of the early 1970s bears testimony to the need for an adequate understanding of the various definitions of geological reserves.

Formation of Mineral Deposits

Analysis shows that a typical section of the earth's crust contains most of the important minerals. These are summarized in table 1. Yet, the location of mineral deposits and mining activity is far from uniform. Table 1 also reveals that, despite their presence in the earth's crust, most minerals have to undergo considerable concentration before they can be mined economically. How does this concentration occur? Briefly, it involves the operation of certain processes on given "source material" under given conditions. Variations in any of these elements involve a range of out-

Table 1. Crustal distribution and enrichment factors of selected metallic minerals

Metal	Symbol	Percentage in crust	Minimum percentage in ore	Enrichment factor	Important ore minerals
Aluminium	Al	8	35	4	Bauxite ($Al_2O_32H_20$)
Iron	Fe	5	20	4	Haemetite (Fe_2O_3) Magnetite (Fe_3O_4)
Titanium	Ti	0.5	5	10	Ilmenite ($FeTiO_3$)
Manganese	Mn	0.1	25	250	Pyrolusite (MnO_2)
Chromium	Cr	0.02	20	1,000	Chromite ($FeCr_2O_4$)
Nickel	Ni	0.008	1	125	Pentlandite [$(Fe, Ni)_9S_8$]
Zinc	Zn	0.008	2.5	312	Sphalerite (ZnS)
Copper	Cu	0.005	0.5	100	Native copper (Cu) Chalcocite (Cu_2S) Chalcopyrite ($CuFeS_2$)
Tin	Sn	0.003	1	330	Cassiterite (SnO_2)
Lead	Pb	0.0016	4	2,600	Galena (PbS)
Molybdenum	Mo	0.001	0.1	100	Molybdenite ($MoS2$)
Tungsten	W	0.0001	0.5	5,000	Wolframite ($FeWO_4$) Scheelite ($CaWO_4$)
Mercury	Hg	0.00001	0.1	10,000	Cinnabar (HgS)
Silver	Ag	0.000007	0.001	1,450	Native Silver (Ag) Argentite (Ag_2S)
Uranium	U	0.000002	.02	1,000	Uraninite (UO_2)
Platinum	Pt	0.0000005	0.003	1,700	Native Platinum (Pt)
Gold	Au	0.0000003	0.001	3,300	Native gold (Au)

Source: Rigby's *Atlas of Earth Resources* (Melbourne: Rigby, 1979), p. 147.

comes, from deposits, to higher than usual concentrations, to no deposits at all.

1. Source materials

There are three basic types of source material from which mineral deposits may be formed: *igneous, sedimentary* and *metamorphic* rocks.

As the earth's interior is molten and the crust consists of material which has cooled, *igneous* rocks can be regarded as *the* source rocks, which originated from the interior of the earth and were pushed through the rocks forming the crust of the earth. They are termed either extrusive or intrusive. Extrusive igneous rocks surfaced as lava flows to produce rocks such as basalt or rhyolite, while intrusive igneous rocks cooled within the crust as igneous masses, resulting in rocks such as granite and porphyry.

Sedimentary rocks are formed from eroded material laid down in freshwater lakes or seas as sediments. These materials later consolidated, producing rocks such as sandstone, shale and limestone. There is considerable variation in the nature of sedimentary rocks, depending on (a) whether they were laid in freshwater or sea water, (b) the nature of sediments and (c) the degree/scale of sedimentation.

Metamorphic or altered rocks are rocks which originated as igneous or sedimentary rocks but whose character was changed by heat and pressure within the crust of the earth. These include rocks such as phyllite, schist and gneiss. Metamorphism is particularly important in certain contexts as a given progression or regularity in mineral formation can be expected from the process. When intrusion takes place the area most immediately and intensely altered is known as "metamorphic aureole". In the aureole, minerals separate out from the original gases and fluids according to their respective temperatures of condensation or solidification according to the principle of "magmatic differentiation".[6] A graded series of minerals results, as summarized in figure 1.

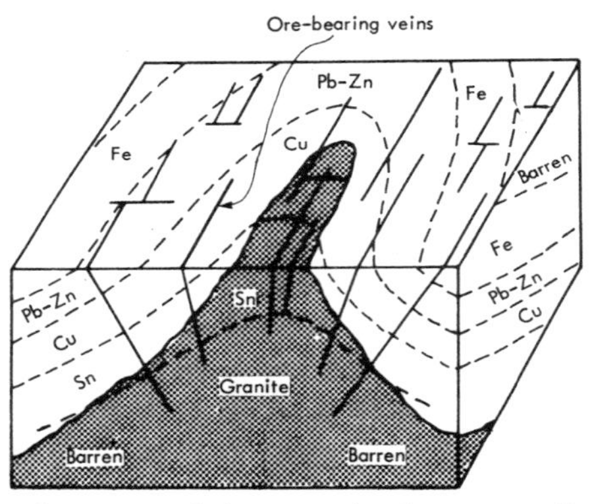

Fig. 1 The arrangement of ore zones relative to granite-slate contact. L. Bauman, *Introduction to Ore Deposits* (Edinburgh: Scottish Academic Press, 1976), p. 50. See table 1 for chemical symbols.

2. Selected processes

A wide range of processes is involved in the formation of mineral deposits. The situation is complicated further by the fact that often a particular mineral may be formed in different contexts through the operation of a series of processes. Thus it has been noted:

> Iron ores are produced on a large scale from concentrations within magmas, as at Kiruna in Sweden, from replacement deposits in other rocks as probably at Bilbao, or from former lake or shallow sea deposits such as the haematite of Krivoi Rog in the USSR or the siderites . . . or limonites . . . such as those of the Jurassic ore fields of Western Europe.[7]

A summary of some of the processes which are involved in the formation of mineral deposits is presented in table 2. Two major types are readily identifiable in the table: surficial processes operating on the surface of the earth and internal processes operating within the crust. Surface processes are further distinguished in terms of (1) physical and (2) chemical and biological processes. Each of these processes operates differently depending on whether the action is taking place on terrestrial or marine environments. Since both the physical surface of the earth and

Table 2. Processes of formation of mineral deposits

Processes	Examples of deposits
I. SURFACE PROCESSES	EXAMPLES OF DEPOSITS
1. **Physical Processes**	
(a) Terrestrial environment	
General sedimentation processes	clays, volcanic ash
Fluvial and eolian processes	gold placers, tin placers and eolian gold placers
Mass-wasting processes	eluvial gold placers
(b) Marine environment	
General sedimentation processes	clays, volcanic ash
Mechanical processes operating along the strand and on the shelf	diamond beach placers, gold beach placers, pebble phosphate, sandstone
2. **Chemical and Biological Processes**	
(a) Terrestrial environment	
Weathering and shallow ground water processes	bauxite, lateritic nickel and iron, residual kaolin
Evaporation processes	salt, potash, lithium, gypsum
Lake, swamp and lagoon processes	coal and lignite, oil shale
Hot spring, fumarolic, and surface volcanic processes	tufa, geyserite, sinter natural steam
(b) Marine environment	
Selective precipitation: accumulation in stratiform bodies or nodules (see evaporation)	phosphate, iron formations, manganese, limestone
Fixation by organisms: accumulation in stratiform bodies or reefs	diatomite, shell reefs
II. INTERNAL PROCESSES	EXAMPLES OF DEPOSITS
1. **Aqueous Processes**	
(a) Ground water processes (cold)	
Selective precipitation and replacement	uranium
(b) Hydrothermal processes (hot)	
Selective precipitation and replacement	copper, gold, molybdenum, copper-gold
2. **Magmatic Processes (Molten Rock)**	
Crystal and melt separations	chromite, diamonds in pipes, anorthosite, feldspar
3. **Heat-Pressure Processes**	
Formation of new minerals in the solid state in response to changes in temperature and pressure (metamorphism)	graphite, anthracite, marble, kyanite, garnet

Source: P.T. Flawn, *Mineral Resources* (New York: Rand McNally, 1966), p. 50.

its climate have varied dramatically over time, surficial processes have played a major role in producing a wide range of mineral deposits.

Internal processes generally involve heat, although the actual manner in which the heat is involved varies. Thus we can have the operation of hot hydrothermal processes, magmatic processes and heat-pressure processes. The most important of these are magmatic processes and metamorphism in the general sense of the term. Their importance stems from the fact that a predictable graded series of ores can normally be expected to be found in areas where these processes have been in operation.

While the distinction between surficial and internal processes is self evident, there is often considerable interaction between the two. Thus an original igneous extrusion may be reworked by surface processes into higher concentration, for example some iron ores; similarly, a deposit originally formed at the surface may be altered by the operation of internal processes, for example chalk to marble.

Elements of Geological History and Structure

Our analysis so far has concentrated on the basic source materials and the processes which result in economic ore deposits. However, to actually understand the economic geology of any particular country, these two elements must be integrated into the geological history and structure of the country concerned because, although deposits for many minerals are continuously being formed, even today, the rate of formation is so slow that it is only over geological time that deposits of significant proportions result.

The standard geological time scale encompasses over three billion years of the earth's history, and has been reproduced in table 3 as frequent reference will be made to it. The intention of this chapter is not to examine the time scale in relation to the evolution of the Australian continent, but rather to highlight those periods which have yielded structures important to an understanding of the economic geology of the continent.

According to geologists the Australian continent is made up of two broad structural units: (1) the older western and central half of the continent or the Australian Shield (or craton) and (2) the younger eastern half of the continent whose existence and development is intimately linked with the Tasman Geosyncline.[8] The Australian Shield is made up of extremely ancient rocks, some of which have been reliably dated to be over three thousand million years old. The Shield, in common with similar areas in all other continents, is highly mineralized. It has provided the greater part of Australia's total production of gold, lead, zinc and uranium. The long geological history and the relative structural stability of the Shield has

Table 3. Geological time scale

Era	Period		Epoch	Absolute Age	Major Orogenies
CAINOZOIC	**Quaternary**		Recent	0–15,000	
			Pleistocene	15,000–3, 000, 000	
	Tertiary		Pliocene	3–7, 000, 000	Kosciusko Uplift
			Miocene	7–26, 000, 000	
			Oligocene	26–38, 000, 000	
			Eocene and Paleocene	38–65, 000, 000	
MESOZOIC	**Cretaceous**			65–136, 000, 000	Maryborough
	Jurassic			136–195, 000, 000	
	Triassic			195–225, 000, 000	
PALAEOZOIC	**Permian**			225–280, 000, 000	Hunter-Bowen
	Carboniferous			280–345, 000, 000	Kanimblan
	Devonian			345–395, 000, 000	Tabberabberan
	Silurian			395–440, 000, 000	Bowning
	Ordovician			440–500, 000, 000	Benambran
	Cambrian			500–570, 000, 000	Tyennan
PRECAMBRIAN	**Proterozoic**	Adelaidean		570–1, 400, 000, 000	Dates of orogenies uncertain
		Carpentarian		1,400–1,800, 000, 000	
		Lower		1,800–2,400, 000, 000	
	Archaean			2,400-3,000, 000, 000 +	

Source: Division of National Mapping, "Geology", *Atlas of Australian Resources* (Canberra: Department of National Development, 1970), p. 14.

meant that processes have had extremely long periods of time in which to effect surface enrichment or reconcentration of original mineral bearing rocks. This has resulted in extremely large scale deposits in the case of some minerals, such as iron ores and nickel.

The areas associated with the Tasman Geosyncline are dramatically different. They are much younger and generally date later than Cambrian. Much of eastern Australia has undergone several orogenies involving lengthy periods of deposition in geosynclinal troughs followed by uplifts. While the geological history of the Tasman Geosyncline is a relatively short one when compared to that of the Shield, there have been major regional variations over time. Thus geologists have identified at least three major regional units: (1) the Lachlan Geosyncline extending from Tasmania, through Victoria to western New South Wales, (2) the New England Geosyncline in northern New South Wales and parts of Queensland and (3) the Tasman Geosyncline in Queensland.

In general terms, orogenic activity was most widespread in the central and southern parts of the Tasman Geosyncline. Most of this activity occurred during the early part of the Tasman Geosyncline, that is, during the Cambrian to Carboniferous period. This regional variation has meant that most mineralization in the south has tended to be associated with igneous activity, while in the northern parts deposition/sedimentation has been the main activity for a long period of time. Another consequence of this activity is that, while the mineralization is quite varied and often of considerable economic significance, the eastern parts of Australia (especially those associated with the Lachlan and the New England sections of the Tasman Geosyncline) have failed to produce major ore deposits such as those found in the western parts of the continent. However, the relative lack of activity in the northern section of the Tasman Geosyncline permitted extensive deposition, especially during the Permian period, producing the extensive black coal deposits of central Queensland.

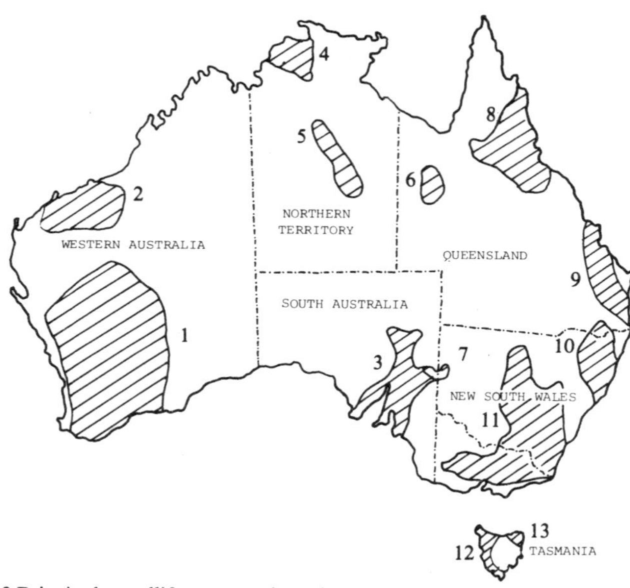

Fig. 2 Principal metalliferous provinces in Australia

1. Central Goldfields-Yilgarn Block
2. Pilbara
3. Stable Shelves
4. Darwin-Maranboy
5. Tennant Creek
6. Mount Isa-Cloncurry
7. Broken Hill
8. Georgetown
9. Gympie
10. New England
11. Victorian Goldfields
12. Lyell
13. Mathinna

Thus, to summarize in broad terms, the older western and the central part of the continent is in many parts highly mineralized, while the eastern third of the continent is made up of much younger rocks which, when combined with structural instability, have resulted in the absence of large scale deposits of metallic ores. While accepting the above continental structure as general background, geologists have attempted to map in more detail those areas where ore deposits are more likely to be found. They have identified and mapped areas where mineral discoveries have been made and where certain mineral associations have been noted or can reasonably be expected. Geologists believe that further study of these areas, which they term metalliferous provinces, could provide clues which might assist in further discoveries. The location of the thirteen major metalliferous provinces in Australia is shown in figure 2. These provinces are:[9]

1. *Central Goldfields-Yilgarn Block.* Mostly basic lavas were deposited 1800-2200 million years ago in a gently subsiding basin. The resulting deposits were thickest in the southern part (for example Hamersley Range). This province in the south-western quarter of Western Australia includes the major gold mining area in Australia today between Meekatharra, Southern Cross, Norseman and Wiluna. Other minerals in this province are iron, nickel, tin, pegmatitic minerals (for example micas, beryl, wolframite), copper and manganese. Chief fields include Kalgoorlie, Coolgardie and Southern Cross. Iron and manganese of the Hamersley Range are located within this province.

2. *Pilbara Province.* The Pilbara Block which constitutes the main structural unit in this province is similar to the Yilgarn Block in character and age, though smaller. (Both blocks contain the oldest rocks in Australia.) The province has extensive deposits of iron, manganese, pegmatitic minerals, tin and copper.

3. *Stable Shelves Province.* This Proterozoic province in South Australia is on the eastern margins of the Shield. The eastern parts of the province were parts of the Adelaidean Geosyncline. The Stable Shelves Province is known for its haematitic iron ore deposits (for example Middleback Ranges) and extensive copper mineralization (Mount Gunson, Wallaroo and Moonta).

4. *Darwin-Maranboy Province.* The structure of this province consists of Lower and Middle Proterozoic sediments and granites on an Archaean basement. Extensive mineralization of the province is related to pegmatite and granite intrusions. Uranium, iron, gold, copper, tin and pegmatitic minerals are the more important deposits at Rum Jungle, Maranboy, Brocks Creek, Alligator Rivers and elsewhere.

5. *Tennant Creek.* This is a Proterozoic province in Northern Territory known for its copper and gold deposits. Minor occurrences of pegmatitic minerals are also reported.

6. *Mount Isa-Cloncurry.* The province is composed of a meridionally trending complex of Archaean, Lower and Middle Proterozoic rocks in north-western Queensland. Granite intrusions have resulted in widespread mineralization. The major copper-lead-zinc deposits at Mount Isa and uranium at Mary Kathleen are among the numerous base metal and uranium deposits in this province.

7. *Broken Hill.* The Broken Hill province is essentially a fault-bounded block of Archaean rocks. Granite intrusion around 1600 million years ago has resulted in the metamorphism of earlier sediments. The well known lead, silver and zinc deposits of Broken Hill are from Precambrian age. Small deposits of gold, uranium and pegmatitic minerals have also been found in the province.

8. *Georgetown Province.* This province in north Queensland has a complex geological history. It is also of interest in that it is one of the few areas in eastern Australia that has rocks from the Precambrian period. These rocks around Georgetown contain gold, copper and lead deposits; the gold at Charters Towers is from considerably younger Devonian rocks. Granitic intrusions of Carboniferous and Permian age introduced important ores of tin, tungsten, molybdenum, copper, lead, silver, zinc and gold. Important mining localities include Croydon, Georgetown, Gilberton, Chillagoe, Mount Garnet and Einasleigh.

9. *Gympie Province.* This province consists of north-north-west trending folds with granite intrusions extending from Marlborough to Brisbane. Ores of copper, gold, silver, mercury and other metals are associated with Permian and Triassic granites. The main area of mining interest is the Upper Devonian deposits of copper-gold-pyrites at Mount Morgan.

10. *New England Province.* This province, embracing the New England area in northern New South Wales and extending north to Stanthorpe in Queensland, is the southern part of the Permo-Triassic Hunter-Bowen orogenic belt and has seen considerable faulting and folding (thrust) activity. Range of ores (gold, copper, tin, molybdenum, mercury, antimony) relates to later Permian to Triassic granites. Chromite deposits resulted from the injection of earlier serpentenites.[10]

11. *Victorian Goldfields-Cobar Province.* In south-eastern Australia extending from Victoria north to Dubbo in New South Wales and north-west to Bourke. In this province periods of intense folding and intrusion of igneous rocks were accompanied by mineralization in Ordovician, Silurian, Devonian and Carboniferous times, when deposits of gold, copper, tin, silver-lead-zinc, tungsten and other metals were introduced. However, the Victorian goldfields and the copper at Cobar are the only deposits of economic significance, despite the history of extensive mineralization in this province.

12. *Lyell Province.* Located in the western part of Tasmania the Lyell Province has a Precambrian basement and Palaeozoic mineralization associated with Palaeozoic intrusions. Most the deposits on the richly mineralized west coast (lead, zinc, copper, silver, gold, tin, iron and other metals) were probably formed in the Middle Palaeozoic. Included in the province are the well known Mount Lyell, Read-Rosebery, Zeehan Dundas, Renison Bell and Savage River mining fields.

13. *Mathinna Province.* This province in north-eastern Tasmania is predominantly a region of tin-tungsten mineralization related to Middle Palaeozoic granites, with gold and other minerals of lesser importance. The Mathinna Province is probably related tectonically to southern Victoria.

Some caveats are in order as the preceding discussion of geological processes and the geological structure of Australia may create some false impressions regarding the importance of metalliferous provinces and the ease with which the role of particular processes in the formation of ore deposits can be identified.

Metalliferous provinces are those areas of the continent which have been found to be highly mineralized. However, extensive tracts of these metalliferous provinces are devoid of any large scale mineral deposits, a fact which has intrigued geologists for some time. Hills has noted:

> Large areas of Precambrian rocks in central Australia are virtually barren; other areas have sparse mineralization . . . The virtual absence of ores of certain elements from otherwise important mineral belts is notable, say in the Adelaide Geosyncline which is devoid of occurrences of tantalum, mercury, tungsten, molybdenum and tin . . . The Central Victorian goldfields are well-nigh devoid of ores other than gold, and the heavily mineralized Mt Isa district . . . has insignificant concentration of gold, tin, tungsten, molybdenum and other elements.[11]

While the metalliferous provinces are of obvious economic importance it does not mean that areas outside them are not of economic importance. The contrary is the case. The areas outside the metalliferous provinces are the sedimentary basins of Australia, in which the main type of mineral found is either of organic origin (for example coal, petroleum, gas, calcareous limestone) or chemical precipitates (limestones, phosphate rocks and evaporites). Thus it is not surprising that the search for oil and gas in Australia has been concentrated in the sedimentary basins. There are twenty well known sedimentary basins spread over the continent. Together they are referred to as the Trans-Australian Platform, which extends considerable distances beyond the continental land mass in places.

The second caveat refers to the role of specific processes in the formation of deposits. Thus, while it is relatively simple to list the whole range of processes which are involved in the formation of mineral deposits, it is somewhat more difficult to identify the role of a particular process in a given context. To take a well known example, there has been considerable debate between two opposed schools of geologists regarding the genesis of the Broken Hill deposits:

> According to the hydrothermal theory, the ore was emplaced by hot fluids that moved up the dip along the crest of folds, replacing two or more stratigraphic horizons; folding and metamorphism are believed to antedate the ore. The alternative hypothesis is that the ores were deposited as chemical sediments and were subsequently folded and recrystallized during metamorphism . . . Both hypotheses are supported by careful geologists who are thoroughly acquainted with the area, and neither hypothesis can answer all the questions.[12]

Indeed Hills has indicated that in many cases far too little is known as yet to state with any certainty the role of specific elements in the formation of many deposits in Australia.[13] He has been careful to point out the role of tectonics and the wide variation in the degree to which tectonic factors are concerned in the formation of ores of different origin. Hills has indicated the need to study palaeo-climatology and geomorphology, as the formation of deposits such as bauxite and lateritic iron ore has little to do with tectonic features.

The Extent and the Distribution of the Resource Base

Australia is particularly well endowed with most of the important mineral resources required for sustaining a technologically advanced society. This is quite apparent from a wide range of data sources. The best known of these assessments is that made regularly by the Bureau of Mineral Resources (BMR), and the following discussion is based on their most recent assessment.[14]

The authors of the BMR report indicate that it is difficult to estimate the magnitude of resources present, as no realistic estimate of identified resources in Australia is yet available for many of the minerals concerned. The difficulty is compounded by the fact that published figures tend to be minimal and ultra conservative. However, in making the estimates in table 4 more realistic, the BMR report has included resources which currently may not be economic, such as shale oil. Furthermore, the estimates are based on the expected life of known reserves at current rates of production.

The basic conclusion that can be drawn from table 4 is clear. Australia, with the notable exception of oil, is well endowed with most important minerals. There are extremely large known reserves of coal, iron ore, nickel, aluminium and phosphorous and more than adequate reserves of uranium, oil shale, manganese, lead, zinc and vanadium. Australia has adequate reserves of natural gas, tin, copper, titanium, zirconium, monazite, magnesium, potassium, cobalt, bismuth, mica,

Table 4. Minerals in Australia, 1978: Origin and reserves

	Distribution	Identified resources	Current raw material imports
Energy Minerals			
Oil	Mainly Bass Strait	Small	About 30% of requirement – crude and refined products
Natural gas	Wide, but long distances to markets	Adequate	–
Uranium	N Australia, WA, SA	Large	–
Coal	Mainly E Australia	Very large	Some high-quality anthracite
Oil shale	E Australia	Large	–
Ferrous			
Iron ore	Well distributed	Very large	–
Nickel	WA, QLD	Very large	–
Chrome	Minor – WA, Victoria	Very small	Bulk of requirements
Manganese	Groote Eylandt, NT	Large (metallurgical)	Battery-grade
Tungsten	King Island, TAS, and QLD Minor – NSW, WA	Adequate	–
Molybdenum	Minor – NSW, QLD, TAS	Very small	Bulk of requirements
Non-ferrous			
Tin	Well distributed – mainly TAS	Adequate	–
Lead	Well distributed – mainly E Australia	Large	–
Zinc	Well distributed – mainly E Australia	Large	–
Copper	Well distributed – mainly E Australia	Adequate	–
Mineral Sands			
Titanium	E and SW coasts	Adequate	–
Zirconium	E and SW coasts	Adequate	–
Monazite	E and SW coasts	Adequate	–
Light Metals			
Aluminium	N and SW Australia	Very large	–
Magnesium	Well distributed (magnesite)	Adequate	About 60% magnesite
Fertiliser/Industrial minerals			
Phosphorus (Phosphate rock)	NW QLD, NT	Very large	All requirements
Potassium	WA	Appear adequate	All requirements
Sulphur	Sulphides well distributed	Elemental nil, sulphide large	50–70% of requirements
Salt	Well distributed	Unlimited	–

Table 4. (cont.) Minerals in Australia, 1978: Origin and reserves

	Distribution	Identified resources	Current raw material imports
Minor Metals			
Vanadium	WA, QLD (oil shale)	Probably large; not developed	–
Bismuth	Mainly NT	Adequate	–
Cobalt	E Australia, WA	Adequate (from nickel ores)	–
Mercury	E Australia	Small but uncertain	–
Mica	Central and WA	Adequate	–
Cadmium	NSW, TAS, QLD	Adequate	–
Antimony	NSW, VIC	Adequate	Very small
Beryllium	NSW, WA	Small but uncertain	–

Source: J. Ward & I.R. McLeod, *Mineral Resources of Australia* (Canberra: Bureau of Mineral Resources, Geology and Geophysics, Record 1980/1, 1980).

Note: Very large – sufficient for more than 100 years
Large – sufficient for 30–100 years
Adequate – sufficient for 15–30 years
Small – sufficient for 5–15 years
Very small – less than 5 years

cadmium and antimony. There are only a few minerals (oil, chrome, molybdenum, mercury and beryllium) where identified reserves are either "small" or "very small"; unfortunately, oil and mercury are critically important.

Table 5 provides additional data on the nature and geographical distribution of the principal mineral deposits, and shows that an understanding of the geological evolution of the continent is of critical importance in understanding the geography of mineral deposits. It is clear that the Precambrian formations of the Australian Shield account for a major proportion of Australia's known mineral deposits. It means that most of the important mineral deposits of this period are in the central and the western two thirds of the continent. In contrast, most of the mineral deposits of the Postcambrian period are in the eastern third of the continent, that is, that part of the continent whose evolution is linked with the history of the Tasman Geosyncline.

It was noted earlier that despite the extensive tectonic activity during the Tasman Geosyncline, the principal mineral deposits listed in table 5 are linked to paleogeography and palaeoclimatology. In other words, the actual formation of the deposits is associated with conditions which permitted deposition, as for example in the case of black coal in central Queensland, or sustained wet periods which allowed the laterization of various nickel, uranium and bauxite deposits. The important mineral deposits which are found in the central and western two thirds of the continent, but which were formed in the Postcambrian period, are

Table 5. Principal Australian mineral deposits

Age* of geological formation in which located	Metal or mineral	State or territory	Locality
Precambrian (over 570)	Copper	QLD	Mt Isa, Gunpowder
		WA	Golden Grove
		NT	Tennant Creek
	Gold	WA	Kalgoorlie, Telfer and other localities
	Iron	SA	Middleback Ranges
		WA	Yampi Sound, Pilbara and Yilgarn regions
	Lead-silver-zinc	NSW	Broken Hill
		NT	McArthur River
		QLD	Mt Isa
	Nickel	WA	Kambalda, Windarra, Scotia, Nepean, Agnew Forrestania, Spargovialle
	Tin (lode)	WA	Greenbushes
	Uranium	NT	Nabarlek, Ranger, Koongarra, Jabiluka
Palaeozoic (235-570)	Black coal	NSW	Hunter Valley, Lithgow, South Coast
		QLD	Bowen Basin, Blair Athol
		WA	Collie
	Copper	NSW	Cobar, Woodlawn
	Copper-gold	TAS	Mt Lyell
	Iron	TAS	Savage River
	Lead-silver-zinc	NSW	Elura
		TAS	Rosebery, Que River
	Phosphate	QLD	Duchess, Lady Annie, Ardmore, Yelvertoft
	Tin (lode)	QLD	Herberton
		NSW	Ardlethan
		TAS	Renison, Luina and north-east of state
	Tungsten	TAS	King Island and north-east of state
Mesozoic (65-235)	Black coal	QLD	Ipswich, Callide
		SA	Leigh Creek
	Manganese	NT	Groote Eylandt
Cainozoic (under 65)	Bauxite	QLD	Weipa, Aurukun
		NT	Gove
		WA	Darling Range
	Brown coal	VIC	Gippsland
	Mineral sands	NSW	North coast
		QLD	South coast
		WA	South-west coast
	Nickeliferous laterite	QLD	Greenvale
	Tin (alluvial)	NSW	Tingha
		QLD	Herberton
		TAS	north-east of state
	Uranium	WA	Yeelirrie

* in million years

Source: *Australia Yearbook 1976–77* (Canberra: Australian Bureau of Statistics, 1978), p. 927.

associated with the climate or the extent of inundation of the continent at the time. Thus there is a clear distinction between the Pre- and Post-cambrian in the manner in which mineralization resulting in economic mineral deposits has occurred in Australia. The following section examines in greater detail those minerals which are economically most important (see chapter 3).

Black Coal

In Australia most black coals[15] are of a bituminous type and the higher grade anthracitic coals are fairly rare. All major black coal deposits were formed during the Permian period when seas inundated low lying areas on the western and eastern seaboards: the Perth, Carnarvon and Canning Basins in the west and the Sydney, Clarence-Moreton and Bowen Basins in the east.

The encroachment of seas was linked with the creation of extensive vegetation-choked swamps which formed the basis for the black coal deposits. The end products of these processes are the massive deposits of Central Queensland and the Sydney Basin. In Central Queensland, the surface nature of many of the deposits and gently dipping coal beds permit large scale open cut mining after the removal of minimal over-burden, while underground mining plays a dominant role in the mining operations in the Sydney Basin.

Smaller deposits of coal are encountered in the West Moreton mining district in Queensland. Outside these east coast regions, the only other Permian deposit of significance is found at Collie in the Perth Basin. Leigh Creek in South Australia has some bituminous coal but these are younger deposits formed during the Triassic. There are extensive indications of Permian coal elsewhere in Australia: under much of the Great Artesian Basin, in North Queensland and in the Canning Basin in Western Australia. However, geological conditions or distance and difficulty of access make them submarginal deposits. These difficulties usually restrict assessment of the resource base to the well known economic deposits. The extent of this base is summarized in table 6.

Iron Ores

Australia's rise to prominence on the world mineral scene is closely associated with the discoveries and the development of extensive iron ore deposits. Reserves of all three iron oxides of commercial interest

Table 6. Australian black coal resources (million tonnes)

	Measured and indicated reserves in situ		Measured and indicated recoverable reserves[a]		Inferred Resources in situ	
State	Bituminous	Sub-bituminous	Bituminous	Sub-bituminous	Bituminous	Sub-bituminous
New South Wales	22,243	500	11,672	450	480,036[b]	10,000
Queensland	24,501	245	14,481	169	22,640	90,260
Western Australia	–	204	–	161	–	2,087
South Australia	–	720	–	720	–	2,300
Tasmania	139	–	69	–	200	–
Australia	46,883	1,669	26,222	1,500	502,876	104,647

Source: *Steaming Coal*, in Australia's Mineral Resources Series, (Canberra: Australian Department of Trade and Resources, 1980).

a. Recoverable reserves consist of the amount of coal that can be physically mined from a reserve at an acceptable cost.

b. Includes some coal unlikely to be economically mined in the near future, for example 130,000 million tonnes below 1000 metres.

(haematite, magnetite and limonite) are massive even when only the higher grades are considered.

The geology of iron ore formation is closely associated with the Precambrian rocks of the Australian Shield. Most iron ore deposits are derived from sedimentary iron ore beds through surface enrichment or reconcentration. Deposits are commonly found in limbs and troughs of synclinal folds. In many areas erosion has inverted the topography so that iron ore deposits have "protected" former valleys and terraces. Consequently, the original land surface has been eroded leaving iron cappings on mesas[16] and terraces. Iron formation on the Shield was closely associated with the prevailing climatic conditions (for example, existing rainfall and existing oxygen levels in the environment). The only known iron ore deposit which departs substantially from the above pattern is the much younger Savage River deposit in north-west Tasmania, formed during the Cambrian period. It also differs in its formation as it is found as a series of massive magnetite[17] lenses in amphibolite. Geologists believe this deposit to be of magmatic hydrothermal origin.

It is difficult to estimate the actual iron ore reserves in Australia as there are no standard criteria for defining submarginal iron ore deposits. As the reserves of high grade ore are extremely large with lifetimes (at current rates of production) of several hundred years it is very unlikely that any great effort will be made to identify the extent of lower grade ores. Part of this difficulty is illustrated in a recent publication which estimated that

> Australia's iron ore reserves total about 35,000 million tonnes. They comprise 17,800 million tonnes of readily usable low-phosphorus ore and 17,200 million tonnes of high-phosphorus ore.
>
> The bulk of these reserves lies in the Pilbara district . . . estimated to contain 24,000 million tonnes of more than 55 per cent iron.
>
> In addition substantial submarginal resources of haematite mineralization grading less than 55 per cent iron occur in association with high grade deposits or as separate deposits. Immense submarginal resources of banded iron formation are also available . . . in the Brockman Formation alone 6,400,000 million tonnes of material grading greater than 30 per cent iron is available for extraction by open cut methods.[18]

It is clear that precise estimates of Australia's iron ore reserves will vary dramatically depending on the criteria used. The term submarginal itself is not a purely technical definition. Thus the Savage River deposits, which are smaller and of lower grade (approximately 35 per cent iron) have been developed yet the larger and high grade (51 per cent iron) deposits of Constance Range in north-west Queensland are unlikely to be developed in the forseeable future. As the bulk of the mined ore is exported, the marginality or otherwise of a deposit is closely tied up with the length of export agreements, prices and so on.

Bauxite

Aluminium is the most commonly occurring mineral, averaging around 8 per cent of the earth's crust. However, a concentration of at least 35 per cent is required before a deposit is of commercial interest. Bauxite, the ore of aluminium, is formed as a result of laterization, a process of deep leaching of soils and decomposed rocks. It operates most successfully in those areas which experience pronounced wet and dry seasons, that is in tropical and semi-arid areas. During the wet period the aluminium is dissolved by carbon dioxide-bearing solutions. In the subsequent dry period the aluminium rises upward and is precipitated at the surface, while other constituents which remain in solution are carried off. This results, over a period of time, in the increasing concentration of alumina. The encrustations and nodules of aluminium, having a yellow-red to intensively red colour, are commonly termed aluminous laterite or bauxite.

Australian bauxite deposits were formed in a similar manner during the Tertiary period, when a tropical and pluvial climate was experienced over a large area of the continent. Thus in Queensland at Weipa and in the Northern Territory at Gove, Cretaceous and Tertiary sediments were laterized. The nature of sediments, the low lying relief and the climate permitted extremely large scale deposits to be formed. The resulting Weipa deposit is probably the largest single occurrence of bauxite in the world. It is conservatively estimated at over 3,000 million tonnes and is up to ten metres thick in places. The deposits are strongly pisolitic[19] in character with an alumina content of more than 50 per cent. The deposits at Gove are similar in character although at an estimated size of 250 million tonnes they are much smaller.

Other major deposits of bauxite are in Western Australia. Currently the largest of these is in the Darling Range in the south-west corner of the state. The deposit at Jarrahdale, which has been formed from the laterization of Precambrian rocks, is estimated at 600 million tonnes, and is of particular interest in that it is a relatively low quality deposit with alumina content around 30 per cent. It is being exploited because its proximity to the settled part of the state has reduced substantially the costs associated with its development and operation.

Other deposits have been found recently in the north-west of Western Australia, for example North Kimberley, Mitchell Plateau and Cape Bougainville deposits. As in the case of iron ore, deposits of high grade bauxite are so immense that little effort is being spent in identifying the scale of submarginal reserves in Australia. At the current rate of production of around 25 million tonnes annually the estimated high grade reserves of 6,200 million tonnes represent a resource base that will last for several hundred years.

Copper

Copper is one of the base metals which is usually found with lead, zinc, gold and silver, due to the fact that these metals are often formed at the same time as separate orebodies. It was noted earlier that during metamorphism a graded series of minerals results (see figure 1). In a granitic intrusion, developing outwards from the aureole, the following sequence is often encountered: tin, copper, lead and zinc. Therefore, it is not surprising that such a link also exists in Australian copper mining operations.

Copper mining has an early history in Australia. The Kapunda mine (South Australia) which began operation in 1842 was the first copper mine. It was followed by others in the Wallaroo-Moonta area. Other major developments at a later date include the Mount Lyell mine (Tasmania), Cobar (New South Wales), Mount Morgan and Mount Isa (Queensland). Of these the three major currently operational mines are Mount Isa, Mount Lyell and Cobar.

Mount Lyell was the chief source of copper in Australia prior to 1953, and its copper occurs with Cambrian rocks at or near their contact with overlying younger conglomerate. Current annual production is around 20,000 tonnes of copper concentrates. Earlier mining operation of high grade copper was by open cut methods but current mining of lower grade ore is underground, which is placing Mount Lyell into a high cost operation situation where it is very sensitive to price fluctuations. Proven estimated reserves include 3.3 million tonnes of 1.4 per cent copper. It is possible that a much higher figure applies to potential reserves, but as the Mount Lyell mine has been supported by government subsidies in the recent past the potential reserves may remain submarginal in the foreseeable future.

Australia's largest and the best known copper mine is the Mount Isa mine in Queensland, where the copper sulphide ores are contained in Proterozoic shale and are of much higher grade than those at Mount Lyell. All mining is underground with current production around 150,000 tonnes of copper concentrates. Primary ore reserves have been estimated at 121 million tonnes of 3.2 per cent copper, while secondary ore has been estimated at 1.5 million tonnes. At current rates of production these reserves should support at least forty years of operation.

Copper was discovered at Cobar in 1869 and mining has been conducted there intermittently since that date. After a period of closure during the 1950s large scale mining operation was resumed in 1965, and current production is approximately 6,000 tonnes of copper concentrates. Reserves of 28 million tonnes have been located as massive copper and copper-zinc sulphide bodies in shear zones. These reserves could support current underground mining operations for another forty to fifty years. Recent intensive exploration in this richly mineralized area indicates a strong possibility of discovering further deposits.

Finally, the Olympic Dam prospect near Roxby Downs station in South Australia is a major recent copper find in Australia. It is a deep occurrence of copper, uranium and gold overlain by a thick cover of barren rock. Reserves have been estimated at 500 million tonnes of copper-uranium ore (compared with 121 million tonnes at Mount Isa). The copper ore grading is between 1 and 2.5 per cent. A new ore body with similar mineralization was found in 1980 at a short distance from the original discovery.

Lead, Zinc and Silver

These base metals are almost always found in close association with copper, although one of Australia's best known base metals mining areas — Broken Hill — is not known for copper mining. Most of Australia's base metals areas are either Precambrian or Cambrian: the lead-silver-zinc deposits of Mount Isa and Broken Hill are Precambrian and those of Rosebery date from the Cambrian. These are the three largest mines.

The Mount Isa mine is the world's largest individual producer of lead and silver and is also a large producer of zinc. With the Broken Hill mines, it produces about three quarters of Australia's lead. Mineralization at Mount Isa occurs within a meridional belt of Proterozoic shale with galena and sphalerite being the main sulphide ore minerals. Annual production is around 2.5 million tonnes of ore composed of 7.2 per cent lead, 6.1 per cent zinc and 203 grams per tonne silver. Reserves of primary ore of similar grade are estimated at 56 million tonnes.

The Broken Hill mines are associated with strongly deformed high grade metamorphosed rocks of Lower Proterozoic of the Willyama complex. Just over 2 million tonnes of ore are processed annually. The grade of ore varies considerably depending on the source. At the North Mine it averages 13 per cent lead, 10.1 per cent zinc and 230 grams per tonne silver. At the South Mine, lead varies between 2 and 8 per cent, zinc between 7 and 10 per cent and silver between 60 and 200 grams per tonne. For these reasons it is difficult to estimate the total reserves at Broken Hill.

The deposits at Rosebery in Western Tasmania are somewhat younger, having being formed in Lower Palaeozoic shale. Annual production of ore is around 650,000 tonnes with the following average grades: 11.8 per cent zinc, 3.7 per cent lead and 134 grams per tonne silver. The estimated reserves are around 8 million tonnes. Compared to Mount Isa and Broken Hill, these deposits have higher grade zinc and silver but lower grade lead.

Petroleum

Petroleum includes both oil and natural gas. Historically Australia has depended on imported crude oil. Discoveries during the 1960s and 1970s have been successful in meeting most domestic needs, but new finds will need to be made if current reserves are not to be exhausted by 1990.

Australia's oil and gas search has been concentrated in the Trans-Australia Platform which covers extensive areas of the continent and extends beyond the landmass in parts. The platform is made up of a series of sedimentary basins which vary dramatically in extent and thickness of sediments. They also vary in terms of their age and the tectonic activity to which they have been subjected. Most exploration interest has focused on basins formed during the Permian-Tertiary period of widespread deposition throughout the continent, when the black coal deposits were laid. Unfortunately, while gas finds have been fairly extensive, the hoped for association with oil has failed to materialize.

Most of Australia's oil and gas production currently comes from the Gippsland and Carnarvon Basins, the former dating from the Tertiary and the latter from the slightly older Mesozoic era. The Surat Basin in Queensland and the Cooper Basin in South Australia are also significant in terms of oil and gas production.

Generally speaking, more success has been met with gas than with oil. Thus the Surat and Cooper Basins have been quite successful in terms of gas production. Current exploration activity is concentrated on the Exmouth Plateau and the Surat Basin, and the objectives of the explorers are quite different in each case. The Exmouth Plateau search is being carried out in the hope of encountering large scale fields of oil and gas. The Surat Basin exploration is based on the probability of encountering many smaller fields, whose development costs and access to the populated centres of south-east Queensland would make them economic.

So far it appears that, while many of the sedimentary basins have suitable reservoirs in which oil could have formed, the vital processes which would have resulted in oil formation did not take place on a large scale. Current exploration activity (other than that in the Surat Basin and the Exmouth Plateau) is concentrated on those basins which are suspected of having fossil reefs in which oil might be trapped. Current reserves of oil are likely to meet the majority of domestic needs till 1990, after which dependence on imports will increase unless further discoveries are made. Recent finds in the Cooper Basin of South Australia and Queensland, in Bass Strait and in Western Australia have enhanced future prospects.

Uranium

Australia's reserves of uranium have been estimated to be as high as 25 per cent of the western world's known deposits. However, despite the scale of reserves, factors such as environmental considerations, Aboriginal land rights issues and especially the non-eventuation of many proposed nuclear fuel programmes overseas have slowed uranium exports from Australia.

Table 7. Major Australian uranium deposits

Deposit	Discovery date	Company-announced resources (tonnes U)*	Average ore grade (% U_3O_8)
Northern Territory			
Ranger	1970	85,000	0.22-0.25
Jabiluka	1971	175,900	0.25-0.39
Nabarlek	1970	12,000	1.84
Koongarra	1970	11,300	0.50
Western Australia			
Yeelirrie	1972	39,900	0.14
Queensland			
Mary Kathleen	1954	5,300	0.12
South Australia			
Beverley	1969	13,500	0.24
Honeymoon	1972	2,100	0.18
Roxby Downs	1975	Not yet proven; crude estimates of 300,000 +	Not available

* These figures generally refer to in situ resources

Source: Australian Department of Trade and Resources, Australia's Mineral Resources: uranium (Canberra: AGPS, 1980), p. 7.

Geologically, the known uranium deposits are associated with the Precambrian formations of the Australian Shield, the most important of which is the Pine Creek geosyncline (which includes the Alligator Rivers province) of Northern Territory. This uranium province accounts for over 80 per cent of known reserves in Australia. Geologists believe that the exploration potential may be five to ten times the known reserves. Other deposits found in Queensland, South Australia and Western Australia are generally either small or of relatively low grade or both (see table 7). The deposits at Yeelirrie in Western Australia are the largest of these proven reserves but they are of relatively low quality. The Yeelirrie deposits are of geological interest as they are secondary deposits formed from weathering of Precambrian granites. Recent finds — claimed to be larger than the Alligator Rivers deposits — have been reported at Roxby Downs in South Australia. However, excess supply in the international uranium industry and the technical problems involved in removing the extensive overburden may prevent the exploitation of the Roxby Downs copper-uranium prospect in the foreseeable future.

Notes

1. While every effort has been made to reduce the usage of geological terms that are not commonly understood by non-geologists, the use of some terms has been unavoidable. Usually these have been explained in the text or in footnotes. Readers encountering difficulty with terms should consult standard geological/earth sciences dictionaries, for example, The American Geological Institute's *Dictionary of Geological Terms* (New York: Anchor Books, 1976) or S.E. Stiegeler, *A Dictionary of Earth Sciences* (London: Macmillan Press, 1976). These dictionaries were the main references used for defining terms in this chapter.
2. P.T. Flawn, *Mineral Resources* (New York: Rand McNally, 1966), p. 1.
3. Ibid., p. 2.
4. E.C. Dapples, *Basic Geology for Science and Engineering* (New York: Wiley, 1959), p. 54.
5. These categories can be briefly described as follows (from Flawn, p. 8):
   Measured reserves are those for which tonnage is computed from dimensions revealed in outcrops, trenches, workings, and drill holes and for which the grade is computed from the results of detailed sampling. The sites for inspection, sampling and measurement are spaced so closely and the geologic character is so well defined that size, shape and mineral content are well established. The computed tonnage and grade are judged to be accurate within limits which are stated, and no such limit is judged to be different from the computed tonnage or grade by more than 20 per cent.
   Indicated reserves are those for which tonnage and grade are computed partly from specific measurements, samples or production data and partly from projection for a reasonable distance on geologic evidence. The sites available for inspection, measurement and sampling are too widely or otherwise inappropriately spaced to permit the mineral bodies to be outlined completely or the grade established throughout.
   Inferred reserves are those for which quantitative estimates are based largely on broad knowledge of the geologic character of the deposit and for which there are few, if any, samples or measurements. The estimates are based on an assumed continuity or repetition of which there are geologic evidence; this evidence may include comparison with deposits of similar type. Bodies that are completely concealed may be included if there is specific geologic evidence of their presence. Estimates of inferred reserves should include a statement of the specific limits within which the inferred minerals may lie.
   This classification is not fully adequate, for an additional concept or category can be used, namely:
   Potential reserves, which are those materials which might become available under future economic conditions and technology, but which are not available because of high costs due to remote location or to inaccessibility due to other factors, difficult extraction conditions resulting from excessive depth or the nature of the ground, small size or the low grade of mineral concentration, or problems of treatment. These materials go beyond the inferred reserves and include a wide variety of marginal, submarginal, or latent mineral materials.
   This classification is used frequently in Australia, although the use and scope of terms is not unanimous. A somewhat different terminology has been recommended recently by the Australian Institute of Mining and Metallurgy and the Australian Mining Industry Council. In this classification the basic distinction is between "recoverable" reserves (those which are expected to be mined) and "in situ" reserves.

The latter category makes no allowance for material not recoverable by established mining practices. The two types are further subdivided into "proven", "probable" and "possible" ores. It is not immediately apparent if this classification is in fact a significant advance on the classification system described earlier. For an application of the latter classification see MIM Holdings Ltd *Annual Report*, 1981, p. 30.

6. *Magmatic differentiation* is the process by which different types of rocks are derived from a single parent magma, or by which different parts of a single molten mass assume different compositions and textures as it solidifies. The term also applies to ores produced by the same process.
7. K. Warren, *Mineral Resources* (Harmondsworth: Penguin, 1973), p. 10.
   Haematite (iron oxide) is the principal ore of iron. Its name refers to the red colour of finely divided or earthy haematite, and most red rocks owe their colour to this mineral. The largest deposits, those of economic value, are of sedimentary origin, the iron having been originally deposited on the floors of shallow seas.
   Limonite is produced by the oxidation and hydration of iron-bearing minerals such as pyrite and magnetite. Limonite is found as a yellow to brown alteration product in all kinds of rocks, as a precipitate in bog iron ore, and as a major constituent of laterite.
   Siderite is usually brown and is a major constituent of bedded ironstones, an important iron ore. It is also found in hydrothermal veins.
8. Geologists have established that in its structural evolution a continent passes through a series of cycles involving uplifts, subsidence and the erosion of resulting surfaces. As a result of these processes, areas emerge with differing levels of stability. Some of these broad structural units achieve stability relatively early and are only slightly affected by later activity. Such stable uplifted areas which form the "backbone" of continents are called cratons or shields. They are generally very old, often being over two billion years old.

   Geosynclines, in contrast, are large generally linear troughs that subsided deeply throughout a long period of time in which a thick succession of stratified sediments and possibly extrusive volcanic rocks accumulated. These sediments have often then been uplifted and formed into mountain chains. This whole process — from crustal sagging, deposition, to uplift — is referred to as orogenesis. The best known and the most extensive geosyncline in Australia is the Tasman Geosyncline which spanned the Cambrian to late Tertiary periods and covered the eastern third of the continent. The major orogenies of this period are listed in table 3.
9. This account of the geological context and the nature of metalliferous provinces has drawn freely on two well known sources: Division of National Mapping, "Mineral Deposits" and "Geology", *The Atlas of Australian Resources* (Canberra: Department of National Development, 1970); and S. Hills, "Tectonic Setting of Australian Ore Deposits", in *Geology of Australian Ore Deposits,* ed. J. McAndrew (Melbourne: The Australasian Institute of Mining and Metallurgy, 1965), pp. 3-12. However, the account here is intentionally presented in a relatively non-technical manner. More technically oriented readers are requested to refer to the above two sources. Somewhat more extensive treatment is presented in the following publication: C.L. Knight, ed., *Economic Geology of Australia and Papua New Guinea,* 3 vols. (Melbourne: The Australasian Institute of Mining and Metallurgy, 1975). Maps showing the distribution of minerals discussed here are presented at the end of Part A.
10. *Serpentine* is a mineral group (typically green or white) characterized by long fibrous crystals. Serpentenite is a rock consisting almost entirely of serpentine minerals derived from the alteration of previously existing olivine and pyroxene.
11. Hills, in McAndrew, *Geology of Australian Ore Deposits,* p. 4.
12. C.F. Park and R.A. MacDiarmid, *Ore Deposits* (San Francisco: Freeman and Co., 1970), p. 311.
13. Hills, in McAndrew, p. 4.
14. J. Ward & I. McLeod, *Mineral Resources of Australia* (Canberra: Bureau of Mineral Resources, 1980).

15. *Coal* is a carbonaceous deposit formed from fossil plant remains. Differences in the kinds of plant materials (type) and in the degree of metamorphism (rank) are characteristic of coal and are used in classification. Coalification proceeds from partially decomposed vegetable matter such as peat, through lignite (brown coal), sub-bituminous coal, bituminous coal, semi-bituminous coal, to anthracite. During this process the percentage of carbon increases and volatiles and moisture are gradually eliminated.

    The background information for black coal and other minerals discussed in this chapter was drawn from the following sources: *The Atlas of Australian Resources*; R. Louthean, ed., *Register of Australian Mining* (Nedlands: Ross Louthean Publishing, 1980); and J. Alexander and R. Hattersley, *Australian Mining, Minerals and Oil* (Sydney: David Ell Press, 1980). The first publication is currently being revised and a new edition of the *Atlas* will be published in 1982. The other two publications are annual issues. Both have extensive and thorough coverage of all minerals and economic activities associated with them. Readers are referred to these publications as they deal with minerals at a level of detail which is outside the scope of the present chapter.
16. *Mesas* are isolated flat topped hills with steep sides most frequently found in old established landscapes based on horizontally bedded strata. Mesas often have almost vertical upper slopes developed on the resistant cap rock, followed by much flatter slopes extending down to a general plain. Iron rich rocks form the cap rock in the Pilbara.
17. *Magnetite* is a black, strongly magnetic ore of iron.
18. Australian Department of Trade and Resources, *Australia's Mineral Resources: iron ore* (Canberra: Australian Government Publishing Service, 1980), pp. 2-3. *Australia's Mineral Resources* is a series which deals with aspects of the most important minerals.
19. *Pisolitic:* refers to material consisting of rounded grains like peas or beans. Such material is often embedded in a clay-like matrix.

2 Historical Aspects

Noela C. Deutscher

The mining industry throughout the world has always been conditioned by certain major forces. The mining of any mineral naturally depends upon the existence of a market for it, and this is affected not only by economic factors but by the state of technology at various periods of history. In addition, speculative or risk capital has to be available to finance prospecting and the eventual exploitation of the minerals found. Some form of reasonably cheap transport is also necessary. In more recent times, another element which has influenced the development of the mineral industry has been public concern and government intervention.

During the greater part of the first century of settlement in Australia, the level of industrial technology throughout the world was such that industrial concerns could make use of only a limited number of minerals, so that early prospectors and miners were interested mainly in the discovery of such well-known minerals as gold, silver, lead, copper, tin, coal and iron ore. From the latter part of the nineteenth century onwards, however, industrial technology grew steadily more sophisticated and the range of minerals required expanded considerably. Two world wars, atomic science and space exploration also acted as forcing agents in this process. Australia has deposits of a vast number of different minerals, but there was no realization at first of the extent and variety of its mineral wealth because the technology existed for using only the more common minerals.

Even when the technology did exist, it was not always possible to mine certain bodies of ore because Australia is a large continent, with most of its mineral wealth located in areas distant from the coast. Consequently, the cost of transport was so high that it was uneconomic to mine some deposits. Until more recent times, this retarded the growth of Australian mining and meant that prospectors tended to concentrate on searching for gold because it was always sufficiently valuable to pay for the cost of development.

The production of any raw material is greatly affected by the level of demand for it. As Australia has never had a large domestic market, much of its mineral production has had to be exported and the mining industry has been particularly subject to fluctuations in world demand and prices, resulting in recurring cycles of "boom" and "bust". Traditionally,

Australia has exported minerals as raw materials to more populous and highly industrialized societies overseas. It is easier to develop complex processing and industrial plants when there is a large domestic market for particular products, and the absence of this has handicapped the mining industry in Australia.

Investing in mining ventures is always risky and requires a supply of speculative capital which is not always easily obtained in a country with a small population. This has resulted, at times, in a shortage of capital and has also entailed considerable dependence upon foreign investors. Both of these factors have caused difficulties for the mining industry.

An important recent development has been the growth of government intervention. There was only a small amount of government regulation during the colonial period up to federation, but since the first world war, and particularly in recent decades, the amount of intervention by both federal and state governments has increased considerably.

If the history of mining in Australia is considered in the light of these determining factors, it could be argued that there were three main developmental phases. First, there was an early period of exploration up to about the 1880s in which prospecting was often a matter of trial and error, technological skills were relatively limited, the range of minerals for which a market existed was restricted, distance and poor transport frequently made it unprofitable to mine some deposits of even those minerals for which technology and markets did exist, there was often a shortage of capital and the process of mining and extracting minerals was conducted on a small scale. It was the era of the small mine and small extractive plant. Secondly, there was an intermediate period after the 1880s in which prospectors became more experienced, technology became more sophisticated and more chemists and scientists were employed. Transport had improved and the development of the railway network made it profitable to develop mines in areas which were more difficult of access. There was also more capital investment in mining, with the result that both mines and processing plants began to be developed on a larger scale. Furthermore, a larger group of professional miners and trained mine managers developed. In the last phase, dating from the 1930s or the second world war, advanced techniques of prospecting, mining and processing were implemented, a greater range of minerals was sought and vastly increased amounts of capital were invested. In this period, industrial technology throughout the world has become extremely sophisticated, markets exist for most known minerals, even if demand for them still fluctuates, and distance and harsh physical conditions are definitely less of a barrier to mining development.

However, if the history of mining in Australia is considered in terms of economic development and prosperity, it could be argued that it fell into four main phases. Firstly, an early experimental period up to about 1880 in which, due to a shortage of trained personnel and lack of capital, mineral discoveries were exploited rather recklessly, with the result that good ore was gouged out of a mine to finance the development of the deposit and the life of the mine suffered. Secondly, there was a period of expansion from about the 1880s onwards, with more investment capital, larger mining companies, overseas experts and more skilled techniques of mining and processing. This period of optimism and expansion continued until about 1907 when the yield from the Western Australian goldfields began to decline. By about 1910, many experts believed that there were few major deposits of minerals left to find in Australia. The base metal industries continued to prosper, especially during the first world war, but collapsed after that and, by the 1920s, the outlook for the mining industry in Australia appeared gloomy. During this third phase, Australian investors began to look elsewhere for profitable mining ventures and skilled mining personnel often accepted positions abroad. During the period after about 1910, a local iron and steel industry was established, together with some large plants for the extraction of zinc, but it was still widely believed that the future of the Australian mining industry was not bright — a view which prevailed until after the second world war. A fourth phase of development began about the late 1940s or early 1950s when improvements in technology and transport produced a demand for minerals which had not previously been intensively sought or exploited. The 1960s saw a dramatic expansion in the production of various minerals, particularly iron ore and black coal, mainly as a result of increased demand from Japan. Mining became, once again, a major element in the economic and social life of the country, providing new wealth and increased importance in world terms and making Australia self-sufficient in almost all minerals, the main exception being petroleum. Although the mining industry is always extremely vulnerable because its prosperity depends upon world demand, which can change so rapidly that prosperity can be replaced by depression in a remarkably short period of time, the future in Australia appears to be promising.

Other possible phases of development could be delineated, especially if only one segment of the mineral industry, such as prospecting, were considered.[1] It is also possible to explore the history of mining in Australia by analyzing major aspects of the industry, such as prospecting, the discovery and production of particular minerals, processing, the investment of capital, cost of transport and legal problems. This is the approach which will be adopted here.

PROSPECTING

The discovery of minerals in Australia proceeded slowly at first because the Aboriginal people had not developed the use of metals and it was not until Europeans began to settle here that some idea of the continent's mineral resources was gained. Even during the early years of European settlement, there were few people in Australia who possessed any scientific knowledge, and those who did were more intent on discovering grazing land or studying the unusual flora and fauna than on looking for minerals. In addition, geology was not a well-developed science at this time. Consequently, the mineral resources of the region tended to be discovered first by miners who had come here as migrants or by pastoralists and rural workers, who were not trained prospectors but did traverse the same ground continually and had the leisure to examine interesting rock formations. However, various factors, including the exodus of Australian colonists to the gold-rushes in California, caused the various colonial authorities to become more interested in supporting geological exploration here and stimulating the discovery of minerals.[2]

As the continent became slightly more settled, as discoveries of some minerals showed that Australia might be likely to be richly endowed in this respect, and as the development of both alluvial and underground mining created a pool of settlers with some knowledge of minerals, there emerged a group of people who spent most of their lives prospecting for minerals. In this sense, they could be called professional prospectors, although only a few had any formal scientific training. Some worked for themselves but others were employed by various interested parties or subsidized by colonial governments. With the passage of time, they became more skilled in examining the country for likely mineral-bearing areas and, as more discoveries were made, they had the advantage of having observed many different mineral fields. Furthermore, as alluvial mining gave way to underground mining, more engineers, assayers, chemists and other trained personnel were employed and this provided prospectors with much-needed advice regarding rock samples, as did the schools of mines which developed from about the 1870s onwards. However, until about the 1930s, or even after the second world war, many prospectors in Australia could still be described as amateurs in that they lacked formal geological training, although they were skilled bushmen and careful observers of the country.[3]

This helps to explain why, for many decades, the minerals discovered here were well-known ones — mainly, gold, silver, lead, copper, tin, iron and coal — which were relatively easily identified by untrained observers. In addition most of the deposits were either alluvial or had a rocky outcrop of ore jutting above the ground. Generally speaking, mineral deposits buried below the ground were not discovered until more sophisticated techniques of prospecting were employed. Both self-trained prospectors and professional geologists employed by the colonial governments also experienced difficulty in discovering pyritic orebodies which were concealed by a layer of ironstone. Nevertheless, the fact that most of the early prospectors were skilled bushmen did make it possible for much mineral exploration to occur in the days before rail, motor and air transport made it easier to prospect in the harsh inland or rugged mountainous areas.[4]

By the 1870s, many of the easier mineral discoveries had been made and it was necessary to extend the search to largely unexplored areas such as the Northern Territory, Western Australia and western Tasmania. Expeditions to these areas had to be better equipped and prospectors had to be able to finance themselves or find backers. This, in itself, led to a somewhat more professional approach. By this time, too, the colonial governments were employing more experts of various types to search for minerals.[5]

A number of rich bodies of ore in Australia were covered by a cap of iron which effectively concealed the rich lodes of mineral beneath. Most prospectors were interested in finding gold, causing them to look for quartz rock and shun ironstone. After the discovery of a number of pyritic orebodies in the 1880s and 1890s, prospectors and geologists

As mines became deeper, it was necessary to raise ore to the surface by means other than manual labour. In many cases, horses were used to turn a type of windlass, known as a whim, which raised material from underground workings. As some mines became more highly capitalized, steam power replaced horses but many small mines continued to use horsepower because it was simple and cheap. This photograph shows a whim in operation at Gympie in Queensland in the 1870s. *(John Oxley Library, Qld.)*

became much more experienced in detecting them, and the use of blasting powder and dynamite to blow rocks apart was also of assistance. However, before prospectors and geologists became used to dealing with this type of deposit, some pyritic orebodies had been discovered by amateurs who had fewer preconceived ideas about where minerals might be found and were willing to try areas which more experienced men ignored.[6]

Enterprising pastoralists and rural workers also continued to play an important role in the discovery of minerals in Australia in the period before the second world war; as well as discovering mineral fields themselves, they helped others to do so by providing such necessary facilities as supplies and water. Sheep or cattle runs in the harsh interior of the continent provided stepping stones for prospectors and the absence of pastoral settlement in the inland regions of the Northern Territory and Western Australia was one of the main reasons why minerals were not discovered there until the latter years of the nineteenth century.[7]

The whole process of prospecting gradually became more scientific, especially after the 1930s when mining companies began to employ geologists to search for new mineral deposits, instead of relying on chance and the work of private prospectors. Pioneering companies in this field included the Western Mining Corporation and Gold Mines of Australia, which were formed to take advantage of geological knowledge in the search for new mineral deposits. About this time, too, the Zinc Corporation of Broken Hill began to use similar methods to search for base metals in western New South Wales and South Australia. The Western Mining Corporation also pioneered the use of aerial surveying, a highly significant development. New methods of prospecting were necessary because earlier prospectors had discovered most of the obvious deposits of well-known mineral ores. What remained to be discovered were deposits of ores for which there had previously been no great demand and "blind" bodies of ore, that is, mineral deposits which were well below the surface, showed no visible sign of their presence and were therefore out of reach of the ordinary prospector. Mining companies which had the capital resources to employ skilled geologists were able to use methods such as underground drilling and exploration of geologically promising areas where other minerals had been found. These methods yielded some significant discoveries, especially in view of the fact that before the second world war large amounts of capital were seldom injected into the quest.[8]

Much new technology was developed as a by-product during the second world war, and the pace of technological change quickened greatly. Prospecting became still more scientific and techniques pioneered in the 1930s were developed much further. Far greater use was made of trained personnel, while techniques such as aerial surveys, aerial mapping and seismic surveys were employed more extensively. More work was also done to compare certain areas in Australia with mineral-bearing land elsewhere — in other words, a process of analogy was used to help select suitable areas in which to prospect for particular minerals. A recent development is the use of satellite photographs to predict areas in which it might be profitable to search. The federal government established the Commonwealth Bureau of Mineral Resources in 1946 and began a systematic programme of research, geological and geophysical surveys, mapping of resources and compilation of statistics, and a number of state governments continued their own geological surveys in a more vigorous manner. As a result, scientists have gained a much better idea of the geological structure and mineral resources of Australia. The federal government also encouraged prospecting by providing tax concessions and other benefits. In the period between the two world wars, some of the large, prosperous mining companies had spent a mere pittance on prospecting but, after the second world war, a greater investment by private companies in the search for minerals was a factor in the discovery of new deposits. It also meant that a more continuous supply of capital was available and that, as a result, there was far less need to resort to the old method of endeavouring to raise money for exploration by floating companies on the stock exchange. Because this technique had succeeded only during mining booms, mining exploration in earlier periods had been an extremely episodic business. The greater availability of capital in the post-war years certainly helped the mining industry to prosper and led to an increase in the known reserves of minerals in Australia.[9]

Although prospecting has become a highly-skilled activity, with most of the major mineral discoveries since the second world war being made by scientists, some finds have continued to be made by fortunate individuals, observant pastoralists or individual prospectors of the old type. Rich deposits of uranium at Rum Jungle and Mary Kathleen, iron ore in the Pilbara and nickel near Kalgoorlie were all discovered by individual prospectors and, in the case of Lang Hancock, by a pastoralist piloting his own plane and observing his property from the air. These men lived in the areas in which they were prospecting, enabling them to observe the country closely and in a leisurely manner.[10]

However, even though some of the mineral discoveries made since the second world war have been the work of private individuals, the fact that geologists and other experts are readily available for consultation and advice can mean that significant clues are followed up even if the prospectors themselves lose interest when the samples which they send to be assayed do not reveal payable quantities of minerals immediately. This was the case, for example, with a nickel field in Western Australia. There is still some role for individual prospectors in Australia, but the facilities of large companies are often needed if their initial discoveries are to be pursued and reach a successful conclusion.[11]

Thus, for the greater part of the history of prospecting in Australia, the majority of prospectors were not highly trained but were hardy men who performed great feats of endurance while searching for minerals in the arid areas of the Australian inland and the rugged terrain of Tasmania. Prospecting is now very much in the hands of scientists and large companies and this trend will intensify, but the earlier individual prospectors did much to reveal the mineral wealth of Australia.

DISCOVERY

The minerals sought, found and mined in Australia in the early decades of settlement were well-known ones for which a ready and often highly profitable market existed, and which could be identified by the naked eye and by relatively untrained observers. The deposits were either alluvial or else protruded above the ground in the form of a rocky outcrop of ore.

Coal was the first mineral discovered and mined in Australia because its presence was easily detected, there was a ready market for it and its importance was widely recognized by British officials due to the fact that Britain had become quite industrialized by the latter part of the eighteenth century. Cliffs of coal were discovered in the Illawarra district south of Sydney in 1796 and rich seams of coal were also found in the Hunter River area, with large deposits being discovered near Maitland in the 1880s. The increasing use of steam-engines in various mining and industrial enterprises, the development of smelting and the building of railways stimulated the growth of the coal industry, especially in New South Wales. Expansion of the black coal industry there was extremely rapid up to the 1890s and, although some mining of coal also occurred in Tasmania, Victoria and Queensland, the main source was New South Wales. Coal not only provided fuel for Australian industrial development but was also exported overseas.[12]

The second mineral to be mined in Australia was *lead,* first discovered near Adelaide in 1841. This success stimulated the search for other minerals and the presence of Cornish miners amongst the migrants who went to South Australia assisted the development of mining both there and in the other Australian colonies, because Cornwall had been an important mining centre since Roman times. Small amounts of silver-lead ores were found in South Australia and elsewhere but it was not until the 1870s and 1880s that significant lodes were discovered, notably at Broken Hill in the Barrier Ranges of New South Wales. A later discovery was Mount Isa which became an important producer of lead, zinc and silver as well as copper.[13]

Copper was the third mineral mined in Australia. Because copper ores advertise their presence by splashes of blue or green colour on rocky outcrops, they are easily seen and identified, even by inexperienced observers, with the result that discoveries of copper in South Australia and elsewhere occurred rapidly from the 1840s onwards. It was not long before copper was being mined in Victoria, New South Wales, Tasmania and Queensland as well, although South Australia remained the main centre for a considerable number of years, at times producing a tenth of all the copper mined in the world. By the 1880s, however, the South Australian mines were in decline and the other colonies became more important as producers of copper. By 1899, for example, Mount Lyell in Tasmania had become the largest copper mine in the British Empire. By the 1950s, Mount Isa had become one of the world's great producers of copper.[14]

From the 1850s until about 1910, mining in Australia was dominated by *gold*. Despite constant rumours that traces of gold had been seen in New South Wales during the early decades of settlement, it was a surprisingly long time before gold was mined in Australia. It has been argued that the colonial authorities, worried about how they would control the convicts if a gold-rush occurred, suppressed the news of earlier discoveries. This may have been a factor in the failure to pursue reports of gold, but the argument is not conclusive. Another reason may have been that none of the early discoveries was sufficiently valuable to excite any great interest. Blainey contends that one of the major reasons why so little notice was taken of these early reports was that, under English law, all deposits of gold and silver became the property of the Crown.[15] This restriction possibly explains why many people were more eager to find copper than gold. However, once alluvial gold was discovered and the news publicized widely by Edward Hargraves, public pressure quickly forced the colonial authorities in New South Wales to clarify the legal position, so this may not have been as great an obstacle as Blainey suggests; although it may explain why immigrant Cornish miners do not seem to have acted upon an English geologist's advice to pan for gold in Australian rivers. Another factor may have been that the early discoverers of gold did not have the skill or machinery to work gold-bearing reefs of quartz rock, which was the type of gold mainly seen before 1851, so it required news of the discovery of alluvial gold to spark a gold-rush. Even after the gold-rushes had actually begun in Victoria, the potential of quartz rock was so little understood that it was used as road metal.[16]

Once the initial discovery of alluvial gold became widely known, the search for it intensified and new discoveries followed in rapid succession. In the 1850s, the success of copper mining was completely overshadowed by the discovery of rich deposits of alluvial gold in Victoria and New South Wales. These discoveries caused a significant influx of free migrants and considerable upheaval in the colonies, the wealth of the Victorian fields becoming celebrated throughout the world. Gold was

This photograph illustrates the process of sluicing often used to wash for minerals such as alluvial deposits of tin. Miners played jets of water from hoses on to cliffs of tin-bearing clay, washing the debris into receptacles which trapped the heavy grains of tin oxide and allowed the sludge to flow away. *(John Oxley Library, Qld.)*

sought and found by both experienced diggers and novices. Discoveries occurred throughout Victoria in the 1850s, New South Wales experienced a revival of discoveries in the 1860s and the search was also extended to the other colonies, including Tasmania. By the end of the 1870s, gold had been found in all colonies except Western Australia, where prospecting was hampered by lack of settlement. By the 1880s, the yield from older gold-fields in eastern Australia had declined but, about that time, there began a series of rich discoveries in Western Australia which, by the 1890s, had become the richest gold-producing region in Australia. By about 1907, however, the yield from even the wealthy Kalgoorlie field had begun to decline, with the result that mining companies and investors began to look overseas to find new mines.[17]

Given the fact that Cornish miners were particularly experienced in mining *tin,* it was a surprisingly long time before significant deposits of this mineral were discovered. Again, it seemed to require the finding of rich alluvial deposits to spark interest in mining tin and this did not occur until the 1870s when rich deposits were found in New South Wales, Queensland and Tasmania. The Australian colonies and the Straits Settlement then became the world's largest producers of tin and, from the 1880s onwards, Tasmania was Australia's largest producer.[18]

In the early years of the twentieth century, commercial processes were developed for the refining of *zinc* from silver-lead ores, and these benefited the mining or processing companies in Broken Hill and Tasmania and the mining industry generally.[19] The decision by the Broken Hill Proprietary Company in 1915 to establish an *iron and steel* industry on a large scale in Australia exerted a profound influence upon the country's industrial development, and also increased the importance of large deposits of iron ore in South Australia.[20]

Another recent development has been the search for *oil*. In the period up to the second world war, oil became an increasingly important source of energy, steadily reducing the demand for coal. The bulk of this oil came from Indonesia, then a Dutch colony, but the strategic implications of this did not seem to cause widespread concern in Australia at the time. There were a few sporadic attempts to find oil here but little was spent on the search. Traces of oil and natural gas were found in Queensland and Victoria but no commercial field was discovered. To find oil in Australia, large amounts of money needed to be spent and geologists employed. It was not until after the second world war that these requirements were met and the search was successful. The possibility of obtaining oil from deposits of shale is another recent development.[21]

The growth of more complex industrial processes increased the range of minerals for which a demand existed throughout the world. *Tungsten* ores, for example, began to be mined in Queensland, Victoria and Tasmania and, in 1905, about half the world's supply of tungsten was produced in Queensland. After the second world war, the mining of *uranium, rutile sands* and *bauxite* became important. The application of open cut technology also created renewed interest in the mining of coal. Many of the minerals discovered in Australia after the second world war were known earlier but, as there had been no great commerical market, their presence and value had not been realized. Thus *manganese oxide,* used for centuries by the Aborigines on Groote Eylandt to colour their bark paintings, was mined when advanced industrial civilizations required it to make special types of steel. The demand for *nickel* and huge quantities of iron ore was largely a post-war development too. In 1902, a government geologist had collected samples of bauxite in the area near the Gulf of Carpentaria, but nothing came of this because it was not until the late forties and early fifties that demand increased to such an extent that new deposits of bauxite were sought throughout the world.[22] W.S. Robinson noted that, on a trip to inspect mines in Queensland in 1906, he had slept near the site of what eventually became the Mary Kathleen uranium mine, but "the value of uranium and the means of locating it were then entirely unknown".[23]

Increased demand for various minerals also meant that both individual prospectors and large companies sometimes re-examined proven mineral-bearing areas where gold or other minerals had been mined previously. Because only well-known minerals had been sought at earlier periods, signs of the presence of other minerals on these fields had been disregarded. Thus, it was not until the 1960s that geologists and mining engineers realized that some former gold-fields in Australia might be base metal fields as well and this led to discoveries such as the finding of nickel at Kambalda, an old gold-field near Kalgoorlie.[24]

Australia has been fortunate in the range of minerals found here and, with the exception of oil, is largely self-sufficient in minerals. Furthermore, with changing world demand and new technology, and harsh conditions in various parts of the continent, Australia's mineral potential has not been fully ascertained even yet.[25]

MINING

The discovery of particular mineral deposits was, of course, only the first step: the question of how to exploit them was often of crucial importance. Due to a range of problems, it was difficult, at times, to bring particular mines into production. Deposits in the rugged west coast area of Tasmania, although rich, were costly to develop; Mount Isa was another rich deposit which was exceptionally difficult to develop due to isolation, the high cost of transport and difficult ore.[26] Successful exploitation of a mineral deposit depends upon a variety of factors including world demand for particular minerals, price structure on the world market, difficulties of terrain, the level of technological competence in both mining and metallurgical spheres, availability of ample capital to finance development and the cost of transport.

Much of the early mining in Australia, especially of gold, was alluvial mining which could be undertaken by untrained men using simple panning equipment. However, the major alluvial deposits of gold and some other minerals were exhausted relatively quickly and it became obvious that underground mining would have to be developed further as the only way to tap most of Australia's mineral wealth. This entailed the growth of a professional class of mine managers and miners such as had emerged already in the New South Wales coal mines and the South Australian copper mines in the period before the gold-rushes.[27]

Of the engineers, mine managers and mine foremen involved in the early years of underground mining in Australia, some individuals had come from Germany and other mining centres, but large numbers were from Cornwall and they provided most of the technology. Their skill was based on canny observation rather than scientific training, but they were good practical miners. Not surprisingly, they took the lead in developing copper mines throughout the Australian colonies, especially in South

Australia. When copper was discovered there in the 1840s, properly organized underground mining began almost immediately. The mines were run on the Cornish tribute system and many of the copper mining towns became replicas of Cornish society. They served the industry well throughout Australia for some years but, by the latter half of the nineteenth century, considerable advances in mining technology had been made in Germany and the United States, and reliance upon the older Cornish managers meant that mining technology in Australia tended to stagnate. It was not until the 1880s that some Australian mining companies, notably the Broken Hill Proprietary Company, made a conscious decision to recruit expert help from mining centres outside Britain.[28]

In the early years of underground mining in Australia, the process was extremely dangerous due to lack of ventilation, the presence of noxious gases, the danger of subsidence and, in many cases, the presence of water in underground workings. The methods used in underground mining were also extremely crude at first, but later became more sophisticated and mechanized. The early Victorian mine managers, for example, were inclined to tear out the rich ore as fast as possible without attempting to preserve the life of the mine. Cornish mine captains, on the other hand, always sought to conserve the ore and thus extend the life of the mine. They also used wooden supports for shafts and underground galleries and avoided a number of the dangerous practices common in the early Victorian mines.[29] As experienced miners from Cornwall, Germany, America, Canada and elsewhere became managers of mines throughout the Australian colonies, as Victorian managers learnt better techniques, and as governments began to implement safety regulations, mining became less dangerous and less foolishly exploitive, although some of the barriers to sound mining practice — such as inexperience, lack of capital or sheer greed — still handicapped the industry.[30]

Professional miners, especially Cornishmen, soon brought to Australia the type of machinery they were accustomed to using in mines elsewhere. In many of the early underground workings in Australia, water was baled out by hand, but it was not long before horses were used to turn whims in most mines, while more advanced concerns were soon using steam-engines to pump water from underground workings and mechanical steam-driven stampers to crush ore. The invention, by a Cornish mine captain, of a mechanical jig for ore-dressing expedited the sorting of poor ore from good. It was not long before attempts were made to use machines to gouge more ore from the rock face and some mine managers experimented with piston rock drills and diamond drills. An experimental rock drill, invented locally, was in use in Bendigo as early as 1869 and thus appears to have preceded American attempts to drill holes in the rock face by mechanical means. Steam-engines were used to haul skips of ore to the surface, replacing human or animal traction. Wire ropes replaced weaker hemp ropes, safety cages transported men underground to work and mechanical blowers supplied fresh air to underground workings. In the early mines, blasting powder was used to shatter rock but this was superseded by dynamite in the 1870s. Open cut techniques were also developed in the 1860s in an effort to obtain the last ore from older mines. Thus, up to the 1870s, some mines in the Australian colonies were relatively highly mechanized, but such development were often handicapped by lack of money because, before the 1880s, most mines were undercapitalized. Undercapitalization also meant that mining machinery was resold repeatedly and moved from location to location to be used again.[31]

As well as machinery, trained mine managers and engineers, a work force of professional miners was necessary. Most of the men who had flocked to the diggings during the gold-rush era drifted into other occupations when the supply of alluvial gold was exhausted, but some decided to turn to underground mining and helped to form a group of professional miners. At first, in order to finance the sinking of deep shafts, groups of miners formed syndicates or co-operatives, but the cost of underground mining was considerable and it was not long before most miners were forced to work for companies. Professional miners from Cornwall and elsewhere in Britain also migrated to Australia and so, by the 1880s, there was a force of reasonably skilled professional miners. However, as mines became deeper and more complex, new methods needed to be introduced.[32]

By about the late 1870s, an infusion of new skills was required urgently in the mining industry. Outdated methods became an increasing problem as the more easily exploited deposits were worked out, and there was a growing realization that the supply of new deposits was not unlimited and that mines must be managed more skilfully. Australian mineowners and companies turned for advice and personnel to Germany and the United States, where mining technology had begun to outpace that in Britain. A group of mineowners commissioned a geologist to study the methods used in the great American mining centres, and his report, written in 1877, showed how backward most Australian mines had become. Some schools of mines were established here in the early 1870s, about a decade after similar schools had been founded in the United States, and the University of Melbourne began to offer lectures in mining engineering in 1874, but it took time for the effect of these developments to be felt.[33]

One of the most significant steps taken at this time was the Broken Hill Proprietary Company's decision in 1886 to hire overseas mining experts to develop the company's mining and smelting operations. The board of directors of BHP was initially composed mainly of pastoralists, but became noted for its willingness to hire good advisers, take gambles and plan on a large scale. Whenever the company faced a difficult situation, the board developed a pattern of hiring new talent or branching out into a new industry, and this was one of the reasons why BHP became so successful.[34] When one of the directors of BHP went overseas in 1886 to hire expert help, he turned to the United States which had become probably the most advanced mining centre in the world and persuaded a leading mining engineer, William H. Patton, to become general manager of BHP at the then lavish salary of £4,000 per year. Through the influence of Patton and other American mining experts hired later by other companies, the Australian mining industry was brought into contact with advanced American and German technology. As various German states were leading centres in the processing of minerals, many German experts were brought here and, until the first world war, a number of German companies held interests in processing plants in Australia.[35]

From the 1880s onwards, more capital was invested in Australian mining and there was a trend towards the development of larger companies and larger mines, many of which were managed by imported experts. A new generation of mine managers — both imported and native — began to implement the best in mining technology. They began

This photograph of an ore-crushing battery not only provides a good view of the machinery but also illustrates how small mines in Australia still operated in a relatively primitive manner even as late as the 1900s. *(John Oxley Library, Qld.)*

to prove reserves in mines in a systematic manner, remove the ore in bulk instead of gouging the eyes out of it, make extensive use of high explosives, employ rock and diamond drills, use steam and later electric power to pump water from underground workings, and adapt to Australian conditions such overseas ideas as the large open cut mines developed in Spain and the square-set method of mining used at the Comstock lode in Nevada. Gradually, with the aid of foreign experts and the development of schools of mining here, a solid basis of mining engineering was established.[36]

However, after the great gold discoveries in Western Australia ceased in the 1890s, few important new deposits of minerals were found until the late 1940s, and many established mining concerns began to decline. The yield from the Western Australian gold-fields had decreased markedly by 1910 and the end of the first world war caused a sharp decline in the profits of the base metal industries. The coal industry, too, was hard hit by competition from cheap imported petroleum and, from about 1927 onwards, lapsed into an extremely depressed state. Mineowners sought desperately to reduce costs, neglected safety precautions and, in many cases, lacked the capital to introduce new machinery and methods. Furthermore, as was the case in Britain, the coal industry here was plagued by severe industrial strife even at times when relations between owners and miners in the metal industries were quite good. As a result, the coal industry declined to such an extent that it was unable to meet even the war needs of the country in the years 1939-45. Because of this stagnation and decline in the mining industry, fewer overseas experts came to Australia and some promising young Australian mining engineers went abroad to work. There were some new developments – modern prospecting techniques, new methods of mining, the exploitation of deposits of brown coal in Victoria by the open cut method, the long struggle to make Mount Isa pay and some renewed development at Broken Hill, Mount Lyell and Kalgoorlie – but the great days of mining in Australia seemed to be over.[37]

After the second world war, however, technical changes created a demand for different types of minerals, causing renewed mining activity here. By the sixties and seventies, a variety of new mining fields – coal, bauxite, iron ore and nickel – had been discovered and brought into production. The cost of developing these fields was so great that Australian companies sought partnerships with large foreign concerns which could provide capital and, if necessary, technological aid and trained personnel as well. This was particularly important in drilling offshore for oil, for example, as Australian firms had little experience of this. Following its accustomed pattern of acquiring expert help whenever necessary, BHP sought an experienced partner amongst the American oil companies and the Esso-BHP combination came into being. New technology and machinery from overseas also played an important part in the development of massive new open cut coal-fields and iron ore deposits. Since the 1950s, larger departments of geology and mining engineering have been developed in Australian universities and, through the involvement of foreign companies, Australia has also had access to foreign technology. This type of interchange can be helpful when a country has only limited resources of population and accumulated wealth, although it is important not to become too reliant upon it. Authorities in Australia could have done more to foster vigorous training programmes at various levels, including replacement of the outdated apprenticeship system.[38]

PROCESSING

The acquisition of more advanced mining technology in Australia was also associated with an increase in the processing of ore here. In the early years, ore was sent overseas to be treated but, as this was extremely expensive, it was not long before mining companies decided to build treatment plants here and import metallurgists and smeltermen from countries such as Wales and Germany where smelting and extractive industries were well established. The Welsh community was so strong in the smelting town of Wallaroo in South Australia, for example, that church services there were conducted in Welsh for forty years.[39] When the first deposits of minerals were discovered, samples of ore had to be sent overseas to be assayed but, after a time, assayers were induced to migrate here and Australians were trained as well. In due course, the foundations of a metallurgical industry were laid, although progress was slow, hampered by lack of trained personnel, lack of capital and the small size of the typical Australian mine.[40]

Small furnaces to smelt lead from the mine at Glen Osmond were built in Adelaide in 1840 and copper smelters were built at the Burra mine in 1849. Other smelting centres soon developed in South Australia, using coal shipped from Newcastle. Despite the lack of coal, South Australia took something of a lead in assaying and in the production of concentrates from various ores, but treatment plants to deal with various minerals, including iron ore and tin, were soon established in the other colonies as well. One of the early leaders in the extractive field was the Port Phillip and Colonial Gold Mining Company which, by the 1860s, operated one of the largest gold treatment plants in the world. Unfortunately, due to undercapitalization and the small size of most Australian mining companies, few followed this lead until the 1880s and progress in metallurgy was slow.[41]

During the early years of mining in Australia, treatment plants were usually built beside the mines because transporting the ore elsewhere —

generally by bullock dray — was so costly. As wood was often used to fuel smelters, it was usually extremely economical to build a smelter near a mine when a good supply was available and, in this way, whole regions were denuded of trees. A smelter serving just one mine was usually relatively small and, at a time when investment capital was often difficult to obtain, this option was attractive.[42] However, it proved very expensive to cart coal and various materials used as fluxes to the smelters so, once supplies of wood were exhausted and railways were constructed, there was a growing tendency to abandon small smelters built near mines and construct large, centralized smelters near coal-fields or near a harbour where coal could be unloaded from colliers. The growth of Newcastle, Port Pirie and Port Kembla as smelting and industrial centres exemplified this trend. Smelting works were also developed in the chief city of each colony because it was usually the centre of the railway network and it was comparatively cheap to rail ore there. Ore could be sent to a smelting centre from a number of mines, whereas a smelter built to serve just one mine had to be abandoned if the ore in that mine were exhausted. This could lead to considerable financial waste.[43] In isolated mining centres, such as Kalgoorlie, Mount Isa and Mount Lyell, treatment plants were still built or maintained beside the mines because the cost of hauling ore to the coast for treatment was so great.[44] Nevertheless, by the 1890s and early 1900s, the trend towards locating large treatment plants on the coast or near coal-fields was becoming noticeable, and plants were becoming larger and more complex. As the small mining company was being replaced by larger concerns from about the 1880s onwards, so was the small treatment plant.[45]

As the mining industry grew, and especially as smelting became concentrated in particular centres, there developed specialized companies which simply treated ores and had nothing to do with mining them. The growth of these companies benefited the mining industry because they experimented with new methods, treated difficult ores and served mines which could not afford to pay their own experts or install their own treatment plants.[46]

In the early days, the treatment processes used were simple and not particularly efficient, with the result that the rate of extraction was not high and minerals other than the one being sought tended to be lost as waste. With the passage of time, chemists throughout the world battled to find more efficient extractive processes. Much progress was made overseas, especially in Germany and the United States, but it was some time before this new technology affected developments in the Australian colonies and the result was that, for a long period, ore was mined here more efficiently than it was treated. The backwardness of the metallurgical industry in Australia may have been connected with the early dominance of Cornish mine managers, because ores from Cornwall were sent to Wales to be treated and Cornish miners often knew little about metallurgy. In addition, the ease with which "kind" ores from a number of the early mines had been treated made many mining men think that metallurgy was unimportant in Australia. The tradition of the small mine and the greed of some owners and shareholders also discouraged the employment of expensive consultants and specialists.[47]

By the 1880s, it was obvious that extractive technology here had to be improved. With many of the easiest deposits of ore exhausted, many mining men were ill-equipped to treat more difficult ores because they lacked scientific training. It became increasingly clear to alert mineowners and companies that extractive technology in the Australian colonies had fallen behind that in other great mining centres such as Germany and the United States. Some schools of mines were established here in the 1870s and metallurgy began to be taught at the Universities of Melbourne and Sydney in the 1880s and 1890s respectively, but it was a considerable time before the effects of this were felt and, in the meantime, many companies began to look abroad for expert help.[48]

One of the leaders of this trend was BHP which, in 1886, sought the services of a leading metallurgist to develop the company's smelting operations. The American chosen, Herman Schlapp, not only assisted BHP, but also gave helpful advice about how other mines in Australia should be developed and acted as a link with the American metallurgical industry, advising on the hiring of other experts by various Australian concerns.[49]

With the increasing size of mines and greater capitalization, there was more willingness to employ skilled metallurgists here and, as a result, extractive plants became more efficient and new methods were tried. The number of mining companies operating at Broken Hill, for example, made both it and Port Pirie important metallurgical centres, and the willingness of companies at Kalgoorlie to finance constant experiments to improve the extraction of gold from the difficult ores mined there meant that, for a time, Kalgoorlie became perhaps the leading gold treatment centre in the world. By the 1890s, there was a ferment of metallurgical activity in the larger mining centres which pushed Australia into the forefront of extractive technology, with many important processes either invented or improved by metallurgists working in Australia. A number of these experts were Americans and Germans but some, such as John Sutherland, had been trained in Australian schools of mines.[50]

At this time, there was fierce competition amongst metallurgists throughout the world to find a cheap commercial process for the extraction of zinc, for which there was a growing demand on the world market and a high price. Broken Hill scientists were to the fore in this battle because, if a commercial process for the extraction of zinc could be found, the huge dumps of tailings at the silver-lead mines there would be worth a fortune. Chemists in London, Melbourne and Broken Hill all made significant discoveries, at approximately the same time, regarding a flotation process for the extraction of zinc but companies at Broken Hill, especially BHP, had the lead and were producing one-fifth of the world's supply of zinc concentrates by the time mining companies in the United States began to work on the problem. In this area, as in some aspects of the treatment of gold, scientists based in Australia led the world.[51]

After about 1900, when the tempo of mineral discoveries in Australia flagged, new developments in processing became possibly the most important aspect of the mining industry here until the renewed boom in discoveries from about the early 1950s onwards. The flotation process for the extraction of zinc was a significant new development and, after a long battle to make the method work, companies such as the Zinc Corporation and BHP did well out of the process. Later, the Electrolytic Zinc Company decided to try a new electrolytic process for producing zinc and, after a costly struggle, established an important new treatment plant in Tasmania in 1921. By the 1960s, this had become the third largest zinc works in the world. Unfortunately, there was less innovation in other areas. The flotation process developed for the extraction of zinc was even more successful when applied to other minerals, especially copper, but American companies were quicker to use this method to replace pyritic smelting of copper, and forged ahead in this field while Australian companies lagged behind.[52]

In the early years of this century, BHP's decision to venture into manufacturing iron and steel was an event of major importance in the history of mining in Australia because it affected the development of secondary industry and also stimulated the mining of iron ore. By 1911, BHP had mined 80 per cent of the ore it was to gain from its leases at Broken Hill and it needed to expand into new ventures if it were to continue to prosper. Once again, the board of directors took a bold step which proved successful: the directors surveyed the highly efficient iron and steel industry in the United States and employed a skilled American engineer to design the new steelworks, thus reaping the benefit of American technology. Newcastle was chosen as the site and the enterprise was planned on a large scale, more than a million pounds sterling being raised in 1913–14 to fund the scheme. The venture eventually became a success and, as a result, BHP diversified its activities, becoming a manufacturing giant which exerted considerable influence upon the industrial history of Australia.[53]

Apart from a brief revival during the 1930s, gold-mining was in decline during the period after about 1910 and the base metal industries experienced grave difficulties after the end of the first world war. However, the profitability of zinc, the growth of the iron and steel industry and the development of new applications of the flotation principle meant that there was continued activity in the metallurgical field. There was also the long struggle to conquer the refractory ores mined at Mount

This photograph illustrates the plant at Mount Morgan in Queensland — a more highly capitalized mine. Even in the case of this mine, however, such a high proportion of the profits was paid in dividends that the managers of the mine were hard put at times to obtain sufficient capital to purchase new machinery, experiment with new techniques and develop the mine properly. *(John Oxley Library, Qld.)*

Isa and, from the 1930s onwards, the giant American Smelting and Refining Company poured money and expert assistance into the operation in an effort to make it pay. Indeed, from about the late 1890s until after the second world war, it could be argued that there was more innovative activity in the metallurgical field than in mining engineering in Australia.[54]

After the second world war, the demand for different types of minerals and some interesting new discoveries sparked renewed interest in Australia's mineral resources. In addition, foreign mining companies were no longer so preoccupied with investing in former European colonies and were seeking new fields of activity. These factors led to new developments in the mining and metallurgical fields — for example, the establishment of alumina plants — and, when necessary, foreign technology was borrowed. At present, Australia seems poised to develop further in the direction of processing minerals instead of being primarily a supplier of raw materials to highly industrialized countries overseas.[55]

Because Australia has always had a small population and limited financial resources, the contribution made by foreign experts who could introduce new technology and ideas in mining engineering and metallurgy has been considerable. The work of some of these experts was of great importance, although not all imported mining men were outstandingly good and a few were inefficient or misguided.[56] On balance, however, the hiring of overseas experts was of great assistance, especially in the 1880s and 1890s, because the skills they brought with them gave new impetus to the mining and metallurgical industries and, in conjunction with the efforts of Australian mining men, enabled Australia to overcome its technological backwardness. The Australian colonies also began doing more to train their own mining engineers and metallurgists. The mining schools at Ballarat and Bendigo, founded in the 1870s, gathered strength and others were established later in centres such as Adelaide, Charters Towers, Zeehan and Coolgardie. Universities in Sydney and Melbourne began to offer some training in mining and metallurgy. The Australasian Institute of Mining Engineers, modelled on the American body founded in 1871, was established in 1893. Australian mining men began to be accepted overseas and were hired by foreign mining companies. Increased federal grants to the universities from the 1950s onwards helped to establish facilities for training Australian mining engineers and metallurgists, but experts employed by foreign companies continue to contribute because Australia has only a limited pool of talented people on which to draw and relatively limited funds to devote to training and research, although more could be done in this regard.[57]

FINANCING

Given that Australia has always had a small population and limited amounts of accumulated wealth, financing mining ventures adequately has presented continuing problems. To obtain the best return from a mine or treatment plant, it is often necessary to pour money into its initial development but, in Australia, there has been a long history of financing the exploitation of mineral deposits by gouging out sufficient rich ore to pay for the cost of development, and this frequently affected the life of the mine. Both Australian and foreign investors have often been speculators, wanting quick capital gains rather than a steady income, so deposits were exploited ruthlessly rather than developed carefully with a view to prolonging their life. Even now, individual Australian investors are quick to join in speculative booms when there is a chance of making capital gains, but the money to finance long-term ventures is not always easily obtained locally.[58]

With the exception of coal in New South Wales and copper in South Australia, most early mining in Australia was of alluvial deposits of gold which required little capital and could be done by individuals. Because so many colonists demanded the right to share in the new wealth, colonial governments decreed at first that the size of each claim or lease should be small. Many miners also wished to remain self-employed and resisted the growth of mining companies. There developed, therefore, a tradition of small mining leases, small mines and prejudice against large, wealthy companies which ultimately affected mining enterprises other than gold. When underground mining of gold became necessary, many former alluvial miners sought to maintain the tradition of self-employment and the small mine by banding together in syndicates to finance the sinking of shafts for deep mining but, as mines went deeper and treatments plants were built, greater financial resources were needed. A few wealthy individuals, such as George Lansell and John Moffatt, risked their entire fortunes to finance the development of certain mines or smelters but, in most cases, companies needed to be formed to provide the necessary capital.[59]

Until about the 1880s, overseas investors were not greatly interested in Australian mining companies, and most shareholders or mine owners were Australian by adoption if not birth. Because they did not provide capital on the scale required for really efficient mining, mines and treatment plants tended to remain relatively small concerns, run on a limited budget.[60] Furthermore, since shareholders demanded quick returns, many mining companies paid out too high a proportion of their profits in dividends and spent too little on development work. These factors affected the way a mine was exploited, its life expectancy and also decisions regarding what deposits it might be possible to develop because, as a general rule, the ore gouged out of the mine initially had to be rich enough to finance developmental costs.[61] With a few exceptions, such as the gold mine at Clunes in Victoria and the tin mine at Mount Bischoff in Tasmania, most mines in Australia before the 1880s were under-capitalized, were not equipped with the latest machinery and did not employ the best mining and metallurgical talent to aid their development. For a number of years, even the legal situation in many colonies (regarding the size of areas able to be mined under the terms of a single lease) favoured the concept of the small mine, thus handicapping efficient mining on a large scale. In addition to the shortage of capital for developmental work, there was a dislike of large mining companies and a distrust of large capitalists, so that, while mining companies in the United States, for example, were forging ahead by the 1870s, mining and processing in Australia lagged behind.[62]

For various reasons, including an alteration in investment patterns in Britain and the development of fast telegraphic communications, British investors began to show more interest in the Australian mining industry in the 1880s and 1890s. There was also some investment by non-British concerns, mainly German, but the bulk of the capital was British. This influx of capital enabled mining and extraction to proceed on a more organized basis instead of by the rather desperate expedients so often adopted in earlier years, although the abundance of available capital also led to some fraud and waste.[63]

The availability of larger amounts of risk capital in Australia in the 1880s and part of the nineties encouraged the growth of large mining companies. The need for better mining methods and costly treatment plants to handle more difficult ores and the development of mineral deposits in more isolated areas also fostered the growth of large companies because only they could cope with the costs and problems involved. With a few exceptions, such as the large Moonta, Wallaroo, Clunes and Mount Bischoff mining companies, the period from the 1840s to the 1880s had been the era of the small mine and extractive plant, but the period from the eighties onwards became the era of the large mine and processing plant.[64] Of the thousands of companies floated before 1880, the majority were small concerns and only three paid more than a million pounds sterling in dividends. However, sixteen of the companies floated in the twenty years after 1880 paid more than a million pounds in dividends, including five companies at Broken Hill alone.[65] A number of these large, profitable companies had an important effect on Australian industry because they became involved in other industrial enterprises in Australia.[66]

The development of large companies and large treatment plants became even more marked after the second world war, because soaring costs meant that only large companies could afford to risk massive amounts of capital and to spend hundreds of millions of dollars on initial exploration and development before any profit could be made. In many ways, the trend towards very large companies has been beneficial because projects are researched thoroughly before mining begins and, in many cases, the mining companies also build the necessary treatment plants, towns and bulk handling facilities, as well as subsidizing railways by virtue of the freight costs paid. However, the cost of providing this infrastructure, especially in isolated and difficult areas, is high. No matter how much research is done prior to the decision to develop a mine or treatment plant, there is still an element of risk in the venture because the world demand for minerals can alter so quickly. Only large companies can weather these storms.[67]

As more large deposits of important minerals were found in Australia after the second world war, an increasing number of overseas concerns became interested in investing in prospecting and in developing deposits here, which has made it easier to find the huge sums of money required. Successful Australian firms have also remained involved but the increased foreign interest has reduced the dominance of British investors. Up to the 1880s, the mining industry was largely owned by Australians. From the 1880s until the first world war, the industry was mainly owned by British interests, with a few relatively large Australian concerns. After that war, Australian shareholders became rather more prominent again and some American interests also became involved, and since the second world war, capital has been provided by a mixture of Australian, American, British, Japanese, German, French, Swiss and Canadian interests.[68]

In the days of quick profits, especially from gold, Australian shareholders often did well as a result of mining booms. When British investors became involved in the mining boom here in the 1880s and 1890s, it appears that they lost much of their investment and that it was the Australian colonies which reaped much of the benefit.[69] However, although it is true that many overseas investors lost their capital or failed to receive a good return upon it, a few individuals and companies did make fortunes, and a percentage of the yield from Australian mines was siphoned off abroad. Nine of the main gold-mining companies in Western Australia, for example, were administered from Britain, with the result that very little of their profit was reinvested in Australia. William Knox D'Arcy, a major shareholder in the Mount Morgan mine, went to live in England and used his profits to finance oil-drilling in Persia, causing the Mount Morgan mine and treatment plant to suffer from lack of developmental capital.[70] Many other Australian mines, however, were financed by foreign capitalists who made no profits, or by Australian firms and syndicates which did make profits which they reinvested in Australia.[71]

As well as the overseas interests, some important Australian capitalists either financed mining ventures themselves or mobilized Australian and overseas capital to do so. The same names keep recurring in connection with new mining developments in Australia, and many of these Australian entrepreneurs — Sylvester Browne, Bowes Kelly, William Orr, William Jamieson, William Knox — had some connection with either the Victorian gold-rushes or the mines at Broken Hill. Other names which run through the earlier history of mining in Australia include William Baillieu, W.S. Robinson and William Corbould.[72] Since the second world war, a new generation of Australian mining magnates has emerged.

The mining industry in Australia has spawned a number of companies which have played an important role in the economic development of the country. BHP, after its entry into the iron and steel industry, emerged as a manufacturing giant, expanding into shipbuilding, cement, chemicals, alloys and other industrial products, and shaping much of Australia's industrial history.[73]

Blainey argues that the wealthy group of gold-mining companies at Kalgoorlie had less effect on the economic history of Australia than the mining companies operating at Broken Hill because they were British-based companies whose profits tended to be returned to Britain rather than reinvested here.[74] In general terms this is true, but there was one important case in which both a Kalgoorlie concern and British capital did

This photograph of the Mount Isa mine, taken about 1931, shows the supply shaft, compressor house and tool-sharpening shop. Although the mine was believed to be a rich deposit, the ore was difficult to treat and the enterprise did not prosper until the giant American mining company, ASARCO, poured millions of pounds into it. For many years, the mine swallowed so much capital for so little return that one of the American firm's executives lamented that his decision to recommend the purchase of shares in Mount Isa was the one great mistake he had made in thirty years with his company. (Copied from *The Queenslander*, 21 May 1931.)

help to create an important force in Australian economic life. Through the activities of the Western Australian gold-mining company, Lake View Consols, British capitalists became linked with a group of Australian mining speculators, based in Melbourne, and led by William Baillieu and others. This created a loose London-Melbourne alliance, the Collins House group of companies, which operated from the early years of this century, tended to dissolve by the late 1930s, but was revived, with some new partners, after the second world war. It was involved in the operations of the Broken Hill North and South mining companies, the Zinc Corporation, the Electrolytic Zinc Company and many other mining companies. The alliance later broadened its interests to include investments in various manufacturing industries, such as the production of chemicals, paint, alloys, light and heavy engineering, timber and paper mills, the search for oil and the formation of the Commonwealth Aircraft Corporation. In the 1920s and part of the 1930s, the Collins House group rivalled the power of BHP and may have played an even more important role in the Australian economy.[75]

With the development of BHP and the Collins House group of companies, Australia moved firmly into the era of the large mine, the large processing plant and the large industrial company, a move which was necessary if expensive new developments were to be financed. The Western Mining Corporation was another company which has had a considerable impact upon mining in Australia, although it was formed at a later date than the others mentioned here and was less diversified industrially. Since the second world war, other important mining concerns have developed here, although they have tended to be linked with multinational firms to ensure access to the huge amounts of capital required.[76]

OTHER FACTORS AFFECTING MINING DEVELOPMENT

The high cost of transport was a major problem during the early years of settlement. Owners of mines were dependent upon bullock and horse teams to transport ore from mines to the coast. This was a costly and difficult process and it was often uneconomic to mine deposits in distant or inaccessible areas.[77] The development of regular coastal shipping services and, even more important, the building of railways, solved many transport problems, lowered freight costs and, in most cases, greatly assisted the mining industry. However, even after the rapid extension of the railway network in the eastern colonies in the 1870s and 1880s, bullock teams, horses, donkeys and camels remained an important means of transport in certain areas, camels being particularly important in the arid regions of Western Australia. The rugged terrain in western Tasmania also continued to hamper development there.[78]

Another significant step in the improvement of communications was the construction of telegraph lines within Australia and the laying in 1872 of a submarine cable linking Australia with Europe. Telegraphic communication spread information about new mining discoveries and expedited sharedealings by local and overseas investors, although it also increased opportunities for fraud.[79] Better roads, the advent of air transport and, more recently, the development of bulk carriers and loading facilities, have improved communications further, with the result that mineral deposits in formerly inaccessible areas can be mined now, provided that world demand is sufficient to justify the cost.[80]

Legal considerations have affected the development of mining in Australia in a number of ways. The question of the ownership of the so-called royal metals – that is, gold and silver – may have retarded the search for them in Australia before the 1850s. At a later stage, during the alluvial gold-rushes, colonial administrators accepted the argument that everyone should have the opportunity to dig for gold, so each digger was allowed to peg only a small claim, often about eight feet square. The credo of the small mine, operated by small capitalist miners, took root and, for a time, handicapped deep mining for gold because of the continual need to sink new shafts instead of mining a large area properly by means of underground galleries and horizontal levels. It was some time before governments considered that the political climate was such that they could increase the size of leases, thus facilitating the development of larger underground gold-mines.[81] Base metal fields were in a different category because, in general, they were not capable of being mined by the alluvial method, so it was recognized that large amounts of capital were needed to develop them. In addition, most of the important base metal fields were developed either on freehold land, to which the owner held the mineral rights, or else were discovered at a later stage after the various colonial governments had increased the size of mineral leases with a view to assisting more efficient mining on a larger scale.[82]

After each of the Australian colonies had been granted self-government, the respective colonial legislatures developed a body of legislation governing such matters as the pegging of claims, the size of leases and the adjudication of disputes. When a federal government came into existence in 1901, it also developed some influence regarding mining, and the situation eventually became extremely complex because the mining industry was affected by both federal and state legislation. Mining companies often used to operate mainly in one colony or state but, particularly since the second world war, the trend has been for companies to diversify and become involved in projects throughout Australia, with the result that their legal advisers face complex problems because they have to consider a variety of state mining laws. The general tendency of the federal government to seek to extend its power, especially since the second world war, has increased problems of jurisdication between the federal and state governments over such issues as the seabed, offshore oil-drilling, refusal of export licences and foreign investment guidelines.[83]

Before 1901, the governments of the various Australian colonies regulated mining law and, after underground mining developed, they also began to exercise some supervision over safety conditions and hours of work.[84] In general terms, however, the mining industry was not greatly affected by government regulation until the entry of the federal government into this area in recent decades. Federal government regulation has had the effect, at times, of delaying or stifling mineral exploration and development, as when it prohibited the export of iron ore. This embargo was imposed in 1938 for strategic reasons but it was maintained long after the second world war, inhibiting the search for deposits of iron ore and causing Lang Hancock and others to keep their discoveries to themselves because such deposits could not be exploited until restrictions imposed by both the federal and the Western Australian governments were repealed.[85] In recent years in particular, the federal government has made definite efforts to assert its authority in relation to mining activities by means of its powers concerning taxation, foreign investment and export licences. In the case of sand mining on Fraser Island, the federal government's refusal to grant the necessary export licences halted mining there completely, while the future of uranium mining will also depend to a considerable extent upon the attitude of the federal government.[86]

Thus, the development of mining in Australia has been shaped by a variety of forces, the most important of which have been analyzed briefly here. After a period of some depression from about 1910 onwards, and of quite severe gloom from the 1920s to the 1950s, the mining industry once again enjoyed a period of prosperity and expansion and the future looks promising. However, the industry is always vulnerable because its prosperity depends upon a fluctuating world demand. The stage has now been reached where more could be done to process and treat minerals here before they are exported, and this could well be the next phase of rapid development in the industry.

Notes

1. See, for example, H.F. King, "History of development of resources of metallic ores", in *Economic Geology of Australia and Papua New Guinea: Metals* (Monograph Series No. 5), ed. C.L. Knight (Melbourne: Australasian Institute of Mining and Metallurgy, 1975), pp. 5-12. For another type of division into phases, see J. Ward and I.R. McLeod, *Mineral Resources of Australia, 1980* (Canberra: AGPS, 1980), pp. 4-11.
2. Geoffrey Blainey, *The Tyranny of Distance: How Distance Shaped Australia's History* (Melbourne: Sun Books, 1966), pp. 2-7, 131-41; Geoffrey Blainey, *The Peaks of Lyell,* 3rd edn. (Melbourne: MUP, 1967), pp. 4-10, 37, 43; Geoffrey Serle, *The Golden Age: A History of the Colony of Victoria* (Melbourne: MUP, 1963), pp. 10, 216-17; Geoffrey Blainey, "Gold and Governors", *Historical Studies* 9 (1961): 339-43; Geoffrey Blainey, *The Rise of Broken Hill,* (Melbourne: Macmillan, 1968), pp. 11-12, 16-19; Geoffrey Blainey, *The Rush that Never Ended: A History of Australian Mining,* 3rd edn. (Melbourne: MUP, 1978), pp. 5-12, 24-25, 34, 65-66, 106-9, 116-18, 128-31, 135, 142-47, 208.
3. Blainey, *Rush,* pp. 44, 138, 147, 162-74, 177-85 208-11, 215-21, 253, 320-22, 324-26, 336-37; Blainey, *Lyell,* pp. 5-37, 57-58; Blainey, *Broken Hill,* pp. 16-19; Blainey, *Mines in the Spinifex: The Story of Mount Isa Mines,* rev. edn. (Sydney: Angus & Robertson, 1970), pp. 11-22, 32-33, 61-72; Serle, *Golden Age,* p. 228; Geoffrey Blainey, "Herbert Hoover's Forgotten Years", *Business Archives and History* 3 (1963): 56; Ian Hore-Lacy, ed., *Broken Hill to Mount Isa: The Mining Odyssey of W.H. Corbould* (Melbourne: Hyland House, 1981), pp. 6-9, 23-26; Robert Duffield, *Rogue Bull: The Story of Lang Hancock, King of the Pilbara* (Sydney: Collins, 1979), pp. 67-70, 88-90, 95-96.
4. King, "Development", in Knight, *Economic Geology,* pp. 5-12; Blainey, *Rush, passim,* esp. pp. 8-9, 28-45, 90-93, 106-9, 116-18, 128-31, 134-47, 161-85, 208-21, 324-25; M.H. Ellis, *A Saga of Coal: The Newcastle Wallsend Coal Company's Centenary Volume* (Sydney: Angus & Robertson, 1969), pp. 3-7; H.G. Raggatt, *Mountains of Ore* (Melbourne: Lansdowne Press, 1968), pp. 319-22; Blainey, *Lyell,* pp. 4-58; Blainey, *Broken Hill,* pp. 11-12, 16-19; F.K. Crowley, *Australia's Western*

Third: A History of Western Australia from the First Settlements to Modern Times (1960; rpt. Melbourne: Heinemann, 1970), pp. 82-90; Blainey, *Mines in the Spinifex,* pp. 11-34, 58-74.

5. Blainey, *Rush,* pp. 90-93, 134-44, 161-85, 208-21, 320-21, 324-25; Blainey, *Lyell,* pp. 5-58; Crowley, *Western Third,* pp. 82-90; K.J. Finucane, "Geological Methods", in *Mining in Western Australia,* ed. Rex T. Prider (Nedlands: Univ. of WA Press, 1979), p. 179.
6. Blainey, *Broken Hill,* pp. 16-19; Blainey, *Lyell,* pp. 4-11, 24-58; Blainey, *Mines in the Spinifex,* pp. 65-79; Blainey, *Rush,* pp. 142-47, 180-81, 212, 215-17, 219-21, 232-44, 324-25.
7. Blainey, *Broken Hill,* pp. 12, 16-24; Blainey, *Mines in the Spinifex,* pp. 1-79; Neill Phillipson, *Man of Iron* (Melbourne: Wren, 1974), pp. 8, 17-19; Duffield, *Rogue Bull,* pp. 88-90; Blainey, *Rush,* pp. 90-93, 134-47, 161-85, 320-22, 324-25; Crowley, *Western Third,* pp. 56-57, 63-64, 67-68, 75, 82-90.
8. King, "Development", in Knight, *Economic Geology,* pp. 5-8; Mark Bersten, "Mining since 1926: Change and Adaptation", *National Bank Monthly Summary* (Dec. 1976): 17-18; W.S. Robinson, *If I Remember Rightly: the Memoirs of W.S. Robinson, 1876-1963,* ed. Geoffrey Blainey (Melbourne: Cheshire, 1969), pp. 61-170, *passim;* Finucane, "Geological Methods", in Prider, *Mining in WA,* pp. 180-84; D.L. Rowston, "Geophysical Methods", in *ibid.*, pp. 186-87; Blainey, *Lyell,* pp. 175-79, 184-94, 272-74, 281-84; Blainey, *Broken Hill,* pp. 156-59; Blainey, *Rush,* pp. 275-82, 290-92, 310, 316, 342-43.
9. King, "Development", in Knight, *Economic Geology,* pp. 7-13; Raggatt, *Mountains;* Blainey, *Rush,* pp. 335-63; Ward and McLeod, *Mineral Resources,* pp. 69-72; Blainey, *Lyell,* pp. 299-305; Blainey, *Mines in the Spinifex,* pp. 201-16; Finucane, "Geological Methods", in Prider, *Mining in WA,* pp. 180, 183-85; Rowston, "Geophysical Methods", in ibid., pp. 187-90; R.H. Mazzuchelli, "Geochemical Methods", in ibid., pp. 191-95; Sir Roderick Carnegie, "Minerals Growth to 2001", Keynote Address, Annual Conference of the Australasian Institute of Mining and Metallurgy, 20 July 1981, rpt. in *Institute of Mining and Metallurgy Bulletin,* No. 455 (1981): 21-23.
10. Raggatt, *Mountains,* pp. 112-15, 142-43, 360-61, 363-64; Phillipson, *Man of Iron,* pp. 70-82, 86; Duffield, *Rogue Bull,* pp. 67-78; Donald W. Barnett, *Minerals and Energy in Australia* (Stanmore: Cassell, 1979), pp. 188-89, 200-1.
11. Raggatt, *Mountains,* pp. 112-15, 142-44, 360-61, 363-64; Phillipson, *Man of Iron,* pp. 8, 17-20, 33-94; Duffield, *Rogue Bull,* pp. 67-87.
12. Ellis, *Saga;* A.G.L. Shaw and G.R. Bruns, *The Australian Coal Industry* (Melbourne: MUP, 1947), pp. 20-21, 56-61, 85-93; Raggatt, *Mountains,* pp. 319-31; Robin Gollan, *The Coalminers of New South Wales: A History of the Union, 1860-1960* (Melbourne: MUP, 1963), pp. 3-14; M.T. Daly, "The Development of the Urban Pattern of Newcastle", in *Urbanization in Australia: The Nineteenth Century,* ed. C.B. Schedvin and J.W. McCarty (1970; rpt. Sydney: Sydney University Press, 1974), pp. 94-103; Edgar Ross, *A History of the Miners' Federation of Australia* (Sydney: The Australasian Coal and Shale Employees' Federation, 1970), pp. 6-9; R.L. Whitmore, *Coal in Queensland, the First Fifty Years: A History of Early Coal Mining in Queensland* (St Lucia: University of Queensland Press, 1981); Blainey, *Tyranny,* pp. 205-22, 228-65, 283-84.
13. Blainey, *Rush,* pp. 106-7, 134-58, 214-15, 217-19, 325-34; Raggatt, *Mountains,* pp. 151-58, 178-87; Barnett, *Minerals and Energy,* pp. 267-68, 271-78; Blainey, *Broken Hill;* Blainey, *Lyell,* pp. 50-56, 65-67; Blainey, *Mines in the Spinifex.*
14. Blainey, *Rush,* pp. 8-9, 105, 108-30, 134-35, 219-47, 283-86, 324-34; Raggatt, *Mountains,* pp. 191-209; Barnett, *Minerals and Energy,* pp. 165-79; Rob Charlton, *The History of Kapunda* (Melbourne: Hawthorn Press, 1971), pp. 8-45; J.B. Austin, *The Mines of South Australia, including also an Account of the Smelting Works in that Colony* (Adelaide: C. Platts et al., 1863); Blainey, *Lyell;* Blainey, *Mines in the Spinifex;* Hore-Lacy, *Broken Hill to Mt Isa.*

15. Blainey, "Gold and Governors", pp. 337-49, esp. p. 346; Blainey, *Rush*, pp. 6, 8, 21. For further information regarding the legal position, see Andrew G. Lang and Michael Crommelin, *Australian Mining and Petroleum Laws: An Introduction* (Chatswood: Butterworths, 1979), pp. 1-2, 11-13.
16. Blainey, *Rush*, pp. 5-29, 36-37; Blainey, "Gold and Governors", pp. 337-49; Serle, *Golden Age*, pp. 10-11; Jay Monaghan, *Australians and the Gold Rush: Californians and Down Under, 1849-1854* (Berkeley: University of California Press, 1966), pp. 165-74.
17. Serle, *Golden Age;* Blainey, *Lyell,* pp. 1-58; Crowley, *Western Third,* esp. pp. 82-90; Hector Holthouse, *Gympie Gold* (Sydney: Angus & Robertson, 1973); G.C. Bolton, *A Thousand Miles Away: A History of North Queensland to 1920* (Canberra: ANU Press, 1963), Blainey, *Rush,* pp. 12-102, 161-220, 232-47, 290-93.
18. Blainey, *Rush,* pp. 130-33, 208-10, 212-14, 290; Blainey, *Lyell,* pp. 12-19, 23, 42-43, 178, 300-5; Raggatt, *Mountains,* pp. 378-81; D.B. Barton, *A History of Tin Mining and Smelting in Cornwall* (Truro, U.K.: D. Bradford Barton, 1967), pp. 154-55.
19. Blainey, *Broken Hill,* pp. 51-55, 65-81, 84, 134-35, 156-59; Blainey, *Lyell,* pp. 240-56, 304; Robinson, *Memoirs;* Blainey, *Rush,* pp. 259-71, 276-82; Barnett, *Minerals and Energy,* pp. 280-82.
20. N.R. Wills, "The Basic Iron and Steel Industry", in *The Economics of Australian Industry: Studies in Environment and Structure,* ed. Alex Hunter (Melbourne: MUP, 1963), pp. 215-46; Neville R. Wills, *Economic Development of the Australian Iron and Steel Industry: An Examination of the Establishment of the Industry, its Development, Present Distribution, Resources and Importance in the Australian Economy* (n.p., June 1948); Helen Hughes, *The Australian Iron and Steel Industry, 1848-1962* (Melbourne: MUP, 1964); Alan Trengove, *'What's Good for Australia . . .!': The Story of BHP* (Stanmore: Cassell, 1975); Paquita Mawson, *A Vision of Steel: The Life of G.D. Delprat, C.B.E., General Manager of B.H.P., 1898-1921* (Melbourne: Cheshire, 1958); Geoffrey Blainey, *The Steel Master: A Life of Essington Lewis* (Melbourne: Macmillan, 1971); Raggatt, *Mountains,* pp. 101-5; Blainey, *Broken Hill,* pp. 65, 77-78, 81-85.
21. Barnett, *Minerals and Energy,* pp. 33-73; Raggatt, *Mountains,* pp. 216-314; C.E.B. Conybeare, "Petroleum, Coal and Uranium: A Case Study in Production, Technology and Consumption", in *The Natural Resources of Australia: Prospects and Problems for Development,* ed. J.A. Sinden (Sydney: Angus & Robertson, 1972), pp. 141-52.
22. For further information regarding more recent developments in relation to the mining of coal, iron ore, uranium, bauxite, tungsten, manganese, nickel, titanium and other minerals derived from mineral sands, see Barnett, *Minerals and Energy,* pp. 75-130, 181-213, 217-65; Raggatt, *Mountains,* pp. 77-150, 317-77; Blainey, *Rush,* pp. 291, 335-64.
23. Robinson, *Memoirs,* p. 47.
24. Raggatt, *Mountains,* pp. 142-49.
25. Ibid., *passim,* esp. pp. 8-33; J.A. Dunn, "Minerals in the Development of Australia", *Economic Papers* 12 (1958): 7-18; Barnett, *Minerals and Energy;* Ward and McLeod, *Mineral Resources,* pp. 11-69; Richard Hattersley and John Alexander, *Australian Mining: Minerals and Oil* (Sydney: David Ell Press, 1980); J.T. Woodcock, ed., *Mining and Metallurgical Practices in Australasia: the Sir Maurice Mawby Memorial Volume,* Monograph Series No. 10 (Melbourne: Australasian Institute of Mining and Metallurgy, 1980); Carnegie, "Minerals Growth to 2001", pp. 21-23.
26. Blainey, *Lyell;* Blainey, *Mines in the Spinifex;* Hore-Lacy, *Broken Hill to Mt Isa,* pp. 184-205.
27. Serle, *Golden Age,* pp. 41 n., 217-29; Ellis, *Saga,* pp. 32, 39, 41; Blainey, *Rush,* pp. 13-51, 57-83, 105-58, 161-202, 208-47; Blainey, *Broken Hill;* Blainey, *Lyell.*
28. For examples of the role of some non-Cornish mining men in the early period of mining in Australia, see Ellis, *Saga,* pp. 39, 41; Blainey, *Rush,* pp. 48, 64-65, 71, 73,

106, 109, 110, 209, 235. For the role of Cornish miners and mining methods, see Oswald Pryor, *Australia's Little Cornwall* (Adelaide: Rigby, 1962); Charlton, *Kapunda,* pp. 8-41, *passim;* Old Colonist, *Colonists, Copper and Corn in the Colony of South Australia, 1850-51,* ed. E.M. Yelland (Melbourne: Hawthorn Press, 1970), pp. 118, 130, 134-35, 137-40, 154-55; Blainey, *Rush,* pp. 7, 9, 48, 77, 79,105-30, 295, 296, 306-7; Blainey, *Mines in the Spinifex,* p. 15. For recruitment of overseas experts, see, for example, Blainey, *Rush,* pp. 78, 154-55, 252.

29. Blainey, *Rush,* pp. 46-48, 64-65, 71, 77-78, 109-10, 121-24, 296-302; Serle, *Golden Age,* pp. 80-81; Blainey, *Broken Hill,* p. 55.
30. See, for example, Blainey, *Rush,* pp. 71, 77-79, 124, 155, 201-2, 209, 221, 236-47 251-52, 273-74, 294-302; Blainey, *Broken Hill,* pp. 55-56; Hore-Lacy, *Broken Hill to Mt Isa,* pp. 126-28.
31. See Blainey, *Rush,* pp. 48, 49, 51, 61, 64-66, 71-73, 76-79, 93-94, 109-11, 113, 118-19, 121-27, 131-32, 176, 209-10, 228-29, 230, 296-98; Serle, *Golden Age,* pp. 218-29; Ellis, *Saga,* pp. 3-134, *passim;* Austin,*Mines of SA;* Pryor, *Little Cornwall,* pp. 34-35, 40-42, 44-46, 53, 72-78; Blainey, *Lyell,* pp. 51, 151, 169, 194; Blainey, *Mines in the Spinifex,* pp. 186-88, 191.
32. Serle, *Golden Age,* pp. 216-29; Ellis, *Saga,* pp. 32, 39, 41, 70-71, 75; Blainey, *Rush,* pp. 9, 46-51, 57-61, 64-73, 76-77, 105-13, 117-29, 294-304.
33. Blainey, *Rush,* pp. 78-79, 154-55, 251-55; H.W. Worner and W.E. Vance, "Mining and Metallurgical Education in Australasia", in Woodcock, *Mining and Metallurgical Practices,* p. 927.
34. Blainey, *Broken Hill,* pp. 18-20, 25, 33-46, 55-56, 62-65, 68-71, 76-79, 81-85; Wills, "The Basic Iron and Steel Industry", in Hunter, *Economics of Australian Industry,* pp. 220-46; Wills, *Economic Development of the Australian Iron and Steel Industry,* pp. 86 ff., *passim;* Hughes, *Australian Iron and Steel Industry,* pp. 55 ff., *passim;* Blainey, *Rush,* pp. 143-57, 220, 252, 259-62, 272-75, 344-45; Blainey, *Lyell,* p. 138.
35. Blainey, *Broken Hill,* pp. 25, 44-45, 57-59, 68, 69, 76; Blainey, *Lyell,* pp. 58-61, 65, 74-76, 244-49; Blainey, *Rush,* pp. 106, 154-57, 200-2, 209, 219-22, 235, 252-55, 260; Hore-Lacy, *Broken Hill to Mt Isa,* pp. 62-79.
36. Blainey, *Rush,* pp. 155, 201-2, 248-56; Blainey, *Broken Hill,* pp. 55-59, 86-104; Blainey, *Lyell,* pp. 73-74, 84, 100, 134-35, 256, 265-67; Blainey, "Hoover", p. 57; Blainey, *Steel Master, passim,* esp. pp. 16-20, 55-56; Roy Bridges, *From Silver to Steel: The Romance of the Broken Hill Proprietary* (Melbourne: George Robertson, 1920), pp. 174-78; Pryor, *Little Cornwall,* pp. 44-45, 144-45; Hore-Lacy, *Broken Hill to Mt Isa;* Worner and Vance, "Mining and Metallurgical Education", in Woodcock, *Mining and Metallurgical Practices,* pp. 927-28.
37. F.R.E. Mauldon, "The decline of mining", *Annals of the American Academy of Political and Social Science* 158 (1931): 66-76; Bersten, "Mining since 1926", pp. 17-18; Ward and McLeod, *Mineral Resources,* pp. 6-8; Shaw and Bruns, *Australian Coal,* pp. 8, 20-21, 40-46, 52-53, 56-93, 136-69, 171; Ellis, *Saga,* pp. 183-237; Blainey, *Broken Hill,* pp. 85 ff., *passim,* esp. pp. 153-59; Blainey, *Lyell, passim,* esp. pp. 240-74, 283-90, 292-93, 299-300, 301-4; Blainey, *Mines in the Spinifex,* pp. 45-57, 65-191; Hore-Lacy, *Broken Hill to Mt Isa,* pp. 141-205; Robinson, *Memoirs;* Blainey, *Rush,* pp. 207, 283-93, 304-17, 324-32.
38. Barnett, *Minerals and Energy;* Woodcock, *Mining and Metallurgical Practices;* Raggatt, *Mountains;* King, "Development", in Knight, *Economic Geology,* pp. 5-12; Bersten, "Mining since 1926", pp. 18-21; Ward and McLeod, *Mineral Resources,* pp. 8-11; Blainey, *Broken Hill,* pp. 153-60; Blainey, *Lyell,* pp. 297-305; Blainey, *Mines in the Spinifex,* pp. 182-200; Blainey, *Rush,* pp. 332-64.
39. Blainey, *Rush,* pp. 71, 106-11, 117-18, 128-29, 131-32, 136, 138-40, 209-35; Pryor, *Little Cornwall,* pp. 84-90, 96-97; Blainey, *Broken Hill,* pp. 24-25; Blainey, *Lyell,* pp. 13, 44-48, 61, 67, 74-76; Barton, *Tin Mining,* p. 155, n. 1.
40. Blainey, *Rush,* pp. 71, 78-79, 108-11, 114, 118, 121, 138-40, 143, 147, 251-53; Hore-Lacy, *Broken Hill to Mt Isa,* pp. 6-32.

41. Blainey, *Rush,* pp. 9, 71, 78-79, 106-7, 110-11, 118, 125-26, 128-29, 131-33, 134-35, 138-40, 143, 209, 235, 251-52; Blainey, *Lyell,* p. 13; Wills, "The Basic Iron and Steel Industry", in Hunter, *The Economics of Australian Industry,* pp. 217-18; Wills, *Economic Development of the Australian Iron and Steel Industry,* pp. 16-33; Hughes, *Australian Iron and Steel Industry,* pp. 1-54.
42. Blainey, *Rush,* pp. 71, 106-8, 110-11, 118, 128-29, 139-40, 148-49, 157, 217, 219, 222, 235, 238-39; Blainey, *Broken Hill,* pp. 25-27; Blainey, *Lyell,* pp. 231-32, 248, n.; Hughes, *Iron and Steel,* pp. 2-9, 15-18.
43. Blainey, *Rush,* pp. 118, 125-26, 128-32, 139-40, 148-49, 157, 209, 219, 255, 262, 272, 274-75, 278-81; Blainey, *Broken Hill,* pp. 25-27, 58, 62-69, 77-81; Hughes, *Iron and Steel,* pp. 2-79, *passim;* Blainey, *Lyell,* pp. 243-46.
44. Hore-Lacy, *Broken Hill to Mt Isa,* pp. 118 ff. *passim;* Blainey, *Lyell,* pp. 42 ff. *passim;* Blainey, *Mines in the Spinifex,* pp. 18-57, 80 ff. *passim;* Blainey, *Rush,* pp. 199-201, 217-30, 235-47, 285-86, 324-34.
45. Blainey, *Rush,* pp. 131-32, 148-49, 157, 199-201, 222, 227-30, 250-56, 259-81; Blainey, *Broken Hill,* pp. 25, 56, 58, 62-81; Blainey, *Lyell,* pp. 74-76, 134-35, 137-61.
46. Blainey, *Rush,* pp. 131, 148, 157, 255.
47. Blainey, *Rush,* pp. 78-79, 110, 140, 235-46, 251-52; Blainey, *Lyell,* p. 276.
48. Blainey, *Rush,* pp. 78-79, 139-40, 251-55; Worner and Vance, "Mining and Metallurgical Education", in Woodcock, *Mining and Metallurgical Practices,* pp. 927-28.
49. Blainey, *Rush,* pp. 154-55, 157, 220-21, 252; Blainey, *Lyell,* pp. 57-62.
50. Blainey, *Broken Hill,* pp. 51-59, 62-81; Blainey, *Lyell,* pp. 74-76, 134-38, 145, 152, 161, 262-64, 276; Blainey, *Rush,* pp. 157, 199-201, 222, 228-30, 238-40, 243-45, 250-56, 259-71; Blainey, "Hoover", p. 58; Blainey, *Steel Master,* p. 56.
51. Blainey, *Broken Hill,* pp. 57-59, 65-77; Blainey, *Rush,* pp. 199-201, 259-81; Robinson, *Memoirs,* pp. 35-44.
52. Blainey, *Broken Hill,* pp. 65-84; Blainey, *Lyell,* pp. 240-69; Blainey, *Rush,* pp. 259-81, 285-86; Blainey, "Hoover", pp. 63-65; Robinson, *Memoirs,* pp. 35-44, 77-90, 93-102, 114-15, 119-25, 206.
53. Blainey, *Broken Hill,* pp. 77-85; Wills, "Basic Iron and Steel Industry", in Hunter, *Economics of Australian Industry,* pp. 215-46; Wills, *Economic Development of Australian Iron and Steel,* pp. 86 ff. *passim;* Hughes, *Iron and Steel,* pp. 55 ff. *passim;* Blainey, *Steel Master,* pp. 41 ff. *passim;* Trengove, *What's Good for Australia,* pp. 86 ff. *passim;* Blainey, *Rush,* pp. 272-75.
54. Blainey, *Broken Hill,* pp. 62-85, 132-59; Blainey, *Lyell,* pp. 74 ff. *passim,* esp. pp. 74-76, 134-61, 179-83, 186-87, 240-304; Blainey, *Mines in the Spinifex,* pp. 45-57, 65-148, 160-200; Hore-Lacy, *Broken Hill to Mt Isa,* pp. 184-205; Wills "Basic Iron and Steel Industry", in Hunter, *Economics of Australian Industry,* pp. 215-44; Wills, *Economic Development of Australian Iron and Steel,* pp. 79-158; Hughes, *Australian Iron and Steel Industry,* pp. 55-150; Blainey, *Steel Master,* pp. 41 ff. *passim;* Robinson, *Memoirs;* Blainey, *Rush,* pp. 199-202, 206-7, 259-93, 309-17, 324-32.
55. Barnett, *Minerals and Energy;* Raggatt, *Mountains;* Ward and McLeod, *Mineral Resources of Australia,* pp. 8 ff. *passim;* Hughes, *Australian Iron and Steel Industry,* pp. 151-92; Blainey, *Mines in the Spinifex,* pp. 192-225; Blainey, Lyell, pp. 297-320; Blainey, *Rush,* pp. 335-59.
56. For some account of the work of able mining engineers and metallurgists such as Patton, Schlapp, Sticht, Diehl, Hoover, Delprat, Baker and Hillman Stevens, see Blainey, *Broken Hill,* pp. 25, 45, 57-58, 68-78, 84, 102-3; Blainey, *Lyell,* pp. 74-76, 134-38, 152, 179-83, 240-64, 276; Blainey, *Mines in the Spinifex,* pp. 144-46, 160-62, 167-68, 179-80, 199-200; Mawson, *Vision of Steel;* Wills, *Economic Development of Australian Iron and Steel,* pp. 91-102; Hughes, *Australian Iron and Steel Industry,* pp. 57, 59, 63-70, 75; Blainey, "Hoover", pp. 53-70; Blainey, *Rush,* pp. 79, 154-55, 157, 200-2, 209, 220-22, 229, 261-62, 273-75, 280, 285-86. For some account of imported mining men who were either inefficient or misguided, see Blainey, *Lyell,* pp. 132-33, 145-52; Blainey, *Rush,* pp. 227-28; Hore-Lacy, *Broken Hill to Mt Isa,* p. 18.

57. Blainey, *Rush,* pp. 200, 251-55, 344; Worner and Vance, "Mining and Metallurgical Education", in Woodcock, *Mining and Metallurgical Practices,* pp. 927-32. For some account of the work of Australian mining engineers and metallurgists, see Hore-Lacy, *Broken Hill to Mt Isa;* Blainey, *Broken Hill,* pp. 51 ff. *passim;* Blainey, *Steel Master, passim,* esp. p. 56; Blainey, *Lyell;* Hughes, *Iron and Steel,* pp. 55 ff. *passim;* Blainey, *Mines in the Spinifex,* p. 200.
58. See Blainey, *Broken Hill,* pp. 16-20, 24-25; Blainey, *Lyell,* pp. 44-45, 57-68, 72, 302-4, 313-20; Blainey, *Rush,* pp. 66, 73-77, 82, 97-102, 108, 138, 140, 143-52, 155, 201-3, 221, 233-47, 284, 360-63; Raggatt, *Mountains,* pp. 51-65; Barnett, *Minerals and Energy,* pp. 299-317.
59. Serle, *Golden Age,* pp. 20, 24, 69, 73-74, 179, 217-29; Crowley, *Western Third,* pp. 82-90, 119-34; Blainey, *Rush,* pp. 17-86, 108-13, 116-26, 130-33, 161-201, 208-10, 215-16, 252, 294-308; Blainey, *Lyell,* pp. 12-41.
60. Blainey, *Rush,* pp. 66, 72-83, 91-94, 97-100, 139-40, 210, 252, 255-56; Serle, *Golden Age,* pp. 220-28; Blainey, *Lyell,* pp. 42-43.
61. Blainey *Broken Hill,* pp. 16-20, 24-25, 33, 55-57, 78; Blainey, *Lyell, passim,* esp. pp. 42-49, 57-78, 168, 301-4; Blainey, *Rush,* pp. 66-83, 108-12, 128-31, 142-55, 199-203, 208-31, 233-47, 274, 284, 357.
62. Blainey, *Rush,* pp. 32, 42-43, 46-51, 57-58, 60, 64-80, 140, 154-55, 251-56; Serle, *Golden Age,* pp. 20, 217-29; Lang and Crommelin, *Mining and Petroleum Laws,* pp. 1-17. For a brief note on the highly capitalized and efficient mines at Clunes and Mount Bischoff, see Blainey, *Rush,* pp. 66-71, 208-9.
64. Blainey, *Rush,* pp. 100-102, 147-58, 199-202, 219-30, 234-56, 259-82, 294-95; Blainey, 250-56, 259-82, 316; Blainey, *passim,* esp. pp. 56-91, 102-72, 187, 240-56; Blainey, *Broken Hill,* pp. 30-31, 46, 49-85.
64. Blainey, *Rush,* pp. 100-2, 147-58, 199-202, 219-30, 234-56, 259-82, 294-95; Blainey, *Broken Hill;* Blainey, *Lyell;* Blainey, *Mines in the Spinifex;* Hore-Lacy, *Broken Hill to Mt Isa.*
65. *Blainey, Rush,* pp. 255-56; Blainey, *Broken Hill,* p. 33; Blainey, *Lyell,* p. 134.
66. Blainey, *Broken Hill,* pp. 33, 62-85, 153-60; Wills, "Basic Iron and Steel Industry", in Hunter, *Economics of Australian Industry,* pp. 215-44; Hughes, *Iron and Steel,* pp. 55 ff. *passim;* Robinson, *Memoirs,* pp. 21 ff. *passim;* Blainey, *Lyell,* pp. 134, 179-83, 313-14; Blainey, *Rush,* pp. 255-56, 272-82.
67. Barnett, *Minerals and Energy, passim,* esp. pp. 1-3, 228, 293-95, 299, 308-10; Raggatt, *Mountains;* Blainey, *Rush,* pp. 335-59.
68. See the references given in notes 60, 63; also Blainey, *Rush,* pp. 272-82, 290, 309-17, 324-59; Blainey, *Lyell,* p. 308; Raggatt, *Mountains;* Barnett, *Minerals and Energy, passim,* esp. pp. 291-93, 299.
69. Blainey, *Rush,* pp. 73-83, 97-102, 143-58, 185-92, 199-207, 217-30, 238-47, 250-56, 259-71, 290; Blainey, *Lyell,* pp. 56 ff. *passim,* esp. pp. 162-72; Blainey, *Broken Hill,* pp. 65-77.
70. Blainey, *Rush,* pp. 236-47, 256.
71. Blainey, *Lyell, passim,* esp. pp. 42-194, 240-56; Blainey, *Broken Hill, passim,* esp. pp. 16-85; Blainey, *Rush,* pp. 79-80, 116-19, 124-25, 131-33, 143-58, 175-76, 181-82, 219-30, 259-82.
72. Blainey, *Broken Hill,* pp. 19, 42-46, 71, 73-74, 82, 83, 136, 149, 156, 158; Blainey, *Lyell,* pp. 53-81, 139-40, 144, 153-63, 176-77, 192, 240-56, 268-69, 275-78; Blainey, *Mines in the Spinifex,* pp. 29-34, 45-57, 64, 80-85, 95-100, 105-19; Hore-Lacy, *Broken Hill to Mt Isa;* Robinson, *Memoirs;* Blainey, *Rush,* pp. 143-58, 175-76, 217-30, 259-82, 306, 316, 325-29.
73. Wills, "Basic Iron and Steel Industry", in Hunter, *Economics of Australian Industry,* pp. 215-46; Hughes, *Iron and Steel,* pp. 55 ff. *passim;* Blainey, *Steel Master;* Trengove, *What's Good for Australia.*
74. Blainey, *Broken Hill,* pp. 46-48, 81-85.
75. Blainey, *Broken Hill,* pp. 72-74, 78-82, 156-60; Robinson, *Memoirs;* Blainey,

"Hoover", pp. 63-66; Blainey, *Lyell,* pp. 254-56; Blainey, *Rush,* pp. 205, 263-68, 275-82.

76. Robinson, *Memoirs, passim,* esp. pp. 151-52; Raggatt, *Mountains, passim;* Barnett, *Mineral and Energy;* Blainey, *Rush,* pp. 281-82, 316, 336-59; Blainey, *Lyell,* pp. 303-4, 318-19.
77. See, for example, Blainey, *Rush,* pp. 110, 114, 128-29, 134-35, 209; Blainey, *Lyell,* p. 13.
78. T.A. Coghlan, *Labour and Industry in Australia: from the First Settlement in 1788 to the Establishment of the Commonwealth in 1901* (1918; rpt. Melbourne: Macmillan, 1969), vol. II, pp. 830-45; Vol. III, pp. 1218-19, 1221-27; Daly, "Newcastle", in Schedvin and McCarty, *Urbanization in Australia,* p. 94; Blainey, *Broken Hill,* pp. 20-27; Blainey, *Lyell,* pp. 18-22, 42, 45-46, 51, 61, 69-78, 92-93, 112-35, 241, 243; Hore-Lacy, *Broken Hill to Mt Isa;* Blainey, *Mines in the Spinifex,* pp. 15-16, 19, 22-27, 30, 34, 37-51, 71-72, 107-13, 176-79; Blainey, *Rush,* pp. 84, 86, 90-96, 118, 134-35, 141, 148-49, 162-94, 213-14, 218, 221-22. For an interesting pictorial record of some of the forms of transport used on gold-fields throughout Australia, see Derrick I. Stone and Sue Mackinnon, *Life on the Australian Goldfields* (Sydney: Methuen, 1976).
79. Blainey, *Tyranny,* pp. 222-27; Brian Fitzpatrick, *The British Empire in Australia: An Economic History, 1834-1939,* 2nd ed. (1949; rpt. Melbourne: Macmillan, 1969), pp. 121, 165-67; Coghlan, *Labour and Industry,* Vol. II, pp. 905-6; Vol. III, pp. 1219-21; Blainey, *Rush,* pp. 90-93, 100, 150, 163, 186-91, 202-6, 222, 238.
80. Blainey, "The Cargo Cult in Mineral Policy", *Economic Record* (Dec. 1968): 474-75; D.J. Hibberd, "Practical Problems in the Development of the Australian Mineral Industry", *Economic Papers* 12 (1958): 23-24; Barnett, *Minerals and Energy,* p. 186; Blainey, *Rush,* pp. 335-59; Blainey, *Mines in the Spinifex,* pp. 153, 182-225.
81. Lang and Crommelin, *Mining and Petroleum Laws,* pp. 1-17; Blainey, "Gold and Governors", pp. 337-40; Serle, *Golden Age,* pp. 20, 73, 219-29; Crowley, *Western Third,* pp. 85, 88, 132-33; Gavin Casey and Ted Mayman, *The Mile that Midas Touched* (Adelaide: Rigby, 1964), pp. 97-107; Blainey, *Rush,* pp. 20-25, 32, 42-43, 46-51, 57-58, 64-80, 195-99, 252, 294-308.
82. Lang and Crommelin, *Mining and Petroleum Laws,* pp. 1-17; Andrew G. Lang, *Manual of the Law and Practice of Mining and Exploration in Australia* (Sydney: Butterworths, 1971), pp. 8-10; Blainey, *Broken Hill,* p. 18; Blainey, *Lyell,* pp. 81, 89; Blainey, *Rush,* pp. 108-9, 112, 116-18, 143-45. For the different situation regarding the ownership of gold, whether located on freehold property or not, see Lang and Crommelin, *Mining and Petroleum Laws,* pp. 12-17; Serle, *Golden Age,* pp. 226-27; Blainey, *Rush,* pp. 66-71. For the Queensland Labor government's introduction of small leases on the base metal field at Mount Isa as late as 1923, see Blainey, *Mines in the Spinifex,* pp. 75-79, 83.
83. See, for example, Lang, *Manual, passim,* esp. pp. 5-6; Lang and Crommelin, *Mining and Petroleum Laws, passim,* esp. pp. 1-54; Serle, *Golden Age,* pp. 219-27; Crowley, *Western Third,* pp. 132-33; Garth Stevenson, *Mineral Resources and Australian Federalism,* Research Monograph No. 17 (Canberra: ANU, 1976), pp. 13-87; K.H. Bailey, "The Constitutional and Legal Framework", in Sinden, *Natural Resources,* pp. 308-29; Barnett, *Minerals and Energy,* pp. 291-321; Blainey, *Rush,* pp. 57-58, 66-70, 98-99, 179-80, 195-98, 296-97, 307-8, 364; Carnegie, "Minerals to 2001", pp. 21-23.
84. Blainey, *Broken Hill,* pp. 86-102, 129-32; Blainey, *Rush,* pp. 57-58, 66-70, 251-52, 294-304; Hore-Lacy, *Broken Hill to Mt Isa,* pp. 126-28.
85. Blainey, "Cargo Cult", pp. 470-79; Raggatt, *Mountains,* pp. 105-15.
86. Barnett, *Minerals and Energy,* pp. 306-25; Stevenson, *Mineral Resources and Federalism;* D.F. Livingstone, "Mineral Policy", in Sinden, *Natural Resources,* pp. 198-227. See also Chapter 9.

3 Economic Structure

W.H. Richmond*

This chapter provides a concise overview of the main economic characteristics of the contemporary Australian mineral sector. The sector as a whole can be regarded as consisting of several more or less distinct industries concerned with the production of different minerals. To some extent the boundaries between these industries are blurred, particularly as an increasing number of mining companies are involved in a wide range of mining and processing activities; but the economic framework in which different minerals are produced and sold varies so markedly that it is useful to adopt an industry-by-industry approach.

The relative importance of different industries has changed over time. For nearly seven decades after the 1850s, gold headed the list of minerals produced in Australia, and was responsible for nearly two-thirds of the total value of mineral production. Subsequently the production of gold became relatively insignificant (though there was a revival in the 1930s and to a lesser extent at the beginning of the 1980s); even during the second half of the nineteenth century the significance of gold gradually diminished as the production of coal, lead, silver, copper and tin expanded. In the inter-war period there was considerable fluctuation in both aggregate mining output and in the relative significance of different minerals. Coal became the dominant industry for a while and then declined in absolute as well as relative terms. In the early post-war decades the sector as a whole underwent a general expansion and diversification, though again there was considerable fluctuation in the value of output of different minerals. The importance of gold steadily declined, coal accounted for around one-quarter to one-third of the value of total output, and copper, silver-lead and zinc together accounted for about one-third. The mid 1960s marked the start of a major mining boom which centred on black coal, iron ore, bauxite and, more recently, oil and gas.

The relative contribution of the major minerals to the total value of mineral output for the years 1977-78, 1978-79 and 1979-80 is summarized in table 8. Black coal, iron ore, bauxite, oil and gas and the base metals,

---

* Jean Bishop provided valuable assistance in compiling the data on which this chapter is based.

Table 8. Minerals produced, Australia, 1977-78 to 1979-80, percentage of total* by value.

Minerals	1977-78	1978-79	1979-80
Copper	3.5	5.2	5.3
Gold	1.9	2.0	3.2
Iron Ore	17.5	15.8	15.5
Lead	5.1	7.0	10.3
Mineral sands	2.3	2.1	2.2
Tin	2.5	2.7	2.6
Uranium	0.5	0.9	1.0
Zinc	2.7	2.7	2.7
Other metallic minerals (including bauxite)	11.3	9.5	10.5
Black coal	35.9	32.4	27.0
Brown coal	1.5	1.6	1.4
Oil and gas	15.3	18.1	18.3
	100.0	100.0	100.0

* Metallic minerals, oil and gas, and coal

Source: Australian Bureau of Statistics, *Mineral Production 1979-80*.

lead, zinc and copper, have come to account for over 85 per cent of total mineral output (excluding non-metallic minerals and construction materials). Accordingly it is these industries which are considered in the following survey, with a brief note on the uranium industry which is discussed in detail in chapter 9.

A considerable amount of mineral processing is undertaken in Australia. While classified by the Australian Bureau of Statistics as part of manufacturing industry, processing operations are commonly regarded as part of the mineral industry, and are of considerable economic significance. Processing is usually closely integrated with mining activities and needs to be understood when analyzing the nature of the market framework in which minerals are produced. The following discussions of the major mining industries thus encompass processing activities where appropriate.

The discussion of each industry follows a similar pattern. The characteristics of producers are summarized — in tabular form where practicable — with particular reference to the number, relative size and nature of ownership of companies. Where relevant, processing as distinct from mining activities is discussed in the same way. The nature of the market in which the mineral is produced and sold is outlined with reference to both domestic and export markets and to the major influences on the determination of price. The economic framework of the industry in the early 1980s is then reviewed, and comments are made on the nature of likely changes in economic structure over the next few years. Important changes which take place between the writing of this chapter and the time of publication will be noted in a postscript following the chapter.[1]

BLACK COAL

Black coal is usually divided into two general categories: *coking* coal, which is used as a feedstock in steel foundries, and *steaming* coal, which is of lower quality and used in power generating plants. Both types have been produced in Australia since the nineteenth century. During the inter-war and early post-war period the industry stagnated and in some years actually declined. However from the 1950s it expanded steadily if unspectacularly to serve the domestic electricity generation and iron and steel industries. In the 1960s, greatly increased export demand for coking coal, particularly from Japan, initiated a dramatic expansion in the industry. This was further boosted in the 1970s as sharp increases in oil prices stimulated the demand for steaming coal as an alternative energy source. Production doubled during the decade of the sixties and doubled again during the seventies. At the beginning of the 1980s black coal is the basis of a growth industry that few would have imagined twenty years earlier.

Accompanying this growth was a marked change in the economic structure of the industry. In the early 1960s the New South Wales industry accounted for nearly 80 per cent of total production; during the next two decades major developments also took place in Queensland, such that by the end of the 1970s the Queensland industry accounted for over 35 per cent of total production and the New South Wales industry just under 60 per cent.[2] It is not surprising therefore that the present industries in the two states (which account for nearly all coal production in Australia) have rather different economic characteristics.

The structure of the industry and details of ownership of major producers are set out in tables 9 (New South Wales) and 10 (Queensland). The New South Wales industry is, in general, older and more established, some companies having been formed nearly a century ago. Most of the coal is mined by underground methods where opportunities for economies of scale are limited. Thus, despite expansion, the industry remains relatively unconcentrated. The Electricity Commission of New South Wales accounts for about 30 per cent of total output; but the next four largest operating companies together produce only about 36 per cent of total output, with the remaining output produced by several smaller companies.[3] By contrast the Queensland industry has, with the exception of some old established mines mainly in the Moreton Basin, developed only since the 1960s. Open cut methods have been used in the majority of cases and significant economies of scale have been possible. Individual operations have thus been very large, and the four largest operating companies account for 84 per cent of total production; the Utah Development Company alone controls nearly two-thirds of total

Table 9. Coal industry, New South Wales, 1978

Operating Company	Major Shareholdings	%	Australian Ownership %	Percentage of NSW Output*
Electricity Commission of NSW (various districts)	(NSW Government instrumentality)		100	30.1
Clutha Development Pty Ltd (various districts)	British Petroleum Co.	100	nil	13.0
Coal and Allied Industries Ltd (various districts)			100	8.9
Australian Iron & Steel Pty Ltd (South Coast District)	BHP	100	90	8.6
Kembla Coal and Coke Pty Ltd	Australian Mining and Smelting	50	59	6.0
	North Broken Hill	30		
	BH South	20		
Austen and Butta Ltd (South Coast and Western District)	Shell Co. of Australia	37.2	51	5.5
	Mitsubishi Chemical Co.	4.6		
	Rivon Investments (Lithgow)	19.2		
	Riange Investments	19.2		
	Marubeni Corp.	4.6		
	Scottish American Investment	2.6		
Buchanan Borehole Collieries Pty Ltd (Singleton-NW and Newcastle Districts)	CSR	92.65	93	3.6
	J. Johnstone	7.35		
The Broken Hill Proprietary Co. Ltd (Newcastle District)			90	3.3
Coalex Pty Ltd (Western District)	Oakbridge Ltd	100	100	2.8
R.W. Miller Holdings Pty Ltd (various districts)	Howard Smith Ltd	34.5	65	2.7
	Ampol Petroleum	33.0		
	Atlantic Richfield Co. of USA	32.5		
The Bellambi Coal Co. Ltd (South Coast District)	Consolidated Gold Fields Australia (CGFA)	68	53	2.6
Peko-Wallsend Ltd (South Maitland and Western Districts)			100	1.8
Bloomfield Collieries Pty Ltd (Newcastle District)			100	1.4
Bayswater Colliery Co. Pty Ltd (Singleton-NW District)			100	1.1
Wambo Mining Corporation Pty Ltd	Hartogen Mining & Investment	45	47	0.8
	Carbonnages de France	25		
	Societe Miniere et Metallurgique de Penarroya	25		
	Austen and Butta	5		

* saleable coal

Table 10. Coal industry, Queensland, 1978

Operating Company	Major Shareholdings	%	Australian Ownership %	Percentage of Queensland Output*
Central Queensland Coal Associates (Mackay District)	Utah Development Co.	76.25	20	50.0
	Mitsubishi Development	12		
	AMP Society	7.75		
	Utah Mining Australia	4		
Utah Development Co. (Blackwater)	Utah International	89.2	11	14.9
	Utah Mining Australia	10.8		
Thiess Bros Pty Ltd (South Blackwater and Callide)	Thiess Holdings	100	84	10.5
Thiess-Dampier-Mitsui Coal Pty Ltd (Kianga-Moura)	Thiess Holdings	22	70	8.7
	Dampier Mining Co.	58		
	Mitsui and Co.	20		
Collinsville Coal Co. Pty Ltd	MIM	100	46	3.5
Queensland Coal Mining Co. Ltd (Blackwater)	BHP	100	90	1.2
Blair Athol Coal Pty Ltd	Conzinc Riotinto Australia (CRA)	62	17	0.5
	Clutha	38		

* saleable coal

production. (There is no direct government or semi-government participation in the industry.)

The New South Wales and Queensland industries also differ in the extent to which they are foreign-owned. In Queensland, the capital requirements of the large-scale operations, together with the need for higher managerial and marketing skills and the larger transport and developmental costs, have resulted in a much higher degree of foreign participation. The degree of Australian ownership (calculated by weighting the volume of coal produced by each company by the proportion of Australian ownership of the company) is around 30 per cent in Queensland, compared to about 75 per cent in New South Wales. In both cases foreign ownership is primarily by overseas mining companies (including oil companies) seeking to diversify their activities. While the expansion of the industry was based very largely on serving the Japanese export market — and in particular the Japanese steel industry — there has been very little direct Japanese investment in the Australian coal industry. This was due partly to government restrictions on overseas investment by Japanese companies before 1971, partly to the need of Japanese steel companies to use their funds to finance expansion of their own plants, and partly to the fact that, because the world coal market is a relatively free one, the pressure to integrate vertically was not great.[4] Foreign firms have entered the established New South Wales industry mainly by the purchase of shares in existing enterprises. In Queensland the industry has been dominated by foreign firms from the beginning of its spectacular growth in the mid-1960s though during the 1970s the degree of foreign ownership decreased with the reorganization of Central Queensland Coal Associates (CQCA).

As one would expect in a growth industry such as coal there have been several changes since 1978. Some producers have greatly expanded production and commissioned new mines, and there have been changes in ownership of some operating companies. CQCA have opened a major new mine in the Mackay district and BHP have started producing for export from their Gregory mine, both in central Queensland. The Thiess company has been taken over by CSR Ltd. However, these developments do not substantially alter the general picture as presented in tables 9 and 10. The general thrust of current and likely future changes is discussed below.

The coal industry serves both a domestic market (mainly for electricity generation and the iron and steel industry) and an export market. The latter has always been significant but has become much more so in the last two decades. Indeed the major expansion of the industry during this time has been based very largely on exports, and in the late 1970s 52 per cent of total production of saleable coal was exported. Developments in the Queensland industry in particular were based on the export market and in the late 1970s more than 70 per cent of saleable coal produced by the Queensland industry was exported. This compares with about 40 per cent in the case of New South Wales.[5] To understand the industry it is therefore necessary to examine both the domestic and export markets.

In New South Wales the domestic market consists almost entirely of the electricity generating and iron and steel industries, and 80-90 per cent or more of coal supplied to these industries is produced by "captive" mines owned by the State Electricity Commission or the steel industry (BHP and Australian Iron and Steel). In Queensland the domestic market consists almost entirely of the electricity generating industry, with the alumina industry being a particularly large consumer. No mines are actually operated by the electricity generating authority. In both states, Coal Boards established by statute have authority to control the price of domestically consumed coal; the manner in which such authority is exercised is not clear, however.

Exports of coal from Australia (which originate entirely from Queensland and New South Wales) are very significant in total world trade in coal; in 1978 Australian exports accounted for nearly 20 per cent of world trade.[6] Reliance on Japanese markets is heavy: there has been a gradual retreat from the near total reliance on Japan in the 1960s, with the development of markets in Western Europe and Asia in recent years, but about 65 to 70 per cent of Australian coal exports still go to Japan.[7] This dependence is of considerable importance for the conduct of the Australian industry, particularly with respect to the determination of prices. Export sales are arranged through contracts which specify (with some flexibility) prices and quantities, and cover varying periods up to ten or fifteen years.[8] In the negotiation of these contracts Japanese coal buyers, particularly the steel manufacturers, have in the past been able to exert considerable bargaining strength. Typically they have co-ordinated their purchasing activities, whereas Australian producers have tended to compete with each other as well as with several international competitors.

Contracts arranged early in 1980 by some coking coal producers, eager to secure contracts in the face of the uncertainty surrounding the Japanese steel industry, drew criticism for the low prices negotiated, and some have seen them as undermining negotiations by other Australian producers.[9] Towards the end of 1980 two major Queensland coking coal hopefuls were also reported to be engaging in keen price competition, in a bid to enter the Japanese steel industry market.[10] On the other hand, the dramatic increases in oil prices in the second half of the 1970s significantly affected the Japanese economy, and led to concerted efforts by government and industry in Japan to switch from oil to coal as an energy source, particularly in the electricity generation and cement industries. This created a potentially huge demand for steaming coal, so significant that some fears were held that the desired supply might not be forthcoming. This gave the Australian producers, especially of steaming coal, more bargaining power. Traditionally seen as a relatively easy mark for tough Japanese cartel negotiators, Australian steaming coal producers, mainly from New South Wales, began to present more of a solid front to Japanese buyers, especially in the case of contracts arranged with Japanese cement manufacturers.[11] One observer of the Japanese industry noted in early 1980 that "for the time being, incredibly, the Australian steaming coal industry appears more organised than the Japanese coal buying industry . . . Japan as a steaming coal market is no Japan Inc".[12]

However, the Japanese buying position was strengthened through developments in 1980. The most significant of these was the formation of the Japan Coal Development Co., which consists of Japan's nine electric utilities and the government-related Electric Power Development Company. The company operates as a joint coal-purchasing organization and co-ordinates the electricity industry's steaming coal buying programme, consulting with other major users of steaming coal, particularly the cement industry.[13] It has been reported that Japanese steel, cement and power industries are working to produce "an overall coal buying master plan" for both steaming and coking coal.[14] Fears have also been expressed that Japanese coal users are trying to orchestrate a situation of world over-supply of steaming coal in order to maintain control over prices.[15] Doubtless the Japanese would like to encourage further the large scale development of coal resources in the United States (especially Alaska), China and South Africa as well as in Australia. The subsequent decline in oil prices and world energy prices generally also weakened the Australian producers' position.

In the context of these developments it was reported that the Australian Government intended bringing into effect what was described as "an export guideline price package" for coal — a report immediately denied — and that it was attempting to form a "coal council" of industry and government representatives to negotiate terms on which Australian coal is exported. There is considerable resistance on both sides to the latter proposal, and it is not clear how significant government influence over the market will be in the near future, nor how the market relationship in general terms will develop.[16]

What is certain, however, is that the industry will undergo substantial expansion in the 1980s which will make black coal even more significant in relation to total mineral production and indeed to the whole economy. The expansion will occur particularly in response to increased domestic and export demand for steaming coal, but major new coking coal projects will also be established. Much of the expansion will occur as a result of increased production by existing producers, in both established and new projects. Some projects will be undertaken by newly-formed consortia, though many of these will include companies already engaged in coal mining activities in Australia; some companies or groups undertaking coal production will be entirely new to the industry; and some changes in ownership of existing producers are likely. What effect this will have on the structure of the industry as depicted in tables 9 and 10 is difficult to say, but two trends can be predicted.

Firstly, there is likely to be an increasing degree of participation by other mining companies, particularly oil companies, diversifying their activities. Shell, BP, Esso and Caltex have all bought into existing or new projects in recent years, as have the local groups H.C. Sleigh and Ampol. Secondly, there will be an increasing amount of direct equity participation in new projects by overseas buyers of both coking and steaming coal. European and Asian (mainly Japanese) steel mills have substantial interests in new coking coal projects in both Queensland and New South Wales. Japanese cement manufacturers and electric power generating companies are also involved in several steaming coal projects. The Japan Coal Development Company has explicitly expressed the intention of purchasing significant equity shareholdings in foreign coal ventures and of developing new mines itself, particularly in Australia.[17] This vertical integration will of course affect the nature of the market relationship and the terms on which export contracts are established.

The level of foreign participation in new projects will be limited by the official policy (which is now being strictly enforced) of restricting foreign equity in resource projects to no more than 50 per cent.

Several projects involving foreign investment have been approved, most of them involving a level of foreign equity near the maximum permitted. On the other hand many projects, particularly in New South Wales, are being undertaken by largely or wholly Australian owned companies. In general it is probable that the overall level of foreign ownership in Queensland will fall slightly and in New South Wales remain much the same.

While much of the increased production will be geared to domestic demand (particularly for steaming coal for use in electricity generation) the industry will remain highly dependent on exports and hence on the state of the world market for coal. In the near future the demand for coking coal is likely to be relatively subdued and the bargaining position of Australian producers is not likely to be strong. The position of producers of steaming coal seems better in the light of the forecast increases in demand from Japan in particular. However, much will depend on the changes in the supply situation in other exporting countries with which Australia competes, and in the price of oil, the main competitive energy source.

IRON ORE

Until the mid 1960s, production of iron ore was geared primarily to serving the needs of the domestic market and was undertaken as part of the vertically integrated iron and steel industry. An embargo placed on the export of iron ore in 1938 was maintained during the post-war years, because it was thought that reserves should be conserved for use by the domestic steel industry.[18] When it became clear at the beginning of the sixties that reserves had been grossly underestimated, the embargo was partially lifted. In 1961 the government of Western Australia, where the major reserves are located, adopted policies to encourage development of the reserves, which brought to light new discoveries and led to the initiation of several large scale projects.[19] As with the concurrent developments in the coal industry, iron ore expansion was based almost entirely on exports, mainly to Japan. Between 1965 and 1970 output increased seven-fold and then doubled again in the next five years. Output then decreased from 92.7 million tonnes in 1975-76 to 68.8 million tonnes in 1979-80.

The structure of the iron ore industry and the details of ownership of the major operating companies are shown in table 11.[20] The industry is highly concentrated, with the three major operating companies accounting for nearly 85 per cent of total output. This may seem surprising in view of the fact that it is relatively easy for new firms to enter the iron ore industry; the technology of iron ore mining is not highly sophisticated and under certain circumstances relatively small deposits can be profitably exploited. However, there are substantial economies of scale to be obtained and clear advantages for very large operations where, as with the Pilbara region in Western Australia, the deposits are remote and basic infrastructure has to be provided.

In this context, and given the very large capital expenditure associated with the projects, it is not surprising to find the degree of foreign participation indicated in table 11. The degree of Australian ownership, calculated by weighting the volume of ore produced by each company by the proportion of Australian ownership of the company, is about 43 per cent. This is represented partly by substantial interests of BHP, which directly or indirectly controls nearly 20 per cent of total output as part of the vertically integrated domestic iron and steel industry; and partly through shareholdings in projects by partly or wholly owned diversified Australian mining companies (including CSR, CRA, MIM and Ampol) and the Australian public.

Foreign equity participation is almost entirely by overseas mining or metals companies. Despite the fact that most exports of iron ore go to Japanese steel mills, there is very little direct involvement of the Japanese steel industry in the Australian iron ore industry. (The combined 35 per

Table 11. Iron ore industry, 1978

Operating Company	Major Shareholdings		%	Australian Ownership %	Percentage of Australian Output
Mount Newman Joint Venture (WA)	Amax Iron Ore Corp.		25	55	35.8
	Pilbara Iron		30		
	Dampier Mining Co.		30		
	Mitsui-C. Itoh		10		
	Seltrust Iron Ore		5		
Hamersley Iron Pty Ltd (WA)	CRA		82.3	34	34.3
	Japanese steel mills		6.2		
Cliffs Robe River Iron Associates (WA)	Cliffs Western Australian Mining Co.		30	23	13.8
	Cleveland Cliffs Iron Co.	53			
	Texasgulf Inc.	35			
	Bank of America	10			
	First Nat. Bank of Chicago	2			
	Mitsui Iron Ore Development		30		
	Robe River		35		
	Cape Lambert Iron Associates		5		
Goldsworthy Mining Ltd (WA)	Consolidated Gold Fields (UK)		46.7	13	6.4
	Utah Development Co.		33.3		
	MIM		20		
Dampier Mining Ltd (WA)	BHP		100	90	4.2
The Broken Hill Proprietary Co. Ltd (SA)				90	2.8
Savage River Mines (Tas.)	Northwest Iron Co.		50	12	2.6
	Pickands Mather	48			
	Cerro Corp.	24			
	Chemical International	3.6			
	Ampol Petroleum, Kathleen Investments, Insurance companies	24.4			
	Dahlia Mining Co.		50		
	Mitsubishi	75			

This giant drag-line and other heavy equipment on the central Queensland coal-fields emphasizes the capital-intensive nature and large scale of operation typical of modern open-cut mining. *(Utah Development Co. photograph.)*

cent interest of three mills in the Cliffs Robe River project is the major exception.) This is due partly to the limited ability of Japanese firms to undertake such investment in the 1960s, and partly to the nature of the world iron ore market which is fairly freely competitive; in this situation there is no particular incentive for backward integration by firms in the steel industry. Foreign participation can thus be characterized as diversification by overseas mining companies rather than integration by steel producers seeking to secure raw material supplies.

The industry is primarily an export industry. The domestic market, which takes about 20 per cent of production, consists of Australia's only steel producer BHP. All of its supplies are obtained from its own mines (or those of its subsidiary Dampier Mining) in South Australia and Western Australia, with about two-thirds coming from the Mount Newman Joint Venture (BHP, through Dampier, has a 30 per cent interest in and manages this project). The world market, on which about 80 per cent of production is sold, is relatively free and competitive. Of several large exporters of iron ore, Australia is the largest, followed closely by Brazil and Canada. Freight costs give Australia some advantage in the Asian market, particularly over Brazil, but some disadvantage selling in the European market. Australian exports go primarily to Japan, and the rapid development of the industry in the 1960s was geared to the spectacular growth of the Japanese steel industry. In the late 1960s and early 1970s Japan was taking over 90 per cent of Australian production; this has now fallen to below 70 per cent as markets have been developed in Europe and Asia (including China), but the Japanese market remains vitally important. Australian exports constitute just under one-half of Japanese imports of iron ore. As with coal, export sales are negotiated by way of long-term contracts, though these provide for adjustments in both contracted quantities and prices over time. The Japanese steel industry is very concentrated and there is a high degree of consultation between the major firms. It has been the practice for negotiations to be done on behalf of the mills by Nippon Steel, the largest of the Japanese producers. The bargaining power of Australian producers has thus been restricted and highly dependent on the fortunes of the Japanese steel industry. The mid and late 1970s were difficult times for the industry, with production in 1980 still below that in the early seventies.[21] As a result the Japanese mills have been taking quantities of iron ore well below base contracted tonnages, and the price received by Australian producers fell in real terms quite considerably during the 1970s.[22] During these years iron ore was in world over-supply and many producers operated at a loss.

At the beginning of the 1980s Australian iron ore producers were still operating in what was essentially a buyers' market, and one in which demand was still relatively subdued. A significant price increase was granted by Japanese steel producers in early 1980, though it has been suggested that this arose mainly through fears that some boost to the profitability of iron ore producers generally was necessary to encourage new investment (in some cases to forestall closure of projects) and thus ensure that a world shortage in the mid 1980s would be avoided. But future projections of the Japanese steel industry indicate only limited growth during the 1980s, particularly in the second half of the decade.[23] On this basis the industry would by 1990 still need less iron ore than the amount for which base contracts have already been signed. Prospects for expansion of the steel industry in China, South Korea and Taiwan — seen as potentially large sources of demand for Australian iron ore — are also subdued. Fears have been expressed that Japanese steel mills, acting together, will increasingly attempt to "play off" Australian and the increasingly significant Brazilian producers against each other, thus strengthening their bargaining position. High, and increasing, freight charges place Brazilian producers at a substantial disadvantage in supplying the Japanese market; nevertheless the iron ore industries in these countries are closely competitive.

Australian government concern that the stronger position of the

Japanese steel producers was resulting in lower than reasonable market prices for Australian iron ore led to the introduction of "export guidelines" in October 1978. These gave the government, through the Department of Trade and Resources, control over negotiations between individual producers and the mills, in an attempt to neutralize the mills' stronger negotiating position. Opposition to these guidelines forced the government in April 1979 to drop iron ore from the list of commodities to which they were applied. A co-operative arrangement has subsequently been established with Brazil to increase the combined bargaining strength of iron ore producers vis-à-vis Japanese steel producers. The arrangement appears to be limited to the exchange of market information and discussions on each country's general perception of the market, and is not an attempt to establish a cartel or in any way control prices. However such an arrangement has the potential to influence the manner in which prices are formed in the market.[24]

The foregoing discussion suggests that there is unlikely to be any major expansion in the Australian iron ore industry in the near future. Additional demand for iron ore could be satisfied by increased output of existing producers, but it is widely considered that the steel industry would prefer to give contracts to a new mine in order to spread its contracts and minimize the risks entailed in strikes, a common feature of the industry. It is thus probable that one new project will be established at some time during the 1980s, and five producers are vying for the opportunity, all in the Pilbara region of Western Australia: Goldsworthy with its Area C (Areas A and B will soon be exhausted); Cliffs Robe River (West Angela project); BHP (Deepdale); CSR (Yandicoogina); and the CRA/Hancock and Wright partnership (Marandoo). Which of these projects will proceed will depend largely on the ability of the developers to negotiate contracts with the Japanese steel mills. The foreign investment guidelines will also be an important determinant, as the government explicitly affirmed in October 1980 that the guidelines would be enforced. It has been reported that all the developers have, in preliminary negotiations, offered substantial equity interests, up to 40 per cent, to Japanese steel mills. Only BHP and CSR, and possibly the CRA partnership, would be able to accommodate such equity.[25]

Thus the structure of the iron ore industry in the 1980s will remain very similar to that presented in table 11. Any new major development will alter the picture in so far as it is likely to involve considerable direct participation by Japanese steel producers. The export market will continue to be of major significance for the industry and will (because of transport costs) become even more concentrated in Asia, with the Japanese market of major significance. Prices will be determined primarily by bargaining between Japanese steel mills on the one hand and Australian producers on the other, with Brazilian producers acting as competitors.

ALUMINIUM

Primary aluminium is produced by refining the mineral bauxite into the metal oxide, alumina, and then smelting alumina into aluminium. Strictly speaking, the refining and smelting processes are classified as part of manufacturing industry and only the initial mining of bauxite (about two to five tonnes of which is required for the ultimate production of one tonne of primary aluminium) is part of the mining industry. However, all three are commonly thought of as part of the mineral industry and, as there is a high degree of vertical integration (with the same companies involved in different stages of production), the industry as a whole is discussed here.

The Australian aluminium industry has developed only recently. The first refinery/smelter was established in the 1950s and used imported bauxite. Subsequently very large deposits of bauxite in Australia were discovered and exploited; several alumina refineries were established during the 1960s and 1970s, and Australian production of alumina increased dramatically from about 1967 onwards. Most was exported but some was used in Australian aluminium smelters established and expanded during the sixties and seventies.

Table 12 summarizes the structure of the industry in 1978. The characteristics of the Australian industry have to be understood in terms of the structure of the world aluminium industry. This is both highly concentrated and highly vertically integrated. Six major producers (Alcoa, Reynolds, Kaiser, Alcan, Pechiney, Alusuisse) account for more than half the total world production of bauxite, alumina and aluminium, and all undertake production at each of these stages.[26] When the industry was established in Australia, it was closely integrated with the world industry and dominated by the major world producers. By the late 1970s the industry was concentrated in the hands of three operating companies at each stage, as shown in table 12. In each case one or more of the major world producers had a substantial or controlling interest. Two of the majors, Alcoa and Kaiser (through Comalco), were involved in all three stages of production; and two, Alusuisse (through Nabalco) and Alcan, were involved in two stages.[27] There is limited Australian equity participation by some Australian mining companies and, to a lesser extent, financial institutions. In 1978 overall Australian ownership — calculated by weighting the Australian ownership of each operating company by the proportion of output it produces — was 32 per cent in bauxite production, 28 per cent in alumina refining and 31 per cent in primary aluminium production.

Bauxite produced in Australia is used primarily as an input in the domestic production of alumina; the remainder is exported. Of the *alumina* produced in Australia, some is used as an input in the domestic

Table 12. Aluminium industry, 1978

Operating Company	Major Shareholdings		%	Australian Ownership %	Percentage of Australian Output (approximate)
BAUXITE					
Alcoa of Australia Ltd (Darling Range, WA)	Aluminium Co. of America		51	43	47
	Western Mining Corporation		20		
	BH South		13.1		
	North Broken Hill		12		
	Australian financial instutitions		3.9		
Comalco Ltd (Weipa, Qld)	Kaiser Aluminium and Chemical Corporation		45	22	34
	CRA		45		
Nabalco Pty Ltd (Gove Joint Venture, NT)	Swiss Aluminium (Aust.)		70	29	19
	Gove Alumina Ltd		30		
	CSR	50			
	Peko-Wallsend	13			
	AMP Society	12			
	MLC Ass. Co.	9			
	Bank NSW	5			
	CBC	5			
	Elder Smith-GM	5			
ALUMINA					
Alcoa of Australia Ltd (Pinjarra and Kwinana, WA)	see above			43	52
Queensland Alumina Ltd (Gladstone, Qld)	Comalco		30.3	7	32
	Kaiser Aluminium and Chemical Corporation		28.3		
	Alcan Aluminium		21.4		
	Pechiney		20.0		
Nabalco Pty Ltd (Gove Joint Venture)	see above			29	16
ALUMINIUM					
Comalco Ltd (Bell Bay, Tasmania)	see above			22	43
Alcoa of Australia Ltd (Point Henry, Vic.)	see above			43	39
Alcan Australia Ltd (Kurri Kurri, Hunter Valley, NSW)	Alcan Aluminium		70	30	18
	AMP Society		10		
	Other Australian financial institutions		20		

production of primary aluminium but most (92 per cent in 1978) is exported. In the case of both bauxite and alumina the majority of "sales" — both domestic and export — are either intracorporate transfers or sales to joint venture partners. This arises from the high degree of vertical integration in the industry both within Australia and in the world as a whole. There is thus a limited free market in bauxite and alumina and the "price" and other terms of "sale" of most Australian bauxite and alumina may be expected to be set by the major producers in the aluminium industry in accordance with their overall corporate objectives, including the minimization of taxation. It may be noted, however, that producers participating in the Australian industry are more reliant than is the case in other major aluminium-producing nations on what are often termed "arms length" sales or sales on the free market, usually negotiated by way of long-term contracts. It has been estimated that (in 1977) approximately 18 per cent of total bauxite output and 40 per cent of alumina was subject to sale by long-term contract or on the spot market (as distinct from being transferred within the one corporation or "sold" to joint venture partners). Alcoa in particular is a significant "arms length" seller of alumina and is responsible for the high figure in the case of alumina.[28] The growth of an industry based to this relatively large extent on arms length sales was possible in the 1960s and early 1970s mainly because of the rapid growth of alumina and primary aluminium industries in Japan at this time. Several long-term contracts were negotiated between Australian and Japanese firms in the manner that such contracts are normally negotiated, by a process of bilateral bargaining. However after the early 1970s several events, foremost among them being the increase in oil prices which caused the Japanese producers considerable difficulty, meant that Australian producers had to accept significant reductions in both quantity and price under the contracts. Some producers, notably Alcoa, have been quite successful in finding other markets. But the general tendency has been for Australian producers to seek a higher degree of integration, a point discussed further below.

Production of *primary aluminium* is less significant than production of bauxite and alumina. The three producers obtain their inputs from their own alumina operation (in the case of Alcan from QAL in which it has a 21 per cent interest). The domestic market uses 70 per cent of production, the remainder is exported, mainly to Japan. In both cases some output is used in further downstream (fabricating) activities of the major producers, either within Australia or overseas; most is sold on the "free market" however.

Several major developments in the industry, mainly involving the production of alumina and primary aluminium, are scheduled for the 1980s. These involve both the expansion of existing refineries and smelters and the construction of new ones. Wholly new projects are summarized in

Table 13. Proposed new projects in the aluminium industry.

Operating Company	Major Shareholdings		%	Australian Ownership %
ALUMINA				
Worsley Alumina Pty Ltd (Worsley, WA)	Reynolds Aust. Aluminium		40	18
	Billiton Aluminium Aust. (Shell)		30	
	BHP		20	
	Kobe Alumina Assoc.		10	
Alcoa of Australia Ltd (Wagerup, WA)	see table 12			43
ALUMINIUM				
Gladstone Aluminium Limited (Gladstone, Qld)	Comalco* (CRA and Kaiser)		30	7
	Kaiser Aluminium and Chemical Corporation		20	
	Sumitomo Light Metal Industries		17	
	Kobe Steel		9.5	
	Mitsubishi Corp.		9.5	
	Yoshida Kogyo K.K.		9.5	
	Sumitomo Aluminium Smelting Co.		4.5	
Alcoa of Australia Limited (Portland, Vic.)	see table 12			43
Tomago Aluminium Pty Limited (Hunter Valley, NSW)	Aluminium Pechiney		35	48
	Gove Aluminium Finance Limited		35	
	CSR	50		
	AMP Soc.	12		
	MLC Ass. Co.	9		
	Peko-Wallsend	13		
	Bank of NSW	5		
	CBC	5		
	Elders-GM	5		
	AMP Society		15	
	Vereignigte Aluminium GmbHAG		12	
	Hunter Douglas Aust.		3	
Alumax Inc. (Hunter Valley, NSW)	Alumax Inc.		45	32
	Alfarl		20	
	Dampier Mining		35	
Alcan Australia Limited (Bundaberg, Qld)	see table 12			30

* Comalco will take a 90 per cent share in the second stage of this project which will lift its interest in the overall project to about 60 per cent.

table 13. Increased bauxite production was planned by both Alcoa and Comalco to supply their Western Australian and Queensland refineries. The new projects were scheduled to come into production in the early to mid 1980s, though the downturn in the world aluminium industry in 1981 led to the deferral of some plans.

The extent of Australian equity in the proposed projects is generally relatively low as can be seen from table 13. Foreign Investment Review Board approval has been granted for the Wagerup and Worsley projects even though the latter has only 20 per cent Australian equity (by BHP). However, the bauxite deposits to serve the Worsley refinery are wholly Australian owned (jointly by BHP and News Ltd). The smelters, as manufacturing projects, are not subject to any specific foreign investment guidelines.[29] The major world producers have substantial equity in all the projects, though a new trend is for minor equity participation by overseas aluminium companies outside the "big six", including several from Japan. Australian participation in new projects, as in the case of existing ones, is partly by mining companies and partly by financial institutions.

These developments are intended to increase the degree of vertical integration in the industry. Comalco, which in the late 1970s relied upon arms length outlets for about one-third of its bauxite production, will direct more bauxite to the expanded QAL refinery and thence to the Gladstone smelter. Both these projects are joint ventures so, although this is not "pure" vertical integration, the effect is the same. Comalco also has interests in overseas joint ventures in both refining and smelting, and so is integrating across national boundaries as well. The expansion of smelting activities by Alcoa will also serve to reduce its reliance on arms length sales of alumina (previously very considerable). Further "downstream" integration will be achieved by development, both within Australia and overseas, of fabricating plants by the major producers; however, a significant proportion of primary aluminium production will be sold on the open market. The development of the Worsley bauxite/alumina project marks the entry of the sixth major world producer (Reynolds) into the Australian aluminium industry.

The increasing degree of integration within the industry means that the free market in bauxite and alumina in particular is becoming very "thin", and the problem of tax minimization through transfer pricing, to suit corporate rather than national objectives, is becoming more acute. Both commodities fall within the ambit of the commonwealth government's mineral exports guideline policy. It is not clear how (or if) this policy has actually been applied. It would seem, from statements made by the Minister for Trade and Resources, that the policy is intended to apply to arms length sales and in particular to those made under long-term contracts. It has been argued, however, that there is not much reason for "export controls" on such sales because the nature of market relationships is such that government intervention would be able to achieve little to secure a better deal for Australian producers.[30] On the other hand there is a very strong case, of a *prima facie* nature at least, for the government to monitor "sales" made as intracorporate transfers (particularly when they are export sales), to ensure that national interests are not submerged in favour of transnational corporate interests. The point applies particularly to bauxite and alumina; there is less of a problem in the case of aluminium, which has a more competitive market and a recognized world market price, and which will become relatively much more significant. Bauxite and alumina will continue to be very significant, and the difficult problem will remain of ensuring that export sales – made primarily as intracorporate transfers or sales to joint venture partners – are arranged at a price which represents a reasonable return to Australian resources employed in the aluminium industry.

In summary, the aluminium industry in Australia is highly concentrated and integrated with the world industry, which is also highly concentrated and integrated. There is a high level of foreign ownership at all stages of production. Major developments will occur in the near future, particularly at the refining and smelting stages, and will increase the degree of integration both within the Australian industry and with the world aluminium industry.

COPPER

Copper occurs in different forms, often in ores containing other minerals (in Australia, gold, lead, zinc, silver, nickel and bismuth). Thus it is sometimes produced in association with, or as a by-product of, other minerals, so not all producers are solely or even primarily copper producers. Copper ores usually undergo some processing at the mine site; concentrates are then smelted (to produce blister copper) and refined. While these processes are strictly part of manufacturing industry, they are discussed here because there is a high degree of vertical integration in the sector.

The first copper production recorded in Australia was from the Kapunda field in South Australia in 1842. From that time until the 1950s several mining and processing operations were started in South Australia, New South Wales, Queensland, Tasmania and the Northern Territory, including the well-known ones at Cobar (New South Wales), Mount Morgan (Queensland) and Mount Lyell (Tasmania). In the early 1950s, the copper lode at Mount Isa in Queensland was found to be of greater potential than previously thought and a large scale expansion programme was instituted. Since this time copper mining and processing

Table 14. Copper mining, 1978

Operating Company	Major Shareholdings	%	Australian Ownership %	Percentage of Australian Output
Mt Isa Mines Limited (Qld)	MIM	100	46	71.1
Mt Lyell Mining and Railway Co. Ltd (Tas.)	Consolidated Gold Fields (Aust.) (CGFA)	56.1	61	8.7
Mt Gunson Mines Pty Ltd (SA)	CSR	100	92	5.6
Cobar Mines Pty Ltd (NSW)	BH South	100	95	2.8
Woodlawn Mines Joint Venture (NSW)	St Joe International Exploration	33.3	9	1.8
	Phelps Dodge Exploration Corp.	33.3		
	Australian Mining & Smelting (CRA)	33.3		
Electrolytic Zinc Co. of Australasia Limited (Tas.)	EZ Industries	100	95	1.8
Peko Mines Limited (NT)	Peko-Wallsend	100	(100)	1.7
Mt Morgan Limited (Qld)	Peko-Wallsend	100	(100)	1.3

Table 15. Copper smelting and refining, 1978

Operating Company	Major Shareholdings	%	Australian Ownership %	Percentage of Australian Output
SMELTING				
Mt Isa Mines Limited (Qld)	MIM		46	87.2
Mt Morgan Limited (Qld)	Peko-Wallsend		(100)	3.0
Electrolytic Refining and Smelting Co. of Aust. Pty Limited (Pt Kembla, NSW)	BH South North Broken Hill	60 40	93	9.8
REFINING				
Electrolytic Refining and Smelting Co. of Aust. Pty Limited (Pt Kembla, NSW)	See above		93	12.0
Copper Refineries Pty Ltd (Townsville, Qld)	MIM	100	46	88.0

in Australia has been dominated by Mount Isa Mines. Some other producers have scaled down operations or ceased production altogether.

Details of major producers are summarized in table 14. Mount Isa Mines accounted for 71 per cent of total mine output of copper in 1978. The Mount Lyell Mining and Railway Co. was the next most important producer with nearly 9 per cent, the remainder being produced by about a dozen other producers. Further processing of ores and concentrates is undertaken in smelters at Mount Isa and Mount Morgan, a smelter/refinery at Port Kembla and a refinery at Townsville. This part of the industry is summarized in table 15, from which the dominance of Mount Isa Mines in processing activities also can readily be seen.

Clearly Mount Isa Mines has a highly integrated operation and is the most important firm in the industry. It is one of many companies now wholly or partly owned by MIM Holdings Ltd, which has as its major shareholder the American Smelting and Refining Company (ASARCO) of the United States. By 1978 ASARCO held 49 per cent of the shares in the company, having declined slightly from a peak of 54 per cent in 1960. Australian residents now hold the majority of the remaining shares, having bought back into the company during its post-war period (in 1947 only 5 per cent of shares were Australian-owned). Mount Lyell is part of the Consolidated Gold Fields group and substantially owned by its UK parent. Other producers are principally Australian-owned, generally by diversified Australian mining companies, one exception being the recently opened Woodlawn Joint Venture. Overall Australian equity in both mining and processing operations is just over 50 per cent.

Of copper produced in Australia, approximately 17 per cent of ores and concentrates, 2.5 per cent of blister and 39 per cent of primary refined copper are exported; thus most ores and concentrates and virtually all blister copper undergo further processing within Australia. Australian producers sell on the relatively free world copper market in which there are many sellers and buyers. The market is characterized by variations in demand and most particularly in supply, which can and do cause considerable fluctuation in the world price which Australian producers must accept.

Several major new projects are in prospect as well as proposals to expand production by existing producers, though the way these proceed will depend to some extent on the fluctuations in the world market. Existing producers Mount Lyell and Peko Mines announced plans to expand their operations, and Mount Morgan is to develop two mines at Mount Chalmers in Queensland. Increased production is planned at Cobar by CRA who now own that mine, and a new project is planned at Teutonic Bore (WA) by Seltrust and MIM (sixty/forty).[31] Together these changes will slightly diminish the dominance of Mount Isa Mines, but will not fundamentally alter the pattern of copper production. Foreign equity participation (generally through diversification activities of foreign mining companies) is substantial in these and other possible developments. In the medium term the massive copper-uranium-gold deposit at Roxby Downs in South Australia, owned by Western Mining and British Petroleum, could become a major producer and, together with possible expansion by Mount Isa Mines, make Australia a more significant supplier to the world market. However, this is likely to remain a free market, subject to short-term fluctuations.

LEAD, ZINC, SILVER

These minerals are commonly found in association with each other and frequently with other minerals, notably copper. Silver is produced as a by-product, chiefly from lead and zinc and copper ores. Lead, zinc and silver have played a major role in the development of the Australian mining industry since the discovery of the large base metal deposits at Mount Isa and Broken Hill (see chapter 2). The Broken Hill mining companies and Mount Isa Mines are still the dominant producers. Despite a decline in relative significance since the mid 1960s, production of lead, silver and zinc still represents a significant proportion of the total value of mineral production (see table 8).

Table 16 gives details of the major producers of lead, silver and zinc and their relative importance, together with details of their ownership. The predominance of Mount Isa Mines and the three long-established Broken Hill companies is obvious. Table 16 also indicates that there is a substantial degree of foreign ownership of the companies involved in the production of these minerals. Foreign participation is entirely the result of diversification by overseas mining companies and has to be understood in terms of the historical development of the Mount Isa and Broken Hill mines.[32] Processing of concentrates is undertaken at five establishments which produce, in different combinations, lead bullion and refined lead, zinc and silver. The details of these establishments are given in table 17,[33] where it can be seen that there is a similar degree of concentration and foreign ownership arising from the vertical integration of operations by the companies involved.

As with all other minerals, production is geared largely to the export market. In the case of *lead,* a relatively small proportion in the form of concentrates is exported, but virtually all lead bullion (produced at Mount Isa and Cockle Creek smelters) and about 60 per cent of refined lead (primary and secondary) goes overseas. A considerable, though variable, proportion of *zinc* (35-55 per cent of total production) is exported as concentrates, and about 60-70 per cent of refined zinc (primary and secondary) goes overseas. In both cases Australian exports constitute an important part of the world market, Australia being the

Table 16. Lead, zinc, silver mining, 1978

Operating Company	Major Shareholdings	%	Australian Ownership %	Percentage of Australian Output	
Mt Isa Mines Ltd (Qld)	MIM	100	46	Pb	37
				Zn	27
				Ag	51
New Broken Hill Consolidated Limited	Australian Mining and Smelting (CRA)	100	27	Pb	18
				Zn	25
				Ag	7
Zinc Corporation Limited	ditto		27	Pb	20
				Zn	16
				Ag	8
North Broken Hill Holdings Ltd (North Mine)			89	Pb	16
				Zn	12
				Ag	14
Minerals Mining and Metallurgy Limited (South Mine)	G. Radford	49	100	Pb	3
				Zn	1
				Ag	2
Cobar Mines Pty Limited (NSW)	BH South	100	95	Pb	1
				Zn	—
				Ag	3
Woodlawn Joint Venture (NSW)	see table 14		9	Pb	<1
				Zn	<1
				Ag	6
Electrolytic Zinc Co. of Aust. Ltd (Tas.)	EZ Industries		95	Pb	6
				Zn	16
				Ag	6

Table 17. Lead, zinc, silver processing, 1978

Operating Company	Major Shareholdings	%	Australian Ownership %	
Broken Hill Associated Smelters Pty Ltd (Port Pirie, SA)	Australian Mining and Smelting (CRA) North Broken Hill	70 30	46	Lead smelter and refinery (sole producer of primary refined lead); electrolytic zinc refinery (14% total output); refined silver produced as by-product of smelting and refining lead concentrates.
Sulphide Corporation Pty Ltd (Cockle Creek, NSW)	Australian Mining and Smelting (CRA)	100	27	Smelting furnace and zinc refluxer; produces lead bullion (17% of lead bullion produced for export) and refined zinc (22% total output).
Mt Isa Mines Ltd (Qld)	MIM	100	46	Lead smelter (83% of lead bullion produced for export).
Electrolytic Zinc Co. of Australasia Ltd (Risdon, Tas.)	EZ Industries		95	Zinc refinery (64% total output).
Electrolytic Refining and Smelting Co. of Australia Pty Ltd (Pt Kembla, NSW)	BH South North Broken Hill	60 40	93	Produces refined silver from copper concentrates; also secondary refined from scrap.

largest exporter in the case of lead and third largest in the case of zinc, and accounting for approximately 20 per cent and 10 per cent of world exports respectively. There is some international vertical integration in the production of these minerals: for example, Mount Isa Mines exports all its lead bullion to a subsidiary company in the UK and a proportion of zinc concentrates exported go to a refinery in the Netherlands which is 50 per cent owned by Australian Mining and Smelting. However, in general there is not a great deal of international vertical integration and a relatively free and competitive world market exists for lead and zinc.

Of the *silver* produced, about 60 per cent is exported in concentrates and bullion. The remainder is refined in Australia and just under half of that is exported. There is a free world market in silver which is subject to considerable fluctuation as a result of speculative activity.

The demand for lead and zinc is expected to grow at only a relatively slow rate and this will limit development in the production of these minerals in Australia. Nevertheless several projects are in train and will lead to increased production in the early 1980s, boosting the production of lead by about 20 per cent and zinc by more than a third. Established producers, North Broken Hill and Australian Mining and Smelting, plan expansions of their Broken Hill mines, Mount Isa Mines are planning development of the Hilton mine just north of Mount Isa by the mid 1980s, and Electrolytic Zinc is developing its Elura deposit in New South Wales. New entrants will be Aberfoyle Ltd, developing the Que River project in Tasmania (lead, zinc, silver, gold, copper), and Seltrust and MIM, developing the Teutonic Bore project (zinc, silver, copper) in Western Australia. However production will remain highly concentrated in the hands of the established Mount Isa and Broken Hill companies. Foreign equity in the Que River and Teutonic Bore projects is substantial, in both cases by foreign mining companies. Thus the general structure of the industry will remain very similar.[34]

No major development in smelting or refining activities appears likely, particularly in the case of zinc where substantial excess capacity is predicted.

PETROLEUM

The production of oil and gas is constituting an increasingly important sector of the Australian mineral industry. Crude oil has only been produced in Australia since the late 1960s. The first commercial field was proven at Moonie in Queensland in the early 1960s, and shortly afterwards larger discoveries were made at Barrow Island in Western Australia and in Bass Strait. In the late 1970s virtually all Australian crude oil was produced by the companies which had discovered these

Table 18. Crude oil production, 1978

Operating Company	Major Shareholdings	%	Australian Ownership %	Percentage of Australian Output*
Esso Exploration and Production Australia Inc. *and* Hematite Petroleum Pty Ltd (Esso/BHP Joint Venture, Bass Strait)	Esso Eastern Inc. (Exxon Corp)	50	45	92.8
	BHP	50		
West Australian Petroleum Pty Ltd (WAPET) (Barrow Island and Dongara, WA)	California Asiatic Oil Co.	28.6	13	6.9
	Texas Overseas Petroleum Inc.	28.6		
	Shell Development (Aust) Ltd	28.6		
	Ampol Exploration Ltd	14.3		
International Oil Pty Ltd (Moonie and Alton, Qld)	Carrick Investments	100	100	0.2

* Measured in barrels

fields, Esso and BHP in the case of Bass Strait and Western Australian Petroleum (WAPET) in the case of Barrow Island. BHP has a 50 per cent interest in the Bass Strait joint venture which is thus nearly 50 per cent Australian owned (allowing for overseas portfolio ownership), but the Australian ownership of the West Australian fields of WAPET, through the one-seventh interest of Ampol, is much lower. Details are set out in table 18. Foreign equity participation in these ventures is entirely by overseas oil companies undertaking diversification.

All crude oil produced in Australia is refined in the country, domestic crude providing about 70 per cent of input to refineries. There are ten major refineries in Australia (excluding lubricating oil refineries) several of which were established before domestic production of crude oil started. Only one of these (Ampol) is substantially Australian owned; of the remainder all but two are wholly owned by foreign oil companies.

Natural gas was first produced in Australia in 1969. By the late 1970s gas was being produced at six fields, though nearly all production (in 1978) came from the three largest fields: Bass Strait (Esso-BHP) — 45 per cent; Cooper Basin, South Australia (equity participation in which is held by several companies under the Cooper Basin Unitisation Agreement, the major partner being Santos Ltd — 36 per cent); and Dongara, Western Australia (WAPET) — 12 per cent. Approximate Australian equity in each of these projects is 45, 82 and 13 per cent respectively, resulting in overall Australian equity of approximately 58 per cent. Foreign equity participation is again entirely by international oil companies.

Major developments likely during the early 1980s are those on the North West Shelf and in the Cooper Basin. The latter will be an extension of the present gas producing project. The North West Shelf Joint Venture is the largest of all the resource projects being established in Australia in the early 1980s, involving an estimated capital expenditure of over $4,000 million, and will vastly increase the scale of gas production in Australia. In both cases the major initial production will be natural gas, though subsequent plans include production of liquid natural gas, condensate and liquid petroleum gas. Considerable quantities of LNG from the North West Shelf project will be exported, mainly to Japan, and negotiations with Japanese companies were in progress in late 1980. Partners in the North West Shelf consortium are Woodside Petroleum Ltd — 50 per cent, BP — 16.7 per cent, California Asiatic Oil Co. — 16.7 per cent, Shell — 8.3 per cent and BHP (through Hematite Petroleum) — 8.3 per cent. The project is approximately 46 per cent Australian owned (through BHP directly and part ownership of Woodside by the Australian public and BHP).

An extraordinarily large amount of exploration for petroleum was being undertaken at the beginning of the 1980s, with some success in all the major exploration areas. For most of the decade, however,

Esso/BHP will continue to be the major producer of crude oil, recent new discoveries in Bass Strait ensuring that present levels of production will be maintained until about 1990. After this time total Australian production will decline unless new oil fields are found either in Bass Strait or elsewhere. Production of oil from shale and through coal liquefaction is a possibility, though unlikely to occur before the end of the 1980s. If it does, the economic structure of the industry could be considerably altered.

URANIUM

Uranium was produced on a relatively small scale in Australia during the late 1950s and 1960s. As a result of predicted changes in the world market for uranium, and Australian government policy, exploration activity was increased in the late 1960s and early 1970s. Several major deposits were found, mainly in the Northern Territory, but development was delayed pending government approval to mine and export uranium. The Mary Kathleen mine in Queensland operated from 1958 until 1963; it was recommissioned in 1976 following government approval (and indeed with the government, through the Atomic Energy Commission (AAEC), as a substantial shareholder), though operations ceased in early 1982. Queensland Mines (Nabarlek, Northern Territory) completed mining of a large quantity of ore in early 1980. Processing operations were commenced in mid 1980 and the first overseas shipment of uranium oxide (yellowcake) was made in November 1980.[35] Energy Resources of Australia (Ranger, Northern Territory) commenced production in early 1982. Details of these and other projects at various stages of development are given in table 19. There are several other projects at less advanced stages of development, the most significant of which is probably the Western Mining/BP project at Roxby Downs in South Australia, based on large copper-uranium-gold deposits.

Commercial development of uranium mines is subject to government approval in relation to environmental questions, Aboriginal land rights and Australian equity participation. Such approval has been granted to owners of the Nabarlek, Ranger, Yeelirrie and Jabiluka deposits; others are still subject to approval in one or more of these respects. The Foreign Investment Review Board guidelines require 75 per cent Australian equity participation in uranium projects, although this requirement was waived when approval was given for the development of the Yeelirrie uranium deposit in Western Australia in 1979. The nominal equity interests in the project are Western Mining — 75 per cent, Esso — 15 per cent, and Urangesellschaff (a German uranium development company)

Table 19. Uranium projects

Operating Company	Major Shareholdings	%	Australian Ownership %	Stage of Development
Mary Kathleen Uranium Ltd (Mary Kathleen, Qld)	CRA Australian Atomic Energy Commission	51 41.6	63	Ceased production early 1982
Queensland Mines Ltd (Nabarlek, NT)	Kathleen Investments	50	(90)	Considerable mining undertaken; processing started mid 1980; first overseas shipment of yellowcake late 1980
Energy Resources of Australia Ltd (Ranger, NT)	EZ Industries Peko-Wallsend Overseas buyers	30.8 30.8 21.4	(80)	Production commenced 1982
Yeelirrie Development Company Pty Ltd (Yeelirrie, WA)	Western Mining Esso Urangesellschaff	75 15 10	see text	Approval granted 1979, production expected to begin about mid 1980s.
Pancontinental Mining Ltd/Getty Oil Development Co. (Jabiluka, NT)	Pancontinental Mining Getty Oil	65 35	(50)	Approval granted early 1982
Denison Mines Ltd (Koongarra, NT)	Denison Mines	100	nil	Development subject to government approval.

– 10 per cent, but under a complex price sharing arrangement, Esso will provide a further 35 per cent of the total cost of the project and will be entitled to take an additional 35 per cent of the product, so the 75 per cent equity requirement was effectively not met. The Government considered that, in view of the benefits of the project to Australia and the difficulty of obtaining greater Australian participation, the project should proceed.[36] The 75 per cent equity requirement was also waived when approval was given later in 1979 for exploration and development of the Roxby Downs copper-uranium-gold deposit with only 51 per cent Australian equity (the project is a joint venture by Western Mining and BP).[37] Further, in September 1980, when giving approval for the purchase of the Koongarra uranium deposit by the Canadian owned Denison Mines, the Treasurer indicated that when development approval was subsequently applied for, approval might be granted for a level of Australian equity lower than 75 per cent (though at least 50 per cent), if a higher level were "clearly unobtainable", and "provided that the project would be of significant economic benefit to Australia and Australian participants would have the major role in determining the policy of the project".[38] The status of some other projects in this respect is still to be determined. There is thus substantial foreign equity in the Australian uranium industry, though it is lower overall than in most of the other mineral industries. Foreign companies participating are in nearly all cases mining companies seeking diversification of their activities. In at least two cases – Denison and Urangesellschaff – they have a substantial and longstanding interest in uranium mining and processing. The Ranger project is interesting in that overseas buyers of the uranium (mainly West German and Japanese electric power utilities) have taken up most of the 25 per cent equity reserved for overseas interests, a pattern which might be repeated in some other projects.

It is uncertain how many further projects will receive government approval and get to production stage, and what the structure of the uranium sector will look like by the mid to late 1980s. Apart from the equity, environmental and land rights questions, approval for additional projects is likely to depend to a large extent also on their commercial prospects. All processed uranium (yellowcake) is exported, and projections of world demand for uranium are becoming progressively less optimistic. Such demand depends almost entirely on the expansion of nuclear power generation, and there are many uncertainties in this regard. Subdued levels of economic activity in major world economies during the 1980s, high interest rates and the relative capital intensity of nuclear power plants, and doubts about the safety and environmental aspects of nuclear power generation, have all been reflected in a marked slow down in reactor construction.[39] What the world market trends will be in the long term remains to be seen.

CONCLUSION

There are some characteristics which are common to each of the major industries reviewed above. Production tends to be concentrated in the hands of a relatively small number of producers; there is a high level of foreign ownership of mining operations; and in virtually all cases the greater part of mineral output, either in a raw or semi-processed form, goes to overseas markets, so each industry is greatly affected by the world demand and supply situation.[40] But for various technical, economic, geographic, political and historical reasons, the economic framework in which each mineral is produced is distinctive in important respects, and it is impossible to generalize too broadly about the mineral sector as a whole.[41] To conclude, some brief comments should be made on two important aspects of the Australian mineral sector which do not clearly emerge from the foregoing review.

The first aspect is foreign *control* of mining operations. The extent of foreign *ownership* has been discussed in some detail: foreign control obviously follows from foreign ownership, but the relationship between the two is not always a simple one and it should not be assumed that a given level of foreign ownership in a particular industry implies the same degree of foreign control.[42] Majority ownership of a company (or indeed minority ownership where there are no other significant shareholders) is sufficient to give control. Thus there can be substantial Australian equity in a mineral project but it can be effectively controlled by an overseas company with a majority shareholding. On the other hand, some mining companies which are subsidiaries of overseas parents, or in which there are substantial foreign shareholdings, are effectively Australian controlled. For example in 1980 Conzinc Riotinto of Australia Ltd (CRA) was 68.2 per cent owned by its UK parent (the Rio Tinto-Zinc Corporation) but claimed itself to be effectively Australian controlled; MIM Holdings Ltd has the US firm ASARCO as a major (49 per cent) shareholder, but has an entirely Australian board of directors and emphasizes that it is not subject to any foreign control. Further there is considerable overseas portfolio ownership of many Australian mining companies (often around 10 per cent of shares are held by overseas residents and in some cases there are considerable holdings by nominee companies which may include overseas residents), but this rarely has any implications for control. However a new form of overseas ownership in mining companies which is becoming increasingly common — the granting of equity in conjunction with the negotiation of sales contracts — can mean that even a relatively small amount of foreign ownership implies a significant element of foreign control. To determine the degree of foreign control in an industry it is necessary to make a case by case examination of the structure of the operating companies or consortia —

after having decided what is deemed to constitute "foreign control". Such studies have generally indicated that the degree of foreign control usually exceeds that of foreign ownership, though there are exceptions, notably that of the iron ore industry. The most recent official data show that for the year 1974–75 the Australian mining industry as a whole was 51.8 per cent foreign owned but 58.9 per cent foreign controlled. The figures for the industries considered above are summarized (where available) in table 20.[43]

Table 20. Foreign ownership and control (according to value added), selected mining industries, 1974–75

	Foreign Ownership %	Foreign Control %
Black coal	62.1	64.1
Iron ore	55.1	34.1
Silver-lead-zinc	49.3	72.9
Crude petroleum (including natural gas and brown coal)	55.8	86.7
Total mining industry	51.8	58.9

Source: Australian Bureau of Statistics, *Foreign Ownership and Control of the Mining Industry, 1973–74 and 1974–75* (Canberra, 1976).

A second aspect which is obscured by an examination of the mineral sector industry by industry is the fact that there are several diversified mining companies which are active in exploring for and mining *several* different minerals. The names of companies such as BHP, CSR, CRA, MIM, CGFA, North Broken Hill, BH South, Western Mining, Peko-Wallsend, EZ Industries and those of several oil companies crop up frequently in the summary tables in this chapter, and many are involved in the mining of other minerals as well. Most are also actively engaged in mineral exploration programmes. Some of these companies are substantially foreign owned and subject to varying degrees of foreign control, but the majority are Australian companies. Most have been in existence for a long time; however, they have generally expanded and diversified their activities considerably in the last two decades. The details of the more important diversified mining companies are summarized in table 21.[44] While there are some companies involved in only one or two industries which are individually more significant than some of these (for example Utah Development Co., the aluminium majors), those listed in table 21 constitute the core of the Australian mining sector.

Table 21. Major diversified mining companies operating in Australia, 1981

Company, and Major Relevant Subsidiary Companies	Ownership[a]	Mineral Production Activities[b]	Interests in New Developments[c]
Ampol Petroleum Ltd Ampol Exploration Ltd (68%)	Australian	Iron ore; oil and gas; uranium; (in addition to oil refining and marketing)	Coal
British Petroleum Co. of Aust. Ltd Clutha Development Pty Ltd	British Petroleum Group	Coal	Coal; base metals; uranium; oil and gas
Broken Hill Proprietary Co. Ltd Dampier Mining Co. Ltd Groote Eylandt Mining Co. Pty Ltd Hematite Petroleum Pty Ltd Mt Newman Mining Co. Pty Ltd	Australian	Coal; iron ore; oil and gas; manganese; gold	Coal; iron ore; bauxite/alumina/aluminium; oil and gas
CRA Ltd Australian Mining & Smelting Ltd Comalco Ltd (45%) Hamersley Holdings Ltd Kembla Coal & Coke Ltd Cobar Mines Ltd	Rio Tinto-Zinc Corporation Ltd (61.1%)	Coal, iron ore; bauxite/alumina/aluminium; copper; silver/lead/zinc; uranium	Coal; iron ore; aluminium; base metals; diamonds
CSR Limited AAR Limited (85%) Buchanan Borehole Collieries Pty Ltd (93%) Delhi International Oil Corp. Gove Alumina Ltd (51%) Thiess Holdings Ltd (93%)	Australian	Coal; iron ore; bauxite/alumina/aluminium; copper; gas; gold	Coal; iron ore; aluminium; oil shale; uranium

Table 21. (cont.) Major diversified mining companies operating in Australia, 1981

Company, and Major Relevant Subsidiary Companies	Ownership[a]	Mineral Production Activities[b]	Interests in New Developments[c]
Esso Exploration and Production Aust. Inc.	Exxon	Oil and gas; (in addition to petroleum refining and marketing)	Coal; copper-zinc; oil and gas; oil shale; uranium
EZ Industries	Australian	Base metals (mining and refining)	Base metals, uranium
MIM Holdings Ltd Collinsville Coal Pty Ltd Mt Isa Mines Ltd	ASARCO (US) (44%)	Coal; iron ore; copper; silver, lead, zinc (mining and processing); nickel	Coal; iron ore; base metals; uranium
North Broken Hill Holdings Ltd	Australian	Base metals	Base metals
Peko-Wallsend Ltd	Australian	Coal; bauxite/alumina; copper; gold; silver, bismuth; mineral sands	Coal; aluminium; copper; uranium
Renison Goldfields Consolidated Ltd Renison Ltd Consolidated Gold Fields Aust. Ltd Mt Lyell Mining and Rly. Co. Ltd Associated Minerals Consolidated Ltd	CGF (UK) (49%)	Copper; mineral sands; coal; tin	Coal
Shell Aust. Ltd.	Shell Oil Group	Coal; oil and gas; (in addition to petroleum refining and marketing)	Coal; bauxite/alumina; oil and gas
Western Mining Corporation Holdings Ltd	Australian	Bauxite/alumina/aluminium; nickel; gold; mineral sands	Alumina/aluminium; copper; gold; uranium

(a) Companies shown as Australian-owned have some overseas shareholders but generally holding less than 10-20 per cent of shares.

(b) Minerals actually being produced by the company, or by a company or consortium in which it has an interest, in 1981.

(c) In most cases the developments are actually under construction and will be producing the minerals in the near future; in addition many of the companies are engaged in exploration programmes for a wide range of minerals.

Notes

1. Details of production and companies (including those relating to ownership) in the following surveys generally relate to the year 1978, the latest for which data were comprehensively available at the time of writing. Since that time there have been several changes. The most important of these, to early 1982, have been noted in the text, and later major changes in the postscript. Readers seeking the most up-to-date details are referred to the following annual publications: John Alexander & Richard Hattersley, *Australian Mining, Minerals and Oil* (Sydney: David Ell Press); *Jobson's Mining Year Book* (Melbourne: Dun & Bradstreet); Ross Louthean, ed., *Register of Australian Mining* (Perth: Ross Louthean Publishing); and the lists of Major Mining and Manufacturing Projects prepared by the Department of Industry and Commerce, Canberra. The Department of Trade and Resources, Canberra, also publishes a series of booklets on mineral resources. Details of industries not covered in this chapter can be found in these sources.

   In accordance with the essential aim of the chapter – to give a readily digestible overview of the main economic characteristics of the major industries – the text and tables have been simplified by giving abbreviated names of companies and in some cases simplified details of ownership. Figures for Australian ownership are approximate or estimates in some cases, and have been rounded to the nearest whole number. Overseas portfolio investment in Australian companies has been taken into account where possible on the basis of information about shareholders with overseas addresses; however, in some cases this information is not readily available, nor has account been taken of foreign ownership through holdings by nominee companies, so there are some inconsistencies in this respect. The summary tables have been compiled from a variety of sources including, in addition to the sources listed above, the *Australian Mineral Industry Annual Review* published by the Bureau of Mineral Resources, various financial and industry journals, company reports and author's calculations.

   The locations of the various deposits and processing plants referred to are indicated in the maps which follow this chapter.
2. Joint Coal Board, *Black Coal in Australia 1978-79* (Sydney, 1980), p. 6. These, and other figures quoted, are for *saleable* coal. Approximately 72 per cent (Qld) and 84 per cent (NSW) of raw coal is saleable after it has been washed to remove "ash".
3. Bureau of Mineral Resources, *Australian Mineral Industry Annual Review (AMIAR)* 1978, (pre-print). The figure for the Electricity Commission includes production from mines owned by the Commission but operated by private companies. Because the Commission uses raw coal in its power stations the figure also tends slightly to overstate its significance as a producer.
4. R.B. McKern, *Multinational Enterprise and Natural Resources* (Sydney: McGraw-Hill, 1976), pp. 67-70.
5. Joint Coal Board, *Black Coal,* pp. 6 and 177. The figures are for the years 1976-77 to 1978-79.
6. *AMIAR,* 1978 (pre-print), p. 39.
7. Joint Coal Board, *Black Coal,* p. 183.
8. For a detailed discussion of the nature of mineral export contracts in general see Ben Smith, "Long-term contracts in the resource goods trade", *Australia, Japan and the Western Pacific Economic Relations,* report presented by Sir John Crawford and Dr Saburo Okita (Canberra: AGPS, 1976), pp. 299-325.
9. *Australian Financial Review (AFR)* 14 February 1980, p. 1; 18 February 1980, p. 1.
10. *AFR,* 10 October 1980, p. 45.
11. *AFR,* 21 February 1980, p. 1.
12. *AFR,* 28 February 1980, p. 22; 29 February 1980, p. 16.
13. *AFR,* 7 February 1980, p. 43.

14. *AFR,* 18 July 1980, p. 19.
15. *AFR,* 29 September 1980, p. 12.
16. *AFR,* 25 March 1980, p. 3; 8 April 1980, p. 1.
17. *AFR,* 7 February 1980, p. 43.
18. Geoffrey Blainey, "The Cargo Cult in Mineral Policy", *Economic Record* 44, no. 108 (December 1968): 470-79.
19. McKern, *Multinational Enterprise,* pp. 52-55.
20. Changes will occur as a result of the CRA's purchase in 1980 of Texasgulf's interests in the Pilbara. The other shareholders in Cliffs WA have a pre-emptive right to purchase any existing shares on offer; they have indicated they wish to exercise that right but to offer them to the Australian public. This would increase Australian equity by over 10 per cent. CRA's purchase will, however, give it a major interest in a possible new project.
21. *AFR,* 3 March 1980, p. 13.
22. *AFR,* 24 January 1980, p. 32; the fall was due in part to revaluations of the Australian dollar against the US dollar (in which contracts were written) in the early 1970s.
23. *AFR,* 4 September 1980, p. 26.
24. *AFR,* 3 September 1979, p. 1.
25. For a review of the likely developments in the industry see *Australian Business,* 4 December 1980, pp. 57-60. Also *AFR* 18 June 1980, p. 25; 17 October 1980, p. 6.
26. The reasons underlying the structure of the world industry are discussed in John A. Stuckey, "Vertical Integration in Aluminium: Theory, Evidence, and Implications for Australia", paper presented to the Eighth Conference of Economists, Melbourne, August 1979; and F.E. Banks, *Bauxite and Aluminium: an Introduction to the Economics of Non-fuel Minerals* (Lexington, Mass.: Lexington Books, 1979).
27. McKern, *Multinational Enterprise,* outlines the background to the establishment of the various projects, pp. 152-56, 219-24.
28. Stuckey, pp. 38-42. The Australian situation is contrasted with a world industry average of 80 or 90 per cent of production being consumed by internal corporate transfers.
29. Foreign Investment Review Board, *Annual Report 1980,* pp. 9-10, 17-18.
30. Stuckey, pp. 46-48; see also chapter 7.
31. *AFR,* (Mining & Oil Review) 19 June 1980, p. 12.
32. See McKern, *Multinational Enterprise,* pp. 79-81.
33. Other establishments produce secondary refined lead and silver using scrap materials.
34. *AFR,* (Mining & Oil Review) 19 June 1980, p. 17.
35. *AFR,* (Mining & Oil Review) 19 June 1980, p. 9; 31 July 1980, p. 23.
36. Foreign Investment Review Board, *Annual Report 1979,* pp. 8-9, 33-35.
37. *AFR,* 10 October 1979, p. 1; Foreign Investment Review Board, *Annual Report 1980,* p. 17.
38. Treasurer, Press Release, no. 122, 14 September 1980.
39. *AFR,* 18 September 1980, pp. 21-23 quoting the Australian Atomic Energy Commission.
40. No significant amounts of petroleum are exported at present, though Australia will soon become a major exporter of liquid natural gas.
41. A large number of other minerals are produced in Australia. They constitute only a small proportion of the total value of mineral output, though in some cases Australian production is significant in terms of world supply. As in the case of the quantitatively more important minerals, the circumstances under which these minerals are produced and marketed are distinctive. Space does not permit discussion of these industries here; information on them can be obtained from the sources cited in note 1.
42. Of course foreign ownership is an important issue in its own right, for it implies that a corresponding proportion of the net profits from mining operations accrues to non-Australians. Foreign *control* raises the additional question of the manner in which the production and marketing of Australian minerals is undertaken and its conformity with the national interest as distinct from that of foreign controlled companies.

43. An enterprise was deemed by the ABS to be foreign controlled if a single foreign investor held at least 25 per cent of the paid up value of voting shares in the enterprise provided that there was no larger holding by an Australian controlled enterprise or Australian resident individual. The figures are calculated on the basis of value added so are not directly comparable with the ownership data cited in the above surveys. Figures for bauxite and copper were not made available.
44. Principal sources for this information from which fuller details can be obtained are the references by Alexander & Hattersley and Louthean, and the Jobson's Year Book cited in note 1, together with company annual reports.

POSTSCRIPT TO CHAPTER 3

Since the above was written there have been some changes in the ownership of operating companies within the industries discussed, the most significant arising from the following developments:

- MIM Holdings announced plans to purchase (both immediately and over 1981-82) additional shares in the US company ASARCO (the major shareholder of MIM) to bring MIM's share in ASARCO to approximately 21 per cent. As part of the proposal ASARCO agreed to make available about 10 per cent of its shareholding in MIM for sale to Australian interests.
- The Consolidated Gold Fields (CGF) group of the UK reorganized its operations. A new holding company, Renison Goldfields Consolidated (RGC), was created, owned 49 per cent by CGF and 51 per cent publicly. RGC owns 100 per cent of (among other companies) CGFA and Mount Lyell. The reorganization will decrease the overseas shareholdings in CGFA, though increase it in the case of Mount Lyell.
- CSR Ltd took over Delhi International Oil Corp. (USA) which has a 17 per cent interest in the Cooper Basin natural gas project.
- By early 1982 some further new black coal projects had come into production or were on the point of doing so, including the large German Greek project in central Queensland, in which Australian equity is just over 50 per cent.
- Production started at the Teutonic Bore copper–silver–zinc mine in Western Australia (BP/MIM, 60/40) and the Que River base metals mine in Tasmania.
- Esso Australia scaled down its interests in alternative energy projects and in particular withdrew from the Yeelirrie uranium project in which it held a 25 per cent interest.

These changes do not fundamentally alter the broad picture presented in the tables and text, though the further "Australianization" of MIM substantially alters some of the figures given for Australian ownership in some industries, particularly copper.

In virtually all industries, market prospects became less bright after about mid 1981; accordingly, prospective developments and thus changes in the structure of industries foreshadowed in the discussion seem likely to be slower in eventuating. In particular several coal projects, particularly in New South Wales, were deferred; the steadily worsening prospects for the world aluminium industry led to the deferral of some of the planned extensions to existing projects and of some of the new projects listed in table 13 (namely the Alcoa refinery in WA, the Alcan smelter in Queensland, the BHP–Alumax smelter in the Hunter Valley (following the withdrawal of Alumax) and the Alcoa smelter at Portland); and a major deferral of work on the North–West Shelf gas project was announced, following difficulties in negotiations with the Japanese power utilities who were expected to be the major buyers.

Fig. 3 Black coal location map

1 Bowen District
2 Mackay District
3 Blair Athol
4 Blackwater District
5 Kianga-Moura District
6 Callide
7 West Moreton District
8 Leigh Creek
9 Singleton-Northwest District
10 South Maitland District
11 Western District
12 Newcastle District
13 Southern District
14 Collie

Fig. 4 Iron ore location map

1 Yampi Sound
(Cockatoo Is., Koolan Is.)
2 Shay Gap, Sunrise Hill
3 Mount Goldsworthy
4 Robe River (Pannawonica)
5 Deepdale
6 Marandoo
7 Yandicoogina
8 Mount Tom Price
9 Area C
10 Mount Whaleback
11 Paraburdoo
12 West Angela
13 Koolyanobbing
14 Iron Knob, Iron Monarch
15 Iron Baron, Iron Prince
16 Savage River

Fig. 5 Aluminium location map

1 Gove
2 Weipa
3 Bundaberg
4 Mitchell Plateau
5 Gladstone
6 Hunter Valley
7 Darling Range (Jarrahdale)
8 Pinjarra
9 Kwinana
10 Wagerup
11 Portland
12 Point Henry
13 Worsley

Fig. 6 Copper location map

1 Dianne
2 Tennant Creek
3 Gunpowder
4 Townsville
5 Mount Isa
6 Mons Cupri
7 Mount Morgan
8 Teutonic Bore
9 Olympic Dam (Roxby Downs)
10 Cobar
11 Mount Gunson
12 Broken Hill Area
13 Woodlawn
14 Port Kembla
15 Rosebery, Hercules
16 Mount Lyell
17 Mount Chalmers

Fig. 7 Lead, zinc and silver location map

1 Mount Isa
2 Elura
3 Cobar
4 Broken Hill
5 Port Pirie
6 Cockle Creek
7 Teutonic Bore
8 Woodlawn
9 Port Kembla
10 Que River
11 Rosebery, Hercules
12 Risdon

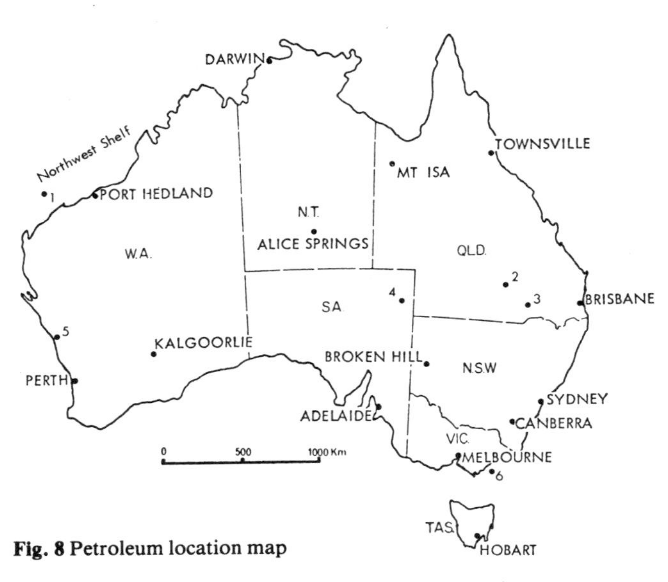

Fig. 8 Petroleum location map

1 Barrow Island Area
2 Roma
3 Moonie/Alton
4 Cooper Basin
5 Dongara
6 Gippsland Shelf (Bass Strait)

Fig. 9 Uranium location map

1 Jabiluka
2 Nabarlek
3 Ranger
4 Koongarra
5 Mary Kathleen
6 Yeelirrie
7 Olympic Dam (Roxby Downs)

Part B

4 Urban Development

J.R. Laverty

The aridity of the continent has shaped the European settlement of Australia and thereby influenced its urban development. Much of the country is desert and large parts of the zones of adequate rainfall, along the east and south-eastern coast and in the south-west, are so mountainous that agriculture is not possible. In the climatically favoured zones the density of rural settlement, as a rule, declines in response to the rainfall pattern with distance from the coast.[1] Not surprisingly Australia's overall population density is a mere two persons per square kilometre and there are only eight persons per square kilometre living in the permanently inhabited areas, the lowest density of the world's seven largest countries. Since the main function of most country towns has been to service their hinterlands, the urban population in rural districts has conformed to the general pattern of coastal concentration. Consequently, by 1971, 84.7 per cent of Australia's population lived within eighty kilometres of the sea.[2]

Australia's size has combined with its aridity to confirm the coastal bias of settlement. In the absence of large, navigable river systems, the Great Dividing Range inhibited inland settlement until railways were constructed after 1850.[3] Indeed, Robinson has argued that the size of the continent and the smallness of the population made the metropolitan concentration of population and economic activity essential to optimal development.[4] Only in the case of mining have geographical factors encouraged urban development which differed basically from the general coastal and metropolitan trend in the Australian settlement pattern, and even the direction and character of mining activity has been modified by the "physical and cultural characteristics of the Australian environment".[5]

Australia's demography and urban system have been more strongly influenced by the timing and nature of colonization than by geographical and environmental factors. For example, the rapidity of Australia's urbanization, illustrated by table 22, has been attributed to its colonization during the age of industrialization.[6] Australia also became one of the most urbanized countries in the world with a greater proportion of its population (64.5 per cent) residing in cities of over a hundred thousand inhabitants than any other country and the highest urbanization-population concentration index number (68).[7]

Table 22. Percentage of the population classified as urban

Year	New South Wales	Victoria	Australia
1861	41.1	40.6	
1871	50.1	54.5	
1881	57.7	56.9	
1891	65.5	65.8	
1901	67.8	64.6	
1921	67.8	62.3	61.1
1933	68.2	65.4	63.8
1947	72.0	71.0	68.0
1954	82.6	81.3	78.7
1961	85.2	84.8	81.9
1966	86.4	85.5	83.2
1971	88.6	87.7	85.6
1976	88.7	87.8	85.9

Source: Figures for 1861–1901 from N. G. Butlin, *Investment in Australian Economic Development 1861–1900* (Cambridge: Cambridge University Press, 1964), p. 184; the remaining figures were taken from Census returns.

McCarty has explained the extreme concentration of Australia's population in the capital cities in terms of the pivotal commercial role the cities played in the colonization process.[8] From the beginning they gave the developing pastoral, agricultural and mining industries access to world markets, and channelled overseas goods, labour, capital and technology into their expanding hinterlands.[9] The nature of these staple industries promoted the growth of the port-cities, while modern transport and communications technology and the fan-like transport systems which evolved in most of the colonies emphasized their commercial/linking role, stimulated their expansion, and increased their dominance.[10] Considerations of security and administrative convenience also made them the logical locales for government. Only in Tasmania and Queensland, where the capital city was eccentrically located, was the primacy of the capital city modified substantially.

Because the capital cities have generally been the major growth nodes, overseas immigration has been a major factor in their development. For example, it promoted the expansion of Sydney and Melbourne in the 1920s and gave a strong impetus to metropolitan growth and economic and social diversification in the post second world war period.[11] But migration within Australia also played a part in shaping the Australian urban system. The growing efficiency of the pastoral and agricultural industries, achieved through improved technology, reduced the need for rural labour, thereby engendering a continuing rural-urban drift of population. Lack of success, the exhaustion of alluvial deposits and mineral ore bodies, and the adoption of technological innovations in mining and metallurgy also sent a fluctuating flow of disappointed diggers and redundant miners from mining camps and towns towards the coast and into the capital cities, where expanding employment opportunities were to be found. Except in the immedite post gold-rush period when many diggers returned directly to the port-cities and towns, there was a stepped internal migration flow from rural areas to local and regional centres to metropolitan areas.[12]

However, the magnitude of this "retreat" to the capital cities and the speed and extent of metropolitan growth should not be allowed to obscure the quite significant expansion of the rural and "other urban" populations before 1900[13] due to rapid mining and agricultural development. Indeed some of the disappointed diggers and redundant miners turned to farming. It was not until the 1930s that rural-urban migration reached a level which led to a declining rural population. Although rural depopulation has continued, Burnley has shown that in recent times internal migration has become more complex, involving two-way flows. After the second world war, employment opportunities in the industrial cities of Newcastle, Wollongong and Geelong attracted both overseas and internal migrants on a sufficiently large scale to moderate the primacy of Sydney and Melbourne. Since 1966 Sydney, Melbourne and Adelaide have lost population to the rest of their states, while Brisbane and Perth have experienced gains as well as attracting an increasing number of people from Sydney and Melbourne. Consequently the continued growth of Sydney and Melbourne has become increasingly dependent on overseas immigration. Of course, Canberra's rapid growth has largely been due to internal migration, much of it from Sydney and Melbourne. A good deal of the migration from these south-eastern metropolitan areas has focused on near metropolitan regions and a limited number of provincial and regional centres.[14] Thus while rural population has declined and metropolitan centralization has continued, non-metropolitan urban places, as a class, have maintained their share of the population, and some individual sub-metropolitan and regional centres have grown substantially.

The expansion of the capital cities and their ability to attract migrants have been due in part to the growth of manufacturing in Australia, since locational advantages arising from their size, their access to overseas labour, capital, technology and raw materials, their location at the hub of the transport and communications systems and the entrepreneurial abilities and political influence of their businessmen ensured that the bulk of manufacturing plants would be established within the confines of the cities.[15] Since 1900 "the industrial specialisation in the metropolitan centres has given them cumulative advantages in securing a rising share of the nation's economic growth".[16] Studies have shown that the concentration of manufacturing employment in the capital cities is greater than the concentration of population and that industrial production is further concentrated in Sydney and Melbourne, which have higher levels of industrial specialization, exhibit a more mature industrial profile and have an extremely high level of commercial leadership.[17]

The international capital flows associated with Australia's economic growth and the exploitation of its natural resources have accentuated metropolitan dominance by concentrating developmental financial transactions and management functions in the capital cities, especially in Sydney and Melbourne. As Stilwell has pointed out, the internationalism of the capitalist system has reinforced the trend towards centralization: "the new imperialism of the multi-national corporation is taking over where British colonialism left off".[18] From the beginning the capital cities were the ideal locations, not only for commercial, financial and government services, but for the tertiary industries in general; as Australia moved towards a post-industrial economy and society after the second world war, the tertiary sector became the prime generator of urban growth and centralization. The capital cities became multi-functional metropoli in which the interdependent sectors of the economy converged.[19] The cumulative effect of these concentrating factors is shown in table 23.

Table 23. Metropolitan population as a percentage of the total population

Year	Sydney NSW	Melbourne Victoria	Brisbane Qld	Adelaide SA	Perth WA	Hobart Tas.	Australia
1861	27.6	25.9	20.1	14.4	–	21.6	–
1871	27.3	28.3	12.5	23.0	20.7	18.8	25.6
1881	29.9	32.8	14.6	37.1	19.6	18.3	32.0
1891	33.9	43.1	23.8	41.6	17.0	22.8	37.1
1901	35.8	41.3	23.7	44.7	36.2*	23.6	36.8
1911	38.2	44.8	23.0	46.1	37.7	20.7	38.0
1921	42.9	50.1	27.8	51.6	47.0	24.5	43.0
1933	47.5	54.5	31.6	53.8	47.3	26.5	46.9
1947	49.7	59.7	36.4	59.2	54.2	29.8	50.7
1954	54.4	62.2	38.1	60.7	54.5	30.8	53.9
1961	55.7	65.3	40.9	60.7	57.0	33.1	56.1
1966	57.8	65.5	43.2	66.7	59.7	32.2	58.1
1971	61.1	71.5	47.5	71.8	68.2	39.2	62.9+
1976	63.3	71.4	47.0	72.4	70.5	40.3	64.7+

* A different basis of calculation used

+ Including Canberra and Darwin

Source: For 1861–91 see Bruce Ryan, "Metropolitan Growth", in *Contemporary Australia: Studies in History, Politics and Economics*, ed. Richard Preston (Durham, N.C.: Duke University Press, 1969), table 2, p. 204; for 1901–76 see Census returns.

There are, however, some significant exceptions to this strong trend towards the centralization of economic activity in the capital cities, which have usually been associated, directly or indirectly, with mineral and metal processing. During the nineteenth century, the introduction of deep-mining and the establishment of ore treatment plants encouraged the growth of manufacturing and induced, or sustained, urban development at centres as widespread as Ballarat, Bendigo, and Castlemaine in Victoria; Wallaroo and Port Pirie in South Australia; Broken Hill, Newcastle and Port Kembla in New South Wales; Mount Morgan and Charters Towers in Queensland; Kalgoorlie in Western Australia; and Zeehan and Mount Lyell in Tasmania. Since 1900, mineral processing plants have also been set up at other sites, including Mount Isa, Bell Bay, Whyalla, Point Henry, Townsville, Gladstone and Kwinana, engendering urban growth at these centres.[20]

The establishment of an integrated iron and steel industry at Newcastle and Wollongong, and of car manufacturing, heavy engineering and petro-chemical plants at Geelong, has given the greatest impetus to the development of manufacturing and to urban growth in non-metropolitan centres. Indeed, Burnley has argued that these developments have been the crucial factor in raising the populations of Newcastle and Wollongong to between 200,000 and 300,000 and Geelong to 120,000, and thereby modifying the extreme metropolitan primacy in New South Wales and Victoria. Burnley has also recognized that the agglomerating effects of the heavy assembling and processing industries and the location of the tertiary sectors of such industries in the metropolitan areas have contributed directly to the growth of Sydney and Melbourne. Stilwell contends that, since these industrial centres lie within the orbit of their capital city, they actually increase metropolitan dominance.[21]

Thus the rapid and extensive urbanization of the population and the centralization of population and economic, political and social activities in the capital cities are the two major features of urban development in Australia. Mining and the processing of minerals are the only activities which have played a continuing, and limited, but significant, role in restraining metropolitan domination and in promoting decentralization.

THE URBAN DIMENSION OF MINING

The development of mining in Australia has gone through three phases. While there may be debate concerning the boundaries between them, these phases can be defined broadly as the period up to 1914, the inter-war interregnum and the post second world war mining boom.[22]

The first period may legitimately be regarded as the golden era of Australian mining. Gold was the glamour metal with power to attract labour and population into the interior. During the decade that followed the gold discoveries in New South Wales and Victoria, the population of Australia grew more than one and a half times, while that of Victoria increased sevenfold. Queensland subsequently also experienced the demographic quickening that gold engendered. The spectacular finds in

Western Australia were largely responsible for the 700 per cent growth in that struggling colony's population between 1885 and 1905.

Gold mining also created more employment, especially during the alluvial phase, and had a much greater value of production than any other mineral mined; indeed, gold dominated mineral production until well into the twentieth century.[23] Nevertheless, the recovery of other minerals also played an important part in the development of mining and the emergence of mining towns and associated centres during this phase. Gold, coal and copper were the three main minerals mined, but silver, lead, zinc and tin were recovered in sufficient quantities to make significant contributions to mineral output and to influence urban development. Coal mining, in particular, emerged as an important aspect of Australian mining and urban growth, especially in New South Wales, during the second half of the nineteenth century as steamships replaced sailing vessels, manufacturing developed, and railways became the country's predominant mode of transport.

The high value/weight ratio of gold and its relatively stable price enticed prospectors into Australia's remotest regions and ensured that alluvial deposits were exploited whenever they were found.[24] But the prospectors also discovered practically all the readily identifiable ore bodies of contemporary commercial value, and those where mining was a viable proposition were exploited. Transport costs posed real problems to prospectors and diggers, even on the central Victorian gold-fields.[25] They placed quite severe limitations on the exploitation of gold reefs and on urban settlement in remote and inhospitable areas, such as the Croydon, Etheridge and Palmer fields of Queensland and the Pilbara, Murchison and Eastern Goldfields of Western Australia. Consequently only the rich and/or extensive reefs in such regions could bear the costs of establishing a productive mine and the associated infrastructure.

Cost considerations were a much more significant restriction on base metal mining. Prospectors and diggers were attracted to alluvial tin-bearing areas since, like alluvial gold, alluvial tin was easily recovered with simple equipment and did not require elaborate and costly treatment. But mineral ores had to be mined and treated by expensive processes. Consequently only the most accessible, rich and extensive reefs were mined. Mining was more likely if the deposit was in a settled area which possessed the infrastructure of a rural community. Despite the difficulties, base metal mines and associated urban centres were established in such remote regions as Cobar and Broken Hill in New South Wales; Cloncurry and Herberton in Queensland; Mount Lyell, Zeehan and Mount Bischoff in Tasmania; and Northampton in Western Australia.

The introduction and development of efficient mining, metallurgical and transportation technologies were of critical importance in overcoming the difficulties posed by distance and a hostile environment. Beginning with the "almost handicraft levels of the early Cornish, Scot and Welsh miners" sophisticated mining and ore treatment techniques were adapted or developed at mining ventures as far apart as Broken Hill, Mount Morgan, Mount Lyell and Kalgoorlie.[26] It was the development of deep mining, facilitated by these advances in technology, that sustained the growth of mining towns and the ports which serviced them.

Although the ease and cheapness of alluvial mining attracted large numbers of diggers to gold and tin fields, this form of mining seldom led to the establishment of significant permanent settlements. Frequently the deposits were small and quickly worked out, as in the Illawarra region of New South Wales where the "Mining towns and gold-fields littered with campers generally, though not invariably, rocketed into existence once a rush began, promptly achieved a pinnacle of population never again recaptured, and then commenced their long, wobbling, dissolution, ultimately expiring altogether or staving off death with new, ignominious functions — a sawmill and eucalyptus distillery at Nerrigundah, rough grazing near Araluen, and a few village facilities at Mogo and Major's Creek".[27] More extensive finds usually produced short-lived boom towns unless their permanency was assured by deep alluvial or reef mining, or by service activities to a surrounding settled region.

The urban dimension of mining is nowhere more apparent than in Victoria where rich gold-fields did much to transform a prosperous pastoral colony into a rapidly urbanizing society. In an analysis of the ebb and flow of urban growth in Victoria, Cloher shows how the establishment and growth of towns accelerated during the gold-rushes of the 1850s, reached a peak in the 1860s as deep mining became common and received a sudden check during the 1870s as gold deposits were exhausted in an increasing number of gold-fields.[28] The decline of gold towns continued in the 1880s even though there was a resurgence of urban growth.

Linge has pointed out that "Of the forty-nine towns with 500 or more inhabitants in 1861, thirty-three owed their origins to gold mining", and that the population of "the gold-fields areas grew by a further forty-two thousand during the 1860s when the addition of sixty-five thousand women and children compensated for a decrease of twenty-three thousand adult males".[29] Butlin found that only three gold towns — Stawell, St Arnaud and Bendigo — sustained their earlier growth during the 1870s before stagnating in the 1880s. The towns which expanded consistently until the end of the 1880s — Melbourne, Warrnambool and Benalla — were not gold towns, nor were the new generation of small towns which emerged in the 1880s.[30] Indeed, from 1870 urban growth centred on Melbourne as gold mining declined, population migrated to the northern colonies, miners shifted into agriculture and the drift of disappointed diggers and others to Melbourne occurred.

The settlements which quickly formed during the rush which followed a gold discovery have been described by Serle in terms reminiscent of an annual fair.[31] It was only after the long-term life of a gold-field had been established that such settlements took on an air of permanency and the trappings of a truly urban community. The introduction of deep mining with its sophisticated machinery and elaborate ore treatment plants set the seal on a town's future. The rapid growth and early social maturation of Ballarat and Bendigo made them the show places of the Victorian gold-fields.[32]

Both of these towns were sheltered from the general decline of mining by the longevity of their deep reefs, the development of engineering, textiles and other forms of manufacturing and their role as service centres for the surrounding agricultural districts. Ballarat's population peaked at fifty thousand about 1870 before declining and then recovering to forty-six thousand in 1891. Bendigo continued to grow until it had some thirty-six thousand inhabitants. Meanwhile, despite an assay into manufacturing, Castlemaine's population had fallen from over thirteen thousand to seven and a half thousand by 1881. At that time only Maryborough, Ararat and St Arnaud had prospects of growth due to surrounding farmlands and a railway junction. Most other gold towns lost population as they "retired into semi-rural slumber from which they failed to awaken during the nineteenth century". However the extent of the decline should not be exaggerated since only the smallest of the gold settlements of Victoria became ghost towns.[33]

Because the deposits were less extensive and more of them were alluvial, the gold towns of New South Wales were generally smaller and less permanent. In Butlin's view, gold-mining in New South Wales did little more than provide a base for towns, such as Deniliquin, Adelong, Gundagai, Orange, Bathurst and Muswellbrook.[34] He might also have added Forbes and Young. Apart from the district service and tourist centre of Gulgong, little is left of the gold settlements north of Bathurst or at Kiandra, apart from a few historical buildings, to remind us of their former size and vigour.

Gold mining was much more important to economic development in Queensland than in New South Wales, but it did not give the impetus to urban growth experienced in Victoria. Although Queensland gold deposits and reefs may possibly have been as rich and extensive as those of Victoria, distance, harsh climate, difficult terrain and the absence of an infrastructure made them more costly and difficult to mine. Moreover, with the notable exception of the Palmer field, alluvial deposits were generally smaller and more scattered and much of the reef gold, especially in North Queensland, was dispersed in shallow shoots. Profits were therefore seldom large enough to finance the systematic prospecting and exploration needed to find the deeper reefs. This pattern of distribution also encouraged the semi-nomadic, fiercely independent miner to resist the introduction of large-scale company mining. Consequently, the gold-fields were seldom effectively exploited and many of the mining settlements were of a transient nature.[35]

Nevertheless, urban expansion in Queensland between 1861 and 1900 "centred around a dozen towns, based on sustained mining activity, coastal agriculture and the development of port towns from which spur railways ran into the interior".[36] It was in gold-fields where companies were formed to mine the deep reefs that substantial mining centres, such as Gympie, Mount Morgan and Charters Towers, and the lesser towns of Ravenswood and Croydon were established. In its heyday Charters Towers was Queensland's second city with a population of twenty-five thousand; the fabulous mine at Mount Morgan spawned a more modest town of ten thousand inhabitants during its most prosperous days. When gold production reached a peak in 1903 and mixed farming was developing in its hinterland, Gympie could boast of fifteen thousand residents. The mines at Croydon and Ravenswood were never extensive enough to support a town of more than five thousand.

As the gold gave out Queensland's gold mining centres decayed. Some, such as Maytown and Ravenswood, became ghost towns. Those located in well-endowed rural districts, like Gympie and Eidsvold, continued as small, but thriving, service centres. Croydon survived as a small inland pastoral service centre. Mount Morgan continued, a shadow of its former self, as a small copper mining town, while Charters Towers, that exemplar of lesser rivals, was saved from an ignominious end by boarding schools, public institutions and the service needs of the surrounding pastoral industry.

In no other colony was distance and a hostile environment such a curb on mining development and urban growth as in Western Australia. That "lightning run of finds" which stretched from the Kimberleys to the Great Australian Bight was matched by the almost equally rapid abandonment of mine and settlement. Many mining centres never got beyond the "canvas and hessian camp" stage before being deserted by diggers rushing to some new discovery. Even settlements on fields with more substantial deposits were often short-lived.[37]

The richer and more extensive reefs of the Eastern Goldfields attracted thousands of miners and soon larger and more permanent towns emerged. By 1901 one third of Western Australia's population lived on the Eastern Fields, more than half of them on the East Coolgardie, or Kalgoorlie, field. Coolgardie was the first boom town, rapidly growing from a camp of tents, pole and hessian humpies and hop-bush and bullock hide huts into a town of galvanized iron hotels, shops, stores and government offices straggling along wide, dusty Bayley Street. By 1895 it had a population of eight thousand and a year later it was linked by rail with Perth. Gradually canvas and corrugated iron were being replaced by jarrah, stone and brick and the town took on an air of permanence. But

Two photographs of Hannan Street, Kalgoorlie, Western Australia, taken in 1895 and 1904, show the remarkably rapid development of the town at this time. *(Copied from the Battye Library Pictorial Collection, 4164P and 9243P.)*

Coolgardie's reef did not run deep and a year later thousands left for other fields and the town slipped into rapid and irreversible decline.[38]

The mantle of "capital" of the Eastern Goldfields had fallen on neighbouring Kalgoorlie, which by 1899 was producing seven times more gold. Standing at one end of the "golden mile" which stretched south to Boulder, Kalgoorlie had the advantage of rich and much more extensive reefs. The golden mile was quickly transformed from "a dirty and very insanitary camp of humpies in a waterless desert of red sand and low scrub", in 1894, into the twin municipalities of Kalgoorlie and Boulder with a population of twenty thousand by 1897: towns complete with electric light, macadamized main street, formed footpaths and side streets and suburbs of white-washed, galvanized iron houses.[39] In a period of eight years the rows of galvanized iron sheds in unformed, dusty Hannan Street were replaced by imposing buildings in brick, stone, cast iron and jarrah wood. A tramway, running down the centre of the street, joined the twin cities and telegraph and electric light poles stood with military precision in a macadamized main street which had once been studded by stunted desert trees.[40]

There was a zest and optimism in the Eastern Goldfields towns of Western Australia. Many of them had amenities far superior to those of larger towns elsewhere, but despite their mushroom growth and precocious sophistication, few survived the exhaustion of the reefs. Only Kalgoorlie and Boulder continued as vital centres until the rejuvenation which followed the nickel boom of recent times. By 1910 ruined stone hotels, piles of empty bottles, neglected graveyards, wrecked vehicles and the ubiquitous poppet heads and rusting mining machinery were the only reminders of the short and eventful existence of scores of Western Australian mining settlements.[41]

With the exception of tin, which was frequently found in alluvial beds, silver and base metals usually required the use of deep-mining methods and the treatment of ores. Consequently, the mines normally spawned urban settlements which varied in size and permanency according to the richness and size of the lodes, the acumen of the managements and the skill of the geologists, technicians, chemists and engineers. World prices for these metals were unstable, unlike gold, and the fluctuations in prices generally forced the closure of the poorly managed and the more marginal mines. The towns dependent on them therefore had checkered histories involving cycles of growth and decay corresponding to the rise and fall of prices. Because the values of metals were much lower than in the case of gold, transport was an important factor in the viability of mines, especially as many of them were in isolated and, sometimes, relatively inaccessible areas. Under the circumstances there was always a temptation for mine managers to rip-out the rich pockets of ore, thus reducing the productive life of the mine and the life of the town to which it gave sustenance. On the other hand, the lower value of the metals led governments to grant larger leases. Mining towns were therefore sometimes company towns, since a single company controlled the whole mineral deposit. Because of the isolation and inaccessible location of many mines, the towns which were established were frequently almost totally dependent on the mine. Consequently when the mine closed the town inevitably had to be abandoned as there was no other activity to sustain it.

Australia's first base metal (silver-lead) mine was opened in South Australia in 1841, for the good reasons that some Cornish miners had come to the colony and the ore body was close to Port Adelaide and easy to develop.[42] It gave a timely boost to the ailing economy, but it was the copper mines at Kapunda and Burra north of Adelaide (1845–78) and on Yorke Peninsula at Moonta and Kadina (1860–1923) which set the colony on its feet and made copper with grazing and agriculture one of the three industries which determined South Australia's pattern of settlement and produced its wealth. The ores were rich enough to justify the introduction of hundreds of Cornish miners and Welsh smeltermen. In the decade following the initial discovery of silver-lead at Glen Osmond in 1841, the population of South Australia rose from fifteen thousand to sixty-four thousand. Towns soon sprang up around the mines. By 1851 Kapunda consisted of some 350 houses and had a population of more than two thousand inhabitants. At its peak in the same year the population of Kooringa and the constellation of mining villages around the Burra mine reached five thousand, half of them living in cave dwellings dug into the creek banks.[43]

The Burra constellation of settlements held the largest urban population outside Adelaide. The mines at both Burra and Kapunda were forced to close in 1851 when the miners rushed to the Victorian gold diggings and did not open again until 1855 and 1856, respectively. The mining settlements were therefore deserted for half a decade. The revival of mining gave them a new, but short, lease of life. Eventually in the mid 1870s the copper ores were exhausted and the mines closed. Kapunda survived as a smaller service centre for the surrounding wheatlands and a producer of agricultural implements; Burra as a small railhead for the pastoralists further inland.

The Wallaroo, Kadina and Moonta mines of Yorke Peninsula were even richer than those at Burra. By 1861 ships were bringing Cornish miners, Welsh smeltermen and Newcastle coal to the region to exploit the rich ores. Soon the smelters at Wallaroo were consuming one tenth of all the coal shipped from Newcastle and the poorer ores were being back-loaded to Newcastle to be smelted there. At their peak of production in 1872–73 the mines at Moonta, Kadina and Wallaroo employed some 2,800 men and boys and the triangle of copper towns had a population of 11,000 at the 1881 Census. Their combined population may have reached 20,000 at the peak of the copper boom. They displaced Burra as South

Australia's second largest urban centre. The Wallaroo mining companies staved off disaster during the copper slump of the 1880s by amalgamating and diversifying into the manufacture of mining and smelting equipment and supplies, chemicals and fertilizers. Poorer ores and the fall in world copper prices at the end of the first world war forced the mines to close in 1923 but the towns continued, much smaller in size, as service centres for the surrounding wheatlands.

The main copper mining centres of South Australia were essentially company towns. The Cornish miners brought with them not only their mining methods and terminology and architecture, but also Cornish company paternalism. Besides providing housing and other urban facilities, the companies supplied a range of welfare services to their employees and their families.[44]

Although both copper and tin were mined in significant quantities in New South Wales, it was in the coal mining centres of Newcastle, Greta, Singleton, Bulli and Wollongong and the silver-lead town of Broken Hill that non-metropolitan urban growth was centred in the 1870s and 1880s. The fortunes of the copper town of Cobar waxed and waned with the rise and fall of world copper prices. At its peak in 1881 it exceeded two and a half thousand. Tin mining in the north of the colony during the 1870s created the short-lived towns of Tingha and Emmaville and gave a timely boost to the district centres of Inverell, Glen Innes and Tenterfield. Exploitation of the rich silver lodes in the far west produced the ephemeral town of Silverton which, in its heyday in 1885, had a population exceeding three thousand.

The far more extensive and rich silver-lead reefs of Broken Hill spawned larger mining companies and attracted many more miners, so that by the beginning of 1888 the town had a population of between six and seven thousand. Three years later it was a small city of twenty thousand inhabitants. Although mining activity and city growth faltered towards the close of the century, technical innovations and the fuller development of the "boomerang lode" boosted Broken Hill's population to over thirty-five thousand, making it the state's third largest city by 1914.

Initially Broken Hill was a rough town; short of water, covered with dust and plagued by typhoid. It would, so it was claimed, "disgrace a Hottentot village".[45] As Broken Hill was not a company town and as the mining moguls lived in Melbourne, the city was still neglected two decades later. Bereft of gardens, sporting facilities and cultural institutions, its suburbs met only basic needs. The unpainted, galvanized iron cottages were unfenced and fronted streets of bare earth.

At first coal mining was largely confined to the Hunter Valley, but subsequently mines were opened in the Illawarra region and at Lithgow. Urban growth in the Hunter Valley centred on Newcastle. Between 1820 and 1851 its size fluctuated around 1,300. The population reached 3,562 in 1861 and rose progressively to 7,024 in 1871 and 8,986 in 1881. By the

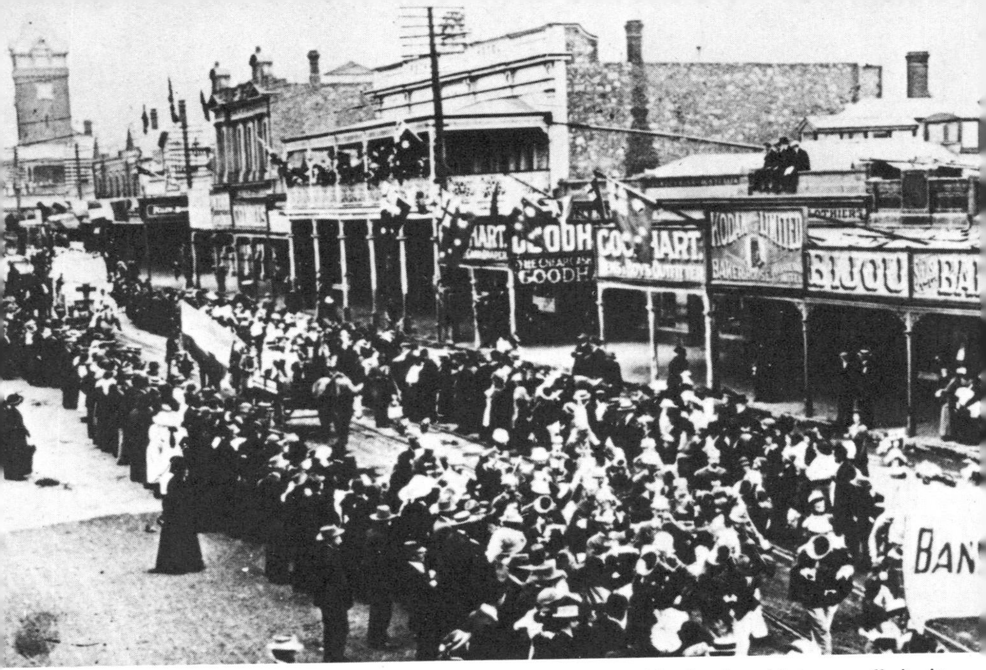

A parade in Argent Street, Broken Hill, NSW, which rapidly developed into a small city in the late nineteenth century. *(Photograph courtesy of Brian Carroll.)*

1880s the central core was surrounded by two rings of mining settlements which had a combined population of more than 15,000. As coal production became less important these suburbs were gradually drawn into the commercial-employment orbit of central Newcastle, so that by 1920 the city had a total population of 88,000. Its growth was due, not only directly to coal mining, but also to the establishment of fuel intensive industries such as copper and tin smelting, brick-making, potteries and foundries, and to its role as regional port for north-eastern New South Wales.[46]

Coal mining in the Illawarra region did not begin to engender urban settlement until the 1860s when villages began to appear at the pitheads or nearby. Eventually a complex of coal towns developed, stretching from Clifton in the north to Port Kembla in the south. Only Bulli had any real sophistication. Wollongong became the service centre and main port for the district, but Port Kembla attracted the coke ovens of Mount Lyell Mines and the smelters and refinery of the Electrolytic Refining and Smelting Company, and showed promise as an industrial centre. Even as late as 1911 Wollongong had a population of only 4,660 and the whole region of 26,452.[47]

Alluvial tin and gold attracted diggers to Tasmania and spawned mining camps as they had done on the mainland, but it was the exploitation of ore-rich reefs by large companies with enough capital to overcome the problems of isolation and inaccessibility which promoted the growth of substantial towns. Tin mining at Mount Bischoff and Ringarooma, silver-lead mining at Rosebery, Mount Heemskirk and Mount Zeehan, gold mining at Lefroy, Beaconsfield and Mathinna, and copper mining at Mount Lyell, all induced urban settlement. Few of these centres reached a substantial size and fewer still survived as significant towns. For a time Beaconsfield was Tasmania's third largest town. The Mount Bischoff mine supported the town of Waratah for more than seventy years. For two decades the Mount Zeehan township had a population which generally fluctuated between 5,000 and 8,000 and could boast of trams, theatres and eighteen hotels. The Mount Lyell Mining Company's smelter centre of Queenstown succeeded Beaconsfield as Tasmania's third town and by 1901 contained 2,900 houses — 1,800 of them bachelor's shacks of one or two rooms — and a main street which offered boisterous hotels, billiard saloons, shops, theatres and share call rooms.[48] By 1914 even the more vital of these mines and towns were in decline.

Few of the base metal mines established in Queensland before 1914 were rich or extensive enough to sustain substantial towns. The tin mining centres of Herberton and Irvinebank grew into prosperous, if conservative, towns. Isolation, a harsh climate, high transport costs and fluctuating metal prices condemned most of the settlements associated with other mines to a checkered, growth-decay existence. The former gold mining town of Mount Morgan, however, continued to prosper after it became a copper mine. The growing demand for coal promoted the growth of Ipswich: by 1910 some 2,000 miners were employed in the district and the town had attracted industries such as the railway workshops, a woollen mill and meatworks. Its population increased from 3,601 in 1861, when coal production was relatively unimportant, to 9,528 in 1911. Although it was also a service centre, much of this growth must be attributed to coal mining.

The establishment and/or development of port towns was one of the important secondary urban influences of mining, especially if an ore treatment plant, such as a smelter, was located at the port. The needs of the copper mines of South Australia prompted the establishment of Port Wakefield and Port Wallaroo and the growth of both was stimulated by the construction of smelters. Port Pirie expanded rapidly after BHP built its smelters there in the late 1880s. The growth of distant Newcastle was boosted not only by the coal trade and the fuel needs of South Australian smelters, but also by smelters constructed in the city itself to take advantage of the economics of back-loading and to treat the tin ores of the New England Tablelands. Wollongong and Port Kembla both benefited from the coal trade and from the employment provided by smelters, coke ovens and other fuel intensive industries established in the region.

Geelong grew faster than Melbourne during the 1850s because it provided readier access to the central gold-fields. Other Victorian ports, such as Port Fairy, Portland and Echuca, also battened on the gold-fields' trade until railways diverted business to Melbourne. Gold, too, gave port development in Queensland a timely impetus. "The ports of Rockhampton, Townsville, Cairns and Cooktown", Blainey asserts, "were made or magnified by mining fields; Brisbane and Maryborough, Port Douglas and Normanton, were enriched and enlarged by gold, and the whole colony was invigorated by the web of commerce that gold spun."[49] In Western Australia a string of ports – Wyndham, Derby, Cossack, Geraldton, Fremantle, Albany and Esperance – were established or developed to service the gold-fields. Burnie, Devonport and Launceston expanded to service the Tasmanian mining fields, though lesser ports such as Trial Harbour and Macquarie Harbour enjoyed a short period of growth and prosperity while they served the mines at Zeehan and Mount Lyell.

The permanency of mining ports, as of mining towns, was tied to the life of the mines they served unless their hinterlands produced other commodities for shipment. Ports, such as Portland in Victoria, Townsville and Cairns in Queensland, Albany and Esperance in Western Australia and Burnie and Devonport in Tasmania, continued to grow after mining declined; Cooktown in Queensland and Cossack in Western Australia soon became derelict after the mines closed. Many others lost much of their newly-won importance as mineral deposits were exhausted or abandoned. Some mining ports, such as Port Douglas, Cardwell and Bowen in Queensland and Trial Harbour in Tasmania, declined after nearby rivals usurped the mining and regional trade. Still others were unable to survive the competition from the capital city.

Perhaps the most important secondary urban effect of mining was the impetus it gave to metropolitan expansion. The position and role of the capital cities in the colonial or state economy and society enabled them to secure the maximum benefit from mining development. This is most clearly seen in the cases of Adelaide, Melbourne and Perth. The mines of South Australia were controlled and largely owned by residents of Adelaide; consequently the massive profits they earned were used to develop and shape the city. The location of smelters in the city shifted employment from the mines at Broken Hill to Adelaide. Furthermore, the transport system and the structure of commerce, finance and manufacturing tended to channel trade and financial transactions associated with mining through Adelaide, and its superior amenities and employment opportunities attracted, not only the wealthy, but also the elderly and redundant miners. Not surprisingly its population grew from 32,810 in 1851 to 192,400 in 1911. It dominated its hinterland like a medieval city state. At no time between 1836 and 1911 did the population of a country town in South Australia reach 10,000; as Hirst has so succinctly put it: "Outside Adelaide there were towns, but no cities".[50]

The growth of Melbourne was the most obvious feature of Victoria's urban development between 1846 and 1891, from a small provincial town of 11,356 to a city of 480,350.[51] Of course, not all of this growth can be attributed to gold mining, but Melbourne achieved its most rapid growth during the golden decade when the population rocketted from 24,100 to 123,000 inhabitants. It was well on its way in 1861 to becoming "Marvellous Melbourne", and the impetus gained in that short period was not lost until the late 1880s. In 1860 a city booster wrote: "The Melbourne of 1852 then but a very inferior English town, unpaved, unlighted, muddy, miserable, dangerous, has become transformed into a great city, as comfortable, as elegant, as luxurious (it is hardly an exaggeration to say it) as any place out of London or Paris . . .".[52] In a more balanced assessment Jill Roe concluded: "As well as making town life more orderly and better controlled, the wealth of the goldrushes made Melbourne more 'up-to-date', more sophisticated, in its appearance."[53]

The city became the Mecca of overseas immigrants, haven for disappointed diggers, colonial port, hub of the transport and communications system, commerical entrepôt, financial leader of the Australian colonies, industrial centre, and a precocious cultural pace-setter. The decentralized urban development occasioned by the gold rushes was rapidly reversed as gold mining declined, as the flow of overseas immigrants and internal migrants to Melbourne continued unabated and as innovations in agricultural, industrial, mining, transport and communications technology enabled Melbourne merchants, financiers and manufacturers to dominate colonial and inter-colonial markets and economic affairs, at the expense of centres such as Ballarat, Castlemaine and Bendigo. By the 1880s gold had ceased to fuel Melbourne's expansion, though Broken Hill silver and lead made a contribution towards the close of the decade. Growth was largely the product of heavy inflows of British migrants and capital. Melbourne stagnated after the boom ended about 1889, but growth subsequently resumed so that the city had a population of 600,280 in 1911.

Mining had, perhaps, the greatest impact on Perth. From a small socially and intellectually isolated community of some five thousand inhabitants living in an over-sized village, Perth grew into a small city of thirty-six thousand people by 1901. It had enjoyed a building boom while the rest of Australia was in the grip of a severe depression. Stone and slate public buildings and timber, shingle and galvanized iron cottages were gradually giving way to red brick and orange tiles. Central Perth's new skyline of elaborate, substantial buildings proclaimed the city's new prosperity. "The city's air of activity and bustle contrasted sharply with the sleepy village of twenty years earlier." Urban settlement had spilled out of the central area into suburbs served by electric trams and supplied with water. It was, however, still a raw town lacking in amenities, such as formed footpaths, kerbs and water-channelling, surfaced roads, adequate drainage and proper sanitation. But its famous parklands were taking shape and, in religion, education and culture, it was progressively assuming the sophistication of a city.[54] It had 111,400 residents in 1911.

The impact of mining on Sydney was modest as compared to Adelaide, Melbourne and Perth. Non-metropolitan urban growth was consistently more vital and extensive than in Victoria, South Australia and Western Australia, and commerce and industry were not centred in the metropolitan area to the same degree. Newcastle emerged as a sub-metropolitan service node and port, as well as a coal-mining and industrial centre, and already the Wollongong urban complex was becoming a location for heavy industry. Moreover, the secondary urban effects of mining at Broken Hill were more strongly felt in Adelaide and Melbourne than in Sydney. Nevertheless, the centralizing forces already discussed were at work so that mining engendered a good deal of metropolitan development; even the mining and industrial activities at Newcastle and Wollongong contributed to Sydney's expansion. Thus mining played its part in raising the city's population from 54,000 in 1851 to 651,800 in 1911.

The growth of Brisbane and Hobart received a comparatively smaller boost from mining because the mines were serviced by more convenient and nearer ports and urban centres. Even the relatively close Gympie gold-field relied on Maryborough rather than Brisbane for transport and supplies. Launceston and Melbourne benefited much more from mining in Tasmania than Hobart. These capital cities nevertheless shared in the general prosperity and economic growth mining engendered, and their natural and acquired advantages ensured that the share was not insignificant. Mining had played its part in raising Brisbane's population from 6,000 in 1851 to 143,500 in 1911, and Hobart's from 25,000 in 1861 to 40,200 in 1911, but it had also promoted the growth of relatively large provincial towns such as Townsville, Cairns, Rockhampton, Maryborough and Launceston, which moderated metropolitan dominance.

The year 1914 marks a convenient transition into the second phase of mining. By 1910 the vigorous wave of prospecting and mining activity, which began in South Australia in the 1840s and was invigorated by the gold discoveries of the 1850s, was spent. The mining revival of the 1890s, based on the "technical awakening" involving more efficient management, better mining methods, more efficient machinery and plant and great advances in metallurgy, could not survive the exhaustion of gold reefs and other ore bodies. Consequently many of Australia's richest gold-fields went into irrevocable decline between 1905 and 1914, and once-thriving mining centres were reduced to ghost towns, faded away into sleepy country villages, or were almost imperceptibly metamorphosed into district or regional service centres of varying significance.

Only those gold mines with substantial reserves of payable ore continued to operate in the second phase of mining. Base metal mines were temporarily protected against the mining slump by high metal prices occasioned by industrialization in Europe, the arms race and the Boer and first world wars. Soon after the war ended in 1918 world metal prices collapsed, destroying the viability of many mines and blighting the future of others. A majority of the copper, tin, silver and lead mines went out of production, and the mining industry was further depressed in the late 1920s by the onset of the great depression, which curtailed industrial activities. The inter-war period was therefore generally one of stagnation for mining in Australia. According to Blainey more than a hundred thousand people were forced to leave the waning mining towns, sacrificing houses, furniture and possessions, and deserting the churches and public halls they had built. Some in North Queensland and Western Australia took up farming; the majority retreated to the capital cities, thus accelerating metropolitan growth and dominance.[55]

There were other reasons for the mining slump. By the second decade of the twentieth century the richer ores in most mines had been worked out and mines only remained profitable because better management, technical improvements and advances in metallurgy enabled companies to exploit the poorer grade ores. In short, only large companies could find the capital for the mine development, plant and equipment needed to mount operations of a scale and nature which could succeed under the new mining and market conditions. It was difficult to attract such amounts of capital when the prospects of achieving adequate profits were so dismal. One of the important consequences was that successful companies were often large "national" corporations with international ramifications, which were able to provide much of the social as well as the physical infrastructure for their mines. The company town thus became a feature of the mining scene in this period.

The struggle to establish a viable mine at Mount Isa illustrates how difficult it was, in a distant, inhospitable region, to raise sufficient capital to exploit even an extensive ore body, and to provide the infrastructure required. It was more than a decade before the mine made its first profit, and the company found that it was necessary to establish a town complete with almost all its amenities in order to attract the labour needed to bring the mine into production. Meanwhile the Mount Lyell Mining Company, like the copper mining companies of South Australia, had concluded that it must assume responsibility for its mining towns and for the welfare of its miners if it was to avoid industrial strife and remain profitable. Similarly, the State Electricity Commission of Victoria developed the town of Yallourn and a good deal of the infrastructure as part of its electricity generating project in the Latrobe Valley. No longer could mining companies expect miners to build their own houses and pave the streets of their town, or speculators to construct railways and

The town of Mount Morgan, Queensland, in the early twentieth century. This photograph is taken from the Town Hall, with the mine and processing plant in the background. *(John Oxley Library, Qld.)*

community utilities, such as gasworks, ice-works, theatres, or even shops.[56] These tasks had become a responsibility the companies themselves could no longer ignore.

One of the bright spots in Australia mining during this phase, and one which did the most to promote permanent urban settlement, was diversification by some large mining companies into metal processing and associated industrial activities which went beyond the recovery and treatment of mineral ores. By far the most important development was the establishment of iron and steel works at Newcastle and Port Kembla. The establishment of BHP's mining-industrial undertaking at Newcastle in 1915 formed the basis for a whole new complex of potential new industries which developed rapidly during and after the war.[57] The reconstruction of the Lithgow steelworks at Port Kembla and its subsequent extension by BHP, improvements to the Newcastle plant and the construction of a steel plant at Whyalla during the 1930s laid the foundation of a competitive, integrated iron and steel industry.

This basic industry promoted the expansion of urban centres, not only in the Newcastle and Wollongong areas where the iron and steel works were located and where the coal used to fuel them was mined, but also induced considerable urban growth at Whyalla and indirectly in a number of other cities and towns, especially Sydney, Melbourne, Geelong and Adelaide. The population of Newcastle grew from 55,380 in 1911 to 127,138 in 1947, and the population of the Wollongong complex from

26,452 to 66,077. Wollongong itself increased fourfold in less than four decades. Whyalla grew from a small port with a population of 150 in 1911 to a significant provincial town of 7,845 in 1947.

The Collins House Group of Broken Hill mining companies also became a large mining-industrial conglomeration, with mines at Broken Hill and Mount Morgan, smelters at Port Pirie, smelters and a refinery at Port Kembla and electrolytic zinc works at Risdon in Tasmania and a diverse collection of industrial undertakings which kept Broken Hill prosperous, sustained Mount Morgan and promoted the growth of mineral processing towns such as Port Pirie, Port Kembla and Risdon. Their mining and industrial activities also contributed to the expansion of Sydney, Melbourne, Geelong and Adelaide and a number of other cities and towns through the linkage effects they forged in the economy.[58]

Rising gold prices triggered a short-lived mining revival in the 1930s which created new towns at Cracow in Queensland and Tennant Creek in the Northern Territory, breathed new life into the old mining towns of Coolgardie, Norseman, Southern Cross, Wiluna, Leonora and Marble Bar in Western Australia and Cobar in New South Wales, and revived the sagging fortunes of others, such as Kalgoorlie, Mount Morgan and several towns in central and north-east Victoria. Most of these resurrected towns died again when gold mining slumped during the second world war, but there were a few significant centres associated with new mining ventures at Ben Lomond in north-eastern Tasmania, King Island in Bass Strait and Mount Isa. By the end of world war two, Mount Isa had a population of some three thousand and was beginning to assume an air of permanence in timber and galvanized iron.

Between 1914 and 1945 Newcastle and Wollongong grew substantially in response to rapid mining and industrial development. They therefore served as a limited counterbalance to the increasing primacy of Sydney, but the growth of Whyalla and Mount Isa and smaller mining centres, and the revival of many other mining towns in the 1930s, did little to counteract the increasing centralization of population and economic activities in the capital cities. During the thirty odd years the proportion of the Australian population living in the capital cities rose from about 39 per cent to 50 per cent.

Despite its lack of vitality during a prolonged period, the mining industry was poised for vigorous expansion at the beginning of the third phase of Australian mining. During the 1950s it entered a new period of growth which has rivalled the golden years 1851–91. For the most part, the contemporary mining boom has been based on what Coghill has called "the new metals and fuels of the twentieth century — aluminium, nickel, uranium, tungsten, titanium, oil and natural gas — but 'old faithfuls' like copper, lead, zinc, tin, iron ore and coal are all experiencing a tremendous revival".[59] Numerous newly discovered mineral deposits and a considerable number of those already known, or being mined, have been exploited on an unprecedented scale.

This new mining boom was not a creature of chance, though, of course, Australia was fortunate in having geological structures which hosted mineral ores. J.M. Powell has identified five factors which combined to bring about the discovery and vigorous exploitation of these numerous deposits:

1. deliberate and liberal government policies to encourage mineral exploration and exploitation;
2. the economic recovery of the United States of America and the rise of Japan as an economic and financial power;
3. the dramatic rise in world demand for minerals and the anticipated depletion of known reserves;
4. the reappraisal of Africa and South America as sources of minerals by American and British mining companies because of their political instability; and
5. Australia's long-standing mineral industry, stable political system and apparent traditional respect for the value of free enterprise.[60]

He believes that overseas mining companies were encouraged to participate actively in mining development in Australia by favourable government policies and the important discoveries made by local companies.

The huge and buoyant demand for minerals during the long period of rising prosperity after 1950 could only be met by very extensive mining, ore treatment and transport operations. Fortunately there were spectacular advances in technology during and after world war two which made these possible. A virtual revolution in mechanical engineering permitted the profitable underground mining of vast amounts of low-grade ore and the extensive open-cut mining of large deposits of bauxite, iron ore, coal and other minerals lying close to the surface. Improvements in transport arising from the same burst of innovation, such as more efficient railways, conveyor belts, pipe lines, bulk-loading facilities and huge ocean bulk carriers, made the movement of unprecedented quantities of mineral ores possibly at greatly reduced transport costs. Finally, advances in ore-treatment technology completed the cycle by increasing the scale and decreasing the costs of producing metals from mineral ores.

These developments had a number of important implications which are reflected in the urban growth associated with mining during the post world war two period. The scale of operations and the sophistication of the mining, ore treatment and transport technology involved made mining a highly capital-intensive industry; that is, vast amounts of money had to be invested in plant, equipment and infrastructure. Geoffrey Blainey has estimated that between 1955 and 1968 the aluminium industry alone had invested about $600 million in development from mines to smelters; Hamersley Iron more than $200 million on equipment and amenities in four years; and the oil industry more than $400 million on exploration and development during the period 1946–66.[61] These estimates merely suggest the scale of the investment: they do not pretend to indicate the total investment in a single segment of the industry, much less that for mining as a whole. It may, perhaps, give perspective to recall that $345 million has been spent on the bauxite treatment plant at Gladstone, and that more than $1,000 million had been invested in the Pilbara region by 1973.[62] Over $5,000 million will be spent before gas from the North West Shelf is marketed. Investment funds of this magnitude cannot be raised in Australia; multinational companies have therefore come to dominate Australia's mining industry.

What is of more importance to Australian urban development is that, despite the staggering levels of production reached, the extent and productive capacity of the plant and machinery have meant that comparatively little labour is required and that the urban settlements associated with these mining activities are relatively small. In short, the workforce at the mines and smelters, and the towns in which they live, are surprisingly small given the output involved. The contrast with pre world war one mining is quite remarkable.

The separation of the mining and ore treatment stages in the production of metals, by locating treatment plants away from the mine at a port and, preferably, near the fuel supply source, was a tendency which emerged during the nineteenth century and became the general rule after world war two. Only in the case of Mount Isa Mines and to a lesser degree of Nabalco at Gove has a substantial part of ore treatment been retained at the mine, though there are proposals to build alumina refineries at Weipa, and mining agreements require the primary treatment of iron ores in the Pilbara. Thus the economics of metal production, the world-wide mineral market and the policies of mining companies have had the effect of limiting the size of mining towns by transferring or locating ore treatment plants, with their associated employment opportunities and urban growth potential, to centres such as Gladstone, Kwinana, Bell Bay, Kurri Kurri, Bluff in New Zealand, Point Henry in Victoria, Newcastle, Port Kembla, and Townsville. As Blainey says: "The importance of a new mining field could no longer be measured by the population which lived near by. Of the mining fields created or extended since 1945, only Mount Isa so far supports a small city; it is populous because it has a higher proportion of married men than most fields, because it concentrates and smelts ores on the field and because it is the largest underground mine in Australian history . . .".[63]

Mining operations usually have national and/or international ramifications because different stages in the production of metals take place in different parts of Australia or the world. Ports and port facilities, such as bulk-handling facilities, have been constructed or extended to handle this unprecedented flow-through of mineral ores and associated products. Heavy investment in plant and equipment has been essential, but, as in the case of Hay Point, few men are required to operate it. Consequently these new and extended port-towns are relatively small, though many of them handle tonnages both of shipping and cargo far in excess of some of the capital city ports. For example, the Western Australian ports of Dampier and Port Hedland exported from thirty to forty million tonnes of cargo in 1975, dwarfing tonnages handled by Fremantle. As early as 1972–73 Gladstone was handling twice as much cargo as Brisbane, and the bauxite port of Weipa and the coal port of Hay Point also shipped more cargo than Brisbane.

With the exception of coal mining in the Sydney Basin, and the possible exception of coal mining in the Bowen Basin in Queensland, most of the new and extended mining operations are in isolated areas, often in most inhospitable environments. They have consequently led to the establishment and/or growth of towns in locations which would not otherwise have attracted urban settlement and have reversed, if only in a minor way, the persistent trends of population concentration towards the eastern, south-eastern and south-western coasts, and especially around the capital cities. The small, but remarkably high rates of population growth in Australia's most isolated areas of the north and far inland have been the most noteworthy exception to those trends.[64]

Not only did this urban expansion give the sparsely settled northern Australia a much higher population growth rate than Australia as a whole, but iron ore, nickel and bauxite mining induced the largest annual population increase registered in Western Australia since 1901 (with the exception of that achieved in the gold era of 1901–11). Indeed, had it not been for the demographic expansion associated with mining, the Perth metropolitan area would have rivalled the population concentrated in and around Adelaide, "where 82.1 per cent of the population reside within 160 km of Adelaide in an area covering only 5.2 per cent of the State". Too much emphasis should not, however, be given to this apparent reversal of long-term trends since the numbers are quite small and, as Holmes points out, the "growth is based almost entirely on resource-extraction projects, notably mining, which have proved notoriously unreliable as a means of long-term support for a population". His summation of the situation in 1971 is clearly on the side of the historical demographic trends:

> Even where aggregate economic and population growth has been stimulated by resource development in areas remote from the metropolis, population growth remains heavily focussed on the metropolis. For instance, between 1966 and 1971 Perth's population increased by 142,000 persons while the rest of the State received an increase of only 41,000.[65]

Holmes divides Australia's most rapidly growing urban centres of 1971 into four categories — mineral centres, resort/retirement centres, dormitory (suburban over-spill) centres, and other or mixed type centres. Burnley adopted a wider classification for 1976: manufacturing towns, service towns, mining and utility towns, professional and administrative towns, and primary production centres.[66] In each case only one category is based primarily on mining though others have also been influenced by it. The new mining towns and centres directly associated with new mining or resource projects are listed in table 24. Not all of these centres are mining towns. Dampier, Port Hedland and Wickham are mineral ore ports. Churchill is also concerned with electricity generation. Exmouth is a base for the communications installation at North West Cape, as well as for the exploitation of oil and gas on the Northwest Continental Shelf. Biloela provides services for the surrounding agricultural district and is concerned with electricity generation, as well as being a coal mining centre. The old gold mining centre of Kambalda has been associated with new nickel mining projects. Practically all are part of the new wave of mining in arid, isolated or tropical areas which has promoted limited urban developments running contrary to the long-term trends of urbanization in Australia.

Table 24. New mining towns and centres

State	Centre	Population 1971	Population 1976
Western Australia	Dampier	3,600	2,700
	Exmouth	4,700	2,300
	Goldsworthy	1,000	1,000
	Kambalda	4,200	4,800
	Karratha	1,800	4,200
	Newman	3,900	4,700
	Paraburdoo	3,000	2,400
	Port Hedland	7,200	11,100
	Roebourne	1,500	1,400
	Shay Gap	— —	900
	Tom Price	3,400	3,200
	Wickham	— —	2,300
Northern Territory	Nhulunbuy	4,400	3,600
	Tennant Creek	1,700	2,200
South Australia	Coober Pedy	1,400	1,900
Victoria	Churchill	2,400	3,500
Tasmania	Savage River	1,200	1,200
Queensland	Biloela	4,000	4,500
	Blackwater	2,000	4,600
	Dysart	— —	1,500
	Moranbah	1,100	4,000
	Moura	1,900	2,700
	Weipa	2,200	1,700

Source: The data are derived from Census returns and from information supplied to the author by the Australian Bureau of Statistics. Figures have been rounded to the nearest hundred. The new mining towns are the subject of more detailed analysis in chapter 5.

It will be obvious from the table that the growth of many of these mining centres and ports had peaked by 1971 and that their populations declined during the next intercensal period. Administrative centres, such as Karratha and Port Hedland, continued to expand, but the main phase of urban growth in the Pilbara region was over. By contrast, the development of new coal mines in Queensland's Bowen Basin induced further growth in the constellation of mining centres in this region. Indeed, the population of new mining towns increased from less than 500 in the early 1960s to over 19,000 in 1976. Mount Isa has also expanded in response to mine development, its population increasing from about 3,000 in 1947 to 26,500 in 1976; during those thirty years the town outgrew company tutelage and is now an open city. Urban development in the Northern Territory associated directly with mining has focused on the new town of Nhulunbuy on Gove Peninsula and the expansion of the old gold mining town of Tennant Creek, but it has little national significance.

New settlements at Savage River and Tuena have been the main additions to mining centres in Tasmania, but renewed mining activities have also sustained or revived the old towns of Rosebery, Zeehan and Queenstown. The expansion of coal mining in the Hunter Valley and the

A new suburb (Healy) of Mount Isa, Queensland, established in the early 1970s, appears similar in design to the new mining towns (see chapter 5). Many of the homes are financed by Mount Isa Mines. The older part of the city and the mine are in the background. (Copied from *Mimag*, March 1973.)

Illawarra region has engendered the only significant growth of mining centres in New South Wales, although new mining ventures have revived Cobar. The vigorous development of brown coal deposits in the Latrobe Valley in Victoria for on-site power stations, as feed stock for gas and chemical production and for fuel generally induced considerable urban development in the towns of Moe, Yallourn, Morwell and Traralgon. But when the promise of substantial industrial growth was not realized this urban expansion lost its momentum, especially after Bass Strait natural gas forced the closure of the coal gasworks and natural gas and petroleum by-products became the feedstock for chemical plants. Even so Moe, Morwell, and Traralgon each had populations of more than 15,000 in 1976 and Churchill of 3,500. Sale expanded rapidly once it became the service centre for the Bass Strait oil and gas fields, but their development had a greater impact on Melbourne.

In recent times, urban development has been particularly strongly associated with the shipment of ores and the processing of minerals. The mineral ports in Western Australia are larger than most of the new mining towns, and the size of mining centres such as Nhulunbuy and Weipa is due in part to their role as ports. As already suggested, ports such as Dampier, Wickham, Weipa and Hay Point are surprisingly small in relation to the tonnages of shipping and cargo handled, and it is only when ore or mineral treatment is also undertaken that the growth and size of mining ports becomes significant. For example, the combination of coal shipment, electricity generation, bauxite refining, alumina smelting and district service centre activities has engendered rapid and extensive urban growth at Gladstone, which had a population of nineteen thousand in 1976. Townsville's growth has also been stimulated by copper and nickel refining. Similarly the growth of the ports of Burnie, Port Latta and Bell Bay in Tasmania and Geelong in Victoria have received a boost from the operation of ore or mineral treatment plants, the handling of ores and mineral concentrates, or the refining and treatment of petroleum products.

This indirect influence of mining on urban growth is seen most clearly, as already indicated, in the rapid growth of the Newcastle and Wollongong industrial complexes during the 1950s and 1960s. Although the rate of growth had already slackened, the estimated population of the larger statistical divisions of these cities stood at 355,700 and 202,800, respectively. By 1979 this had risen to 379,800 and 224,000, representing increases of 6.77 and 10.45 per cent. Whyalla in South Australia also grew rapidly into a city of 33,000 by 1976, as an iron and steel industry was developed and associated metal fabricating activities, such as shipbuilding, were introduced. Much of Geelong's growth in the post second world war period has been largely due to activities based on mining, such as mineral processing, oil refining, chemical production and metal fabrication. However in this case vigorous expansion persisted for longer, so that the estimated population of the Geelong statistical division has increased by 14.9 per cent from 122,800 in 1971 to 141,100 in 1979.

Thus mining and activities based on mining have induced quite significant growth in centres outside the capital cities, but have generally had a greater impact on the expansion of the metropolitan areas through the agglomerating effects they have produced. The growth of Newcastle, Wollongong and Geelong can be regarded as an essential aspect of the metropolitan development of Sydney and Melbourne, and the extensive mining developments in Western Australia and Queensland have obviously given an extra impetus to the growth of Perth and Brisbane. Burnley has argued that the exploitation of the Pilbara iron ore deposits was brought about by the development of the iron and steel industry in eastern Australia and that this mining activity, in turn, strongly influenced the growth of Perth. He says that "Perth's rapid growth in the 1960s and continued strong growth throughout the 1970s even in a time of national economic recession, can at least partly be attributed to the very substantial mineral discoveries in its Western Australian hinterland." Certainly the government used its control over mineral deposits to persuade the companies to establish iron and steel, aluminium and oil refining plants at Kwinana in the Perth area in order to promote the city's development.[67]

Brisbane's rapid growth in the 1960s and especially during the 1970s must also be partly attributed to the development of base metal mining at Mount Isa, bauxite mining at Weipa and treatment at Gladstone, and coal mining in the Bowen Basin. During the second half of the 1960s Brisbane's growth rate was second only to that of Perth among the state capital cities, a position it has retained during the 1970s as table 25 shows.

Table 25. Growth in estimated population of Australian capital cities (statistical divisions) 1971 and 1979

City	1971	1979	Increase	Percentage Increase
Sydney	2,977,300	3,093,300	116,000	3.90
Melbourne	2,575,400	2,739,700	224,300	8.91
Brisbane	891,100	1,015,200	124,100	13.92
Adelaide	850,700	933,300	82,600	9.70
Perth	711,800	833,600	171,800	24.13
Hobart	153,100	168,500	15,400	0.10
Canberra	160,800	241,300	80,500	50.06

Source: Extracted from table 8, p. 9, *Australian Demographic Statistics Quarterly*, December 1979 and March 1980 (Canberra: Australian Bureau of Statistics, 1980), Catalogue no. 3101.0.

Thus mining since world war two has produced a small, but significant exception to the trend towards urban concentration in Australia. This has been more than outweighed by the contribution it has made to metropolitan growth and to the development of the ports and ore treatment centres in the more populous parts of Australia. As in the past, the life of the new remote mining towns and ore ports will depend on the size and quality of the mineral deposits which sustain them and on the world prices of the minerals they produce. Fortunately the size of many of the deposits and the world demand for minerals appear to guarantee the vitality of a sizable proportion of these towns for many years. If the dreams of on-site ore and mineral processing are realized some of them may grow into sizable industrial centres.

Notes

1. The exceptions are usually due to the construction of irrigation schemes. See J.H. Holmes, "Population", in *Australia: a geography,* ed. D.N. Jeans (Sydney: Sydney University Press, 1977), pp. 344-50.
2. Holmes, "Population", pp. 332-34.
3. I.H. Burnley, *The Australian Urban System: Growth, Change and Differentiation* (Melbourne: Longman Cheshire, 1980), p. 47; see also Geoffrey Blainey, *The Tyranny of Distance* (Melbourne: Sun Books, 1966), pp. 136-37.
4. Albert J. Robinson, "Regionalism and Urbanization in Australia: a Note on Locational Emphasis in the Australian Economy", *Economic Geography* 39, 2 (1963): 154-55; see also Holmes, "Population", pp. 336-38.
5. C. Duncan, "Mineral Resources and Mining Industries", in Jeans, *Australia,* pp. 435, 438.
6. See Adna Ferrin Weber, *The Growth of Cities in the Nineteenth Century, A Study in Statistics* (Ithaca, New York: Cornell University Press, 1899, reprinted 1963), p. 1; Kingsley Davis, "The Urbanization of the Human Population", in *Cities: A Scientific American Book,* ed. Gerard Peel & others (Harmondsworth: Penguin reissue, 1967), p. 16; N.G. Butlin, *Investment in Australian Economic Development 1861–1900* (Cambridge: Cambridge University Press, 1964), pp. 6, 181.
7. Holmes, "Population", pp. 338-39; Kingsley Davis and A. Casis, "Urbanization in Latin America", quoted in P. Solznick, ed., *Sociology* (New York: Row Paterson, 1958), p. 416.
8. Weber, *Growth of Cities,* p. 138; see also, G.R. Parkin, "Australian Cities", *The Colonial Century* (March 1891), p. 690; and J.W. McCarty, "Australian Capital Cities in the Nineteenth Century", in Urbanization in Australia: The Nineteenth Century, *Australian Economic History Review* X (1970).
9. J.H. Holmes, "The Urban System", in Jeans, *Australia,* p. 414.
10. J.M. Powell, ed., *Urban and Industrial Australia: Readings in Human Geography* (Melbourne: Sorrett Publishing, 1974), p. 13; Frank J.B. Stilwell, *Australian Urban and Regional Development* (Sydney: Australia & New Zealand Book Co., 1974), p. 63; Burnley, *Urban System,* pp. 48-49.
11. Burnley, *Urban System,* p. 67; Stilwell, *Urban and Regional Development,* pp. 30, 35.
12. Stillwell, *Urban and Regional Development,* p. 32.
13. Burnley, *Urban System,* pp. 40, 43; Butlin, "The Shape of the Australian Economy, 1861–1900", *Economic Record* xxxiv, 67 (1958): 21.
14. See Burnley, *Urban System,* pp. 73-92.

15. See G.J.R. Linge, "Manufacturing", in Jeans, *Australia,* pp. 477-78 and G.J.R. Linge, "Governments and the Location of Secondary Industry in Australia", in Powell, *Urban and Industrial Australia,* pp. 27-35.
16. Stilwell, *Urban and Regional Development,* p. 64.
17. G.J.R. Linge, "Governments and the Location of Secondary Industry in Australia", *Economic Geography* LXIII (1967): 59; Linge, "Manufacturing", p. 482; G.J.R. Linge, "The Location of Manufacturing in Australia", in Alex Hunter (ed.), *The Economics of Australian Industry,* (Melbourne: Melbourne University Press, 1963), pp. 18-64.
18. Stilwell, *Urban and Regional Development,* pp. 64-65; Holmes, "Population", p. 342.
19. Burnley, *Urban System,* p. 130.
20. Burnley, *Urban System,* pp. 108-9; Stilwell, *Urban and Regional Development,* pp. 44-45.
21. Burnley, *Urban System,* pp. 111-2; Stilwell, *Urban and Regional Development,* pp. 16-18.
22. Alex Kerr, "The economic and social significance of Western Australian mining developments", in *Mining in Western Australia,* ed. Rex T. Prider (Nedlands: University of Western Australia Press), pp. 294-95.
23. Duncan, "Mineral Resources", p. 439.
24. H.G. Raggatt, *Mountains of Ore* (Melbourne: Lansdowne Press, 1968), p. 3.
25. Blainey, *Tyranny,* p. 141.
26. Duncan, "Mineral Resources", p. 442.
27. B. Ryan, "The South Coast of New South Wales: A Regional Survey of Urban Growth", *Business Archives and History* 6, 2 (1966): 146-47.
28. D. Urlich Cloher, "A perspective on Australian urbanization", in *Australian Space Australian Time: Geographical Perspectives,* ed. J.M. Powell and M. Williams (Melbourne: Oxford University Press, 1975), chapter 4.
29. G.J.R. Linge, "The forging of an industrial nation: manufacturing in Australia 1788–1913", in *Australian Space Australian Time: Geographical Perspectives,* ed. J.M. Powell and M. Williams (Melbourne, Oxford University Press, 1975), p. 158.
30. Butlin, *Investment,* p. 186.
31. Geoffrey Serle, *The Golden Age: a history of the colony of Victoria 1851–1861* (Melbourne: Melbourne University Press, 1963), p. 218.
32. See Geoffrey Serle, *The Rush to be Rich: a history of the Colony of Victoria, 1883–1889* (Melbourne: Melbourne University Press, 1971), pp. 2-3; Anthony Trollope, *Australia and New Zealand,* 2 vols. (London: Chapman & Hall, 1873), pp. 406-7, 418.
33. Serle, *Rush to be Rich,* p. 3; Butlin, *Investment,* p. 187; Powell, *Urban and Industrial Australia,* p. 13.
34. Butlin, *Investment,* p. 189.
35. See G.C. Bolton, *A Thousand Miles Away: A History of North Queensland to 1920* (Brisbane: Jacaranda Press, 1963), pp. 66, 121, 130, 269-70; Geoffrey Blainey, *The Rush That Never Ended: A History of Australian Mining,* 2nd ed., (Melbourne: Melbourne University Press, 1969), p. 240; *The Cooktown Herald,* November 1875, 17 April 1978.
36. Butlin, *Investment,* p. 192.
37. F.K. Crowley, *Australia's Western Third: A History of Western Australia from the first settlements to modern times* (London: Macmillan, 1960), p. 121.
38. Ibid., p. 123.
39. Ibid., p. 124.
40. See comparative photographs in Brian Carroll, *Australia's Mines and Miners: An illustrated history of Australian mining* (Melbourne: Macmillan, 1977), p. 56.
41. Blainey, *Rush That Never Ended,* pp. 193-94, 207.
42. Ibid., p. 106.

43. J.B. Hirst, *Adelaide and the Country 1870–1917: Their Social and Political Relationship* (Melbourne: Melbourne University Press, 1973), p. 8; Douglas Pike, *Paradise of Dissent: South Australian 1829–1857* (Melbourne: Longmans, Green, 1957), p. 335; see also Philip Payton, *Pictorial History of Australia's Little Cornwall* (Adelaide: Rigby, 1978), pp. 10-23; Ian Auhl and Denis Marfleet, *Australia's Earliest Mining Era: South Australia 1841–1851, Paintings by S.T. Gill* (Adelaide: Rigby, 1975), p. 10; E.M. Yelland, ed., *Colonists, Copper and Corn in the Colony of South Australia 1850–51 by Old Colonist* (Melbourne: Hawthorne Press, 1970), pp. 137-38, 150-53; Philip Cox & Wesley Stacey, *Historic Towns of Australia* (Melbourne: Lansdowne Press, 1973), p. 160; Pike, *Paradise of Dissent*, p. 336; see also Blainey, *Rush That Never Ended*, p. 111.
44. Blainey, *Rush That Never Ended*, pp. 111, 306.
45. Ibid., pp. 152, 306.
46. See M.T. Daly, "The Development of the Urban Pattern of Newcastle", Urbanization in Australia, *Australian Economic History Review* X (1970).
47. See Ross Robinson, ed. *Urban Illawarra* (Melbourne: Sorrett, 1977).
48. Blainey, *Rush That Never Ended*, pp. 222-23.
49. Ibid., p. 87.
50. Hirst, *Adelaide and the Country*, p. 5.
51. See Cloher, "Australian Urbanization".
52. Sir Archibald Michie, *Readings in Melbourne; with an Essay on the Resources and Prospects of Victoria for the Emigrant and Uneasy Classes* (London, 1879), p. 39.
53. Jill Roe, *Marvellous Melbourne: The Emergence of an Australian City*, (Sydney: Hicks Smith & Sons, 1974), p. 54.
54. Crowley, *Australia's Western Third*, pp. 142, 146-7.
55. See Blainey, *Rush That Never Ended*, pp. 283, 286, 288.
56. Ibid., pp. 306, 307, 328-29.
57. W.A. Sinclair, "Capital Formation", in *Australian Economic Development in the Twentieth Century*, ed. Colin Forster (London: Allen & Unwin, 1970), p. 34.
58. Blainey, *Rush That Never Ended*, pp. 280-81.
59. Ian Coghill, *Australia's Mineral Wealth* (Melbourne: Sorrett Publishing, 1971), p. 12.
60. Powell, *Urban and Industrial Australia*, pp. 18-19.
61. Blainey, *Rush That Never Ended*, pp. 356-57.
62. John Bach, *A Maritime History of Australia* (West Melbourne: Thomas Nelson, 1976), p. 420.
63. Blainey, *Rush That Never Ended*, p. 357.
64. J.H. Holmes, *Population Distribution and Growth*, 2nd ed., Atlas of Australian Resources, Second series, Geography Section, Division of National Maping (Canberra: Department of Minerals and Energy, 1975), p. 3.
65. Holmes, *Population Distribution and Growth*, p. 6, 7, 9.
66. Holmes, "The Urban System", pp. 418-29; Burnley, *Urban System*, pp. 112-18.
67. Burnley, *Urban System*, p. 53, 55, 106-7. See also Stilwell, *Urban and Regional Development*, p. 25.

5 The New Mining Towns — "Outback Suburbias"?

P.C. Sharma

The visitor to any of the new mining towns often finds himself disoriented in the town. The sight of planned shopping centres, mothers with toddlers in strollers, the marked absence of the elderly and the general absence of adult males during working hours, together with uniformly laid out housing blocks and the general "suburban look", are somewhat at odds with the knowledge that the particular town is "in the middle of nowhere". The term "outback suburbia"[1] encapsulates the visual images presented by the new mining towns and is also the appropriate point of entry for a study of their human geography.

In a sense there is nothing new in the current research interest in mining communities. In chapter 4 we were introduced to a rich literature that exists on mining communities. Much of this material has been produced by historians; contributions by planners, sociologists, psychologists and geographers are rare. Yet it is the latter group that features so prominently in the current research literature on mining communities. Why is this so? The latter half of the nineteenth century saw the major pushing back of the frontier on the Australian continent. This expansion was short-lived and, by the early years of the twentieth century, the beginnings of rural population decline were already evident. As the frontier was progressively mastered and farming began to exhibit the features it has today, the "great retreat" of population began. But the growth of the mining communities represents an exception to this general trend.

Examination of the census data for the last decade shows that the great retreat of the rural population has continued. The only exception is provided by the rapidly growing mining communities in Western Australia, the Northern Territory and Queensland. Considerable research has been conducted on the following aspects of these communities:

- *Growth rates:* the new towns exhibit (currently at least) significant rates of population growth when much of rural Australia is stagnating or declining in terms of population numbers. While obviously the growth is related to mining development there is interest in its impact and permanence.
- *Remote communities:* an outstanding characteristic of the vast majority of the new towns is their isolation or remoteness.
- *Urban planning:* the mining towns are also of interest from a

planning point of view as most of them have been planned "from the ground up". Despite the mixed record of success, the planning of the new mining towns represents the main attempt at the manipulation of the built environment in Australia during the last decade.

- *Tropical development:* the location of the new mining towns within tropical Australia makes them of particular significance; for, while the extreme form of the "populate or perish" argument is heard less frequently these days, there is considerable interest in these communities from the point of view of expanding settlement in these areas. Despite some well known failures, even today "northern development" remains a partisan and emotional issue in Australia.
- *Company towns:* there is a long standing interest in the study of closed company towns in mining and in other industries. These towns create a situation where it is possible for the employers to exercise extensive control over their workers during work and non-work situations. Interaction within this socio-political context has received attention from serious research workers as well as from writers of popular literature.

Clearly, not all these approaches or interests can be encompassed in the present chapter. It focuses, in the widest sense of the term, on the architectural aspects of these communities, as these are the most visible manifestations of the human geography of current mining activity in Australia. Selected socio-demographic characteristics of these communities are also examined in some detail. Perhaps some *caveats* are in order, for it is easy to over-interpret the importance of the new mining towns when seen against the background of the "resources boom". Despite their growth rates the great majority of these towns are unlikely to grow beyond a population of five thousand. Furthermore, despite the high purchasing power of their inhabitants, the towns are unlikely to have the economic and social vitality of "normal" rural centres with similar populations. They are not much more than dormitory settlements, essentially suburbs without cities. In fact it is probable that, despite the obvious growth of mining towns, the demographic impacts of the "resources boom" are unlikely to alter dramatically the settlement geography and the overall population distribution of Australia.[2]

In what follows the evidence obtained through various social surveys in these towns is reviewed and analyzed. Generally speaking these surveys have focused on the relationship between aspects of the built environment and human behaviour, addressing themselves to the broader issue of the social well-being of residents in remote communities. The review draws heavily on the CSIRO Remote Communities/Mining Towns project and is supported by the author's work in the Bowen Basin of Queensland.[3] The chapter is in two parts. The first is a general over-

Fig. 10. Location of selected Australian mining towns

1 Moranbah
2 Dysart
3 Middlemount
4 Blackwater
5 Moura
6 Broken Hill
7 Norseman
8 Kambalda
9 Kalgoorlie
10 Paraburdoo
11 Tom Price
12 Newman
13 Karratha
14 Dampier
15 Port Hedland
16 Goldsworthy
17 Shay Gap
18 Kununurra
19 Katherine
20 Tennant Creek
21 Alice Springs
22 Nhulunbuy
23 Weipa
24 Mt Isa
25 Mary Kathleen

view of the new mining communities and examines in some detail the physical and the social fabric of these communities. The second examines the impact of these mining towns on their surrounding regions. Figure 10 shows the location of selected mining towns in Australia.

THE NEW MINING TOWNS: AN OVERVIEW

Types of Mining Towns

While the new mining towns have an overall similarity of character there are important differences between them. Indeed, it would be unusual if the type of mining operations, climate and geographic location did not produce important differences in the types of mining towns encountered in Australia. No systematic classification of mining towns is known to have been attempted in Australia. One recent publication has used the distinction between "new" and "old" mining towns but, as the concern of this chapter is with new mining towns only, this classificiation is of little use.[4]

Examination of literature and general evidence suggests that the new mining towns can be tentatively classified into two general types, the Pilbara type and the Bowen Basin type.

Pilbara type: These towns occur in the Pilbara region of Western Australia, an area of arid tropical climate, noted for its geographic isolation. Considerable distances are involved in travel to settlements of any significance, and there is heavy dependence on air links for communication and routine travel. A variant of this type is found in the humid tropics in the northern tip of Australia (Weipa, Gove). These towns are generally located in the sparsely settled parts of the continent, although in places such as Gove significant Aboriginal populations may be present.

Bowen Basin type: These towns occur mainly in the Bowen Basin of Central Queensland, although towns of this type are encountered elsewhere in Australia, for example in the Victorian coal-fields. These towns are not as isolated as their counterparts in the Pilbara. They are also located in areas with a reasonably long history of rural settlement and different impacts can be expected from those experienced in the Pilbara.

These distinctions are real and must be kept in mind in the following discussion of mining towns. They often provide the underlying explanation for variations that are noted in the new mining towns.

Elements of the Physical Fabric of the New Mining Towns

The interest in the physical aspects of the new mining towns stems from several sources. The location itself warrants consideration, for alternative sites were almost always available and certain consequences clearly follow the siting decision. Also, the fact that these towns were often located in remote tropical regions spurred research into two distinct but related areas of study. The first concentrated on the broader aspects such as layout and administration, seeking answers to the perennial question of the relationship between the built environment and human behaviour.[5] The fact that these towns were in remote areas meant that certain "laboratory type" conditions existed.

The second group of studies concentrated on the narrowly defined architectural aspects, such as building design, assessing its general suitability for the harsh environment. While some of these investigations are still in progress, it can be seen that a wide range of issues relating to the architecture of the new mining towns has been investigated over the last decade. Indeed, the topic was of sufficient interest to have a special issue of *Architecture in Australia* and the *New Towns in Isolated Settings* international seminar devoted to it.[6]

The analysis in this segment will not be as wide-encompassing. It will be restricted to an examination in some detail of the location, town planning principles, housing and house types and administration of the new mining towns. These aspects provide the basis for critical enquiry into the human geography of the new mining towns.

Location

Most of the new mining towns are located within tropical Australia and usually at considerable distances from the settled parts of the continent. Indeed with some licence it may be claimed that the pattern of mineral distribution is almost the opposite of the pattern of population distribution. Thus while the bulk of the Australian population is located in the south-eastern corner of the continent the bulk of its mineral resources is located in the northern and western sections of the continent. While there are more general issues which may be discussed as possible consequences of this "mismatch" between the two distributions or patterns, the following section deals with the more specific local issues.

There is some debate regarding the appropriateness of the actual location of many of these mining towns.[7] This debate stems from the fact that the absence of any overall governmental regional planning and the existence of a "development-at-any-cost" mentality has meant that mining companies have had a "free run" as to the location of the new townships. Consequently, most of the new townships have been developed close to the minesite and/or at the coastal shipping ports, the latter location reflecting the export based nature of recent mining developments. There has been considerable criticism of the *laissez-faire* developments that have taken place, especially in Western Australia.

> Mineral producers argue that there can be no control over the location of ore deposits, and use this fact as the basis for their insistence on the establishment of towns in close proximity to the minesite. While there can be no denying that the location of some ore bodies is so remote as to justify the existence of a small isolated settlement, there are many instances where separate development was not warranted. For example, in the vicinity of Dampier in the Pilbara, there are five separate small settlements within a radius of 30 kilometres and the two towns of Tom Price and Paraburdoo, each serve ore deposits which are only 60 kilometres apart.[8]

In addition, in the West Australian case they argue that in some cases it would have been possible for the new town to be located on the coast rather than at the inland minesite. A coastal location, it has been argued, would have made life somewhat more pleasant in this arid region.

Similar criticism, perhaps with less force, can also be made of developments in the Bowen Basin, although in this case the development of mining towns has been less haphazard. Thus Blackwater, an existing town, has been enlarged to serve several mining companies, Dysart serves two mines and Moranbah will serve three when the Riverside mine becomes operational. However, even in the Bowen Basin some overall planning could have produced better results. For example, better roads could have made it quite feasible to amalgamate Moranbah and Dysart into a single town of some twenty to twenty-five thousand people. Alternatively, it would have been possible to develop Dysart and Middlemount as a single centre with a population of around twenty thousand.

The geographic isolation of most recent mining developments has meant that the companies involved have, in the main, borne infrastructure development costs. While this has involved them in considerable costs it has also created the impression in the public mind that the states are "getting something for nothing". The argument for governmental intervention in the siting of mining towns stems from two relatively independent considerations. The first of these relates to the question of regional planning: it is argued that the mining projects and the associated towns presented an excellent opportunity for the regional development of the areas involved. In such a situation regional infrastructure could have been developed in terms of wider objectives rather than to meet a particular mining company's requirements. The second consideration concerns the towns themselves: it has been argued that many of the problems of these mining towns stem from their small size and that a combined town (as a consequence of centralized developments) would have resulted in towns of larger size. The advocates of centralized development claim that the larger towns would have increased levels of choice for residents in a variety of areas, reduced the influence of mining companies overall, reduced population turnover and reduced infrastructure costs.[9] It is unfortunate that neither of these arguments has been investigated in any detail.

Town planning principles

The new mining towns represent the main efforts at planning or manipulating the urban environment in Australia over the last decade (excluding the plans for the capital cities and the growth centres during the short period 1972–75 when a Labor federal government was in power). As the new mining towns in most cases have been developed "from the ground up", there were considerable opportunities for the testing or application of town planning principles. However, despite an abundance of general principles, there existed no acceptable working models of mining towns which could be copied, and mining companies were forced to develop their towns on a "trial and error" basis. Thus the early efforts differed little from the Australian suburbia which enshrines many conventional town planning principles, including segregation of land uses, provision of certain facilities at the neighbourhood level and regulation of vehicular traffic.

The slipping into suburban type planning is perhaps an unwitting recognition of the dormitory nature of most new mining towns. The label "outback suburbia" is appropriate! As planners followed conventional planning principles it is not surprising that the end results could be mistaken for conventional suburbias. However, it is worth looking at some of the interesting efforts of the mining companies and their planners.

The first experiment was in South Hedland, where the Radburn concept of planning was introduced. The basic tenet of the scheme is that

> . . . the vehicular and pedestrian flows be separated or where they are in close proximity, the road design must limit vehicle speeds. . . . The concept is particularly suited to northern mining towns because of their high level of car ownership and the high proportion of population too young to drive, bringing about acute driver pedestrian conflict.[10]

Apparently this mining town design fell far short of the hoped for success in South Hedland, and this has been attributed to two things. Firstly, modification of some of the basic principles to vehicular traffic control meant that residents experienced only a quasi-Radburn system. Secondly, it appears that as most residents were having to cope with a difficult new environment the unusual traffic/block layout became the focus and surrogate for other less clearly identifiable concerns.[11] The planners, somewhat prematurely, regarded the South Hedland experiment as a failure and the model has not been applied elsewhere.

Another early experiment in mining town design was attempted at Shay Gap, where a variant of typical urban medium density housing was developed. This resulted in a tightly packed township occupying a small area, and early assessments suggest limited success in this case as well. These limitations, it seems, are associated with problems of the built structures themselves as well as with human relations in such harsh environments. This experiment is also unlikely to be repeated.

Paradoxically, it appears that the relatively conventional designs are more successful. This can be illustrated with examples from Queensland, although equally successful towns can be found elsewhere. With the bauxite mining township of Weipa, the broad objective of the mining company (Comalco) was to create a community which would attract and retain people with a view to establishing a permanent population.[12] The company recognized that an isolated mining development required total planning, and instructed its planners to experiment with a mixture of layout and housing types within a relatively conventional framework. Apart from its overall success with housing design, the company has

This vertical aerial photograph of Dysart, Central Queensland in 1980, displays many typical features of new mining towns. Clearly visible are the centralized shopping and other amenities. The caravan park on the western edge and the executive residences (darker roof) on the north-eastern edge of the town stand out against the typical housing. *(Photograph courtesy of Utah Development Co.)*

done extremely well with landscaping, creating a township with an air of permanence about it in a relatively short space of time.

The new mining towns in the Bowen Basin provide excellent examples of the evolution of mining town development. The earliest town to develop as a result of the mining boom of the sixties was Blackwater, where the new mining town component was grafted on to the existing township, producing a somewhat bleak-looking result. This impression was created largely by the indiscriminate removal of the existing poor tree cover, by the flat terrain and by the scattered layout of the township. Absence of a clear town centre tends to add to the visual confusion. The next town, Moranbah, was a major planning improvement. The overall attractiveness of Moranbah comes from its general design, from the gently undulating site as well as from the serious effort made in planting gardens to give an established look to the place as soon as possible.

The two remaining Bowen Basin townships of Dysart and Middlemount reflect a further refinement in mining town design. In Dysart the judicious use of bricks instead of concrete blocks in town utilities and services has created a "quality" atmosphere. Dysart's planners managed to create an atmosphere of permanence even while the town was being established. The new township of Middlemount is similar in many respects but goes further in integrating the single men's quarters within the township, rather than locating them at the minesite as has generally been the case both in the Bowen Basin and in the Pilbara. Thus it is clear that there has been a general evolution of mining town design and the incorporation of developments which clearly reflect some of the lessons learnt from earlier experience. The overall assessment to date suggests a mixed result. Clearly, there is a need to conduct studies which evaluate the success or otherwise of the town planning policies which have been incorporated in the new mining towns.

Housing and house types

To the casual observer there is a monotonous uniformity about most suburban housing, and the same impression is obtained in the outback suburbias. However, as in the case of big city counterparts, closer inspection reveals the existence of major housing distinctions in these towns. Most homes are owned by the mining company, with the state (through the Housing Commission) owning the remaining homes. The high cost of construction, the often short stays and doubts regarding the future viability of mining operations severely discourage home ownership. In any case in the "closed" company town, where all land is owned by the company, private ownership is generally not possible.[13] One survey recorded the following percentages for company ownership of homes: Nhulunbuy 99 per cent, Newman 94 per cent, Tom Price 99 per cent,

Port Hedland 66 per cent and Kambalda 89 per cent.[14] In the Bowen Basin the situation is similar although the company ownership figure tends to be lower: Blackwater 78 per cent, Moranbah 75 per cent and Dysart 67 per cent.[15] Non-company home ownership tends to be higher in "open towns" and the Bowen Basin townships reflect this.

The State Housing Commission is usually the second major landlord in mining towns. In the Pilbara there is often visible separation between mining company housing and State Housing Commission housing. This, it has been claimed, has important consequences for mining communities. In the first instance, Housing Commission homes are generally of inferior quality, as there is no compelling need for the state government to provide attractive housing for its personnel. Its system of transfers for staff to any part of the state means that there is an element of compulsion in the move, and the state's obligation is merely to provide "adequate" accommodation for its personnel. The mining companies can exercise no such compulsion on personnel they wish to recruit. Consequently superior housing itself becomes an incentive item for staff recruitment purposes. Furthermore, for the occupants of mining company houses there are additional benefits in terms of housing-related subsidies with regard to rentals and general running costs. Thus airconditioning, water subsidies and nominal rentals tend to produce a qualitative difference between company and non-company housing.

The distinction in housing has led to the creation of a socio-economic distinction between the mining company-employed "haves" and the state-employed "have nots" in the Pilbara.[16] In the Bowen Basin the housing distinctions do exist but they are less sharp, perhaps due to the fact that the company housing is generally of a slightly lower quality than that encountered in the Pilbara, and the Housing Commission housing is not visibly inferior to that provided by the mining companies. Additionally, the climate is not as harsh and only minimal airconditioning is provided in most company houses, so the distinction between company and state housing is a relatively small one.

There are further housing distinctions created by the quality, size and location of the homes of company executives and other staff. Almost all mining towns have their equivalents of "Nob Hill" where the select residences of senior staff are located. There are also distinctions at other levels in the company hierarchy. The general distinction between "staff" and "wages" personnel is reflected in housing as well, and often the staff category is divided into further levels and the relevant housing types identified.

While the housing distinctions are real enough their importance appears to be over emphasized, as such distinctions are similar to those encountered in any Australian city or town. However, the one-employer-company-town aspect of these communities makes housing a very real index of local social stratification. In the Bowen Basin at least, it is the policy of the mining companies to reduce these disparities between different types of housing by changing the two standards involved: the standard of housing of the Housing Commission is being raised and that of the mining company is being lowered slightly. It is precisely what was recommended for the Pilbara towns by Brealey and Newton.[17]

Administration

"Closed" and "open" are the terms used to describe the two basic types of administration practised in most of the new mining towns in Australia. The first is the fully company owned and operated "closed" town, while the second category contains various types of "open" towns. In the former, the mining companies control entry into the town by linking employment with company-owned accommodation, and termination of employment with the company invariably results in a forced departure from the township. The latter category of towns are open to the extent that, in theory at least, it is possible for individual households and commercial establishments to move in and out of the town at will, or to change employment, residence and so on within the township. Despite the theoretical distinction between the two types, modifications in actual operations lead to a blurring of distinction in reality.

In many of the new mining town developments, the closed company town seemed the only option open to the mining companies. The relative isolation, the need to attract personnel, and the general inability of the state (or federal) governments to provide the necessary infrastructure meant that the mining companies had to set up their own settlements and services as the first step in their mine development. The closed towns also meant that mining companies had a tight control over the initial period of substantial capital investment and thus avoided holdups in the accommodation for the mine workforce.

Most mining companies claim that they would prefer to relinquish the direct control of these towns. However, despite this long-term objective, circumstances are not changing rapidly enough to bring about its realization in the foreseeable future. The main source of change would come either from government or from individuals. The view of governments, in general, seems to be that the present arrangements are satisfactory; their reluctance to get involved is understandable as such involvement would almost certainly mean increased expenditure on the part of the government concerned. With regard to individuals the most likely source of change is to come from a stable population and home ownership. The available evidence shows that even with relatively stable populations the level of home ownership is not likely to rise due to a series of factors, the main one being the availability of rental accommodation at nominal cost. This factor, perhaps more than anything else, will work against the opening up of closed towns.

The "control" of the workforce (a frequently suspected aspect of closed company towns) may in fact turn out to be counterproductive, as evidence suggests that attitudes towards the employer are more unfavourable in closed towns than they are in the open towns. Thus it has been suggested that, as far as closed towns are concerned, it would be in the company's interest, as well as that of the community, if all town administration were vested in elected bodies at the earliest opportunity.[18] It is, however, necessary to avoid any misconceptions regarding "instant solutions" as far as the nature of town administrations are concerned. It does seem that it would be best to move away from closed company towns, where frustrations are said to build up because it is not possible to express dissatisfaction with conditions in the township, since the employer is the administrative authority for the town. This situation is avoided in open towns. However, the reality is that in these small single-resource single-employer communities the change from closed to open town may be merely a "cosmetic change", because the company presence is still felt when most of the participants in the administrative agency are employed by the same company. The mining company may also be able to maintain the dominant presence in other ways. Moranbah is an open town but the Utah Development Company still has veto rights on any development in the town, and any benefits that accrue to residents are the result of Utah's action rather than that of the Belyando Shire Council, under whose jurisdiction the township falls. The residents of Moranbah, not being ratepayers, are represented on the Shire council by their employer, the Utah Development Company.

Thus, while the goal of opening all closed towns is a desirable one, the reality of employer presence and dominance seems unlikely to change. The objective would be more likely to be met if there were some regional planning leading to centralized multi-company towns of larger size, which would allow a more genuine participation on the part of the residents. Additionally, some modifications to local government acts may be involved if the current franchise restriction to ratepayers only is continued.

Aspects of Socio-Demographic Characteristics of New Mining Towns

The socio-demographic characteristics of the new mining towns which have created considerable interest among social scientists are: male dominance, atypical age structure, presence of overseas-born migrants in large numbers, single employer workforce and high population turnover. These features hardly appear to be in line with the vivid imagery of the mining towns and their inhabitants as discussed in chapters 4 and 10. However, while the diggers and their rip-roaring activities gave the old mining towns the social character and atmosphere that has been captured by literary writers and historians, the planned new mining towns in contrast do not offer the same fare to the investigating social scientists. The planned character of the new mining towns leaves little scope for the creation of rip-roaring townships. Instead we have outback suburbias where there is serious concern with the socio-demographic characteristics of the population and with the identification of those aspects of behaviour which would result in the creation of "integrated, warm communities . . . which would lessen if not eliminate the tensions inherent in relatively small isolated mining communities".[19] The five distinctive characteristics listed earlier will now be examined.

Male dominance

Extensive research by social scientists indicates that frontier communities tend to be male dominant, indeed these communities may be exclusively male during their early phases. With the development of the frontier community there is an increase in the representation of females, but usually the male bias persists. A similar situation exists in mining communities.

Several factors make the male dominance in mining communities inevitable. In the beginning, during the town and mine construction phase, mining communities are essentially all male communities. The existence of "males only" jobs and the absence of any female (and family) accommodation is a severe deterrent to the presence of females during this phase of development of mining communities. There is greater female representation when married workers move in with their families. However, as many mining companies have a certain proportion of accommodation reserved for single male workers, and as there is usually little incentive for single females to be present in these communities, male dominance becomes a permanent feature of mining communities. This male dominance has important consequences as it is linked directly to the age structure and the population turnover in these communities.

Examination of the 1976 Population and Housing Census data indicates strong support for these generalizations. The sex ratios of selected mining towns in Australia are shown in table 26. It can be seen that the greatest imbalance is found in the Pilbara. The low figure for Karratha is due to the fact that the single men's accommodation for the town is located in Wickham and Dampier, which has the effect of raising their ratios even higher. The ratios for Roebourne and those for the Eastern Goldfields townships support the observation that the longer

Table 26. Pattern of sex ratios and age structure in selected mining towns, 1976

	Sex Ratios*	Age Structure						
		0-4 %	5-14 %	15-19 %	20-34 %	35-49 %	50-64 %	65+ %
Pilbara								
Dampier	190	8.7	15.3	4.7	45.9	19.7	5.2	0.5
Wickham	194	11.4	14.7	0.2	43.0	22.6	7.5	0.6
Karratha	116	16.2	21.8	5.4	36.2	15.4	3.6	1.4
Roebourne	120	14.0	20.2	6.2	29.5	16.3	9.6	4.2
Arnhem								
Gove	130	14.5	20.6	4.0	36.3	19.0	4.8	0.8
Balance rural	101	15.5	27.1	10.0	24.0	13.6	7.5	2.3
Eastern Goldfields								
Kambalda	124	14.6	25.0	6.8	32.8	15.3	5.0	0.5
Kalgoorlie-Boulder	111	10.1	19.7	9.8	23.7	15.6	12.1	9.0
Coolgardie	104	12.8	22.9	6.4	22.9	15.2	10.6	9.2
Bowen Basin								
Moura	116	14.8	25.9	6.1	31.0	15.5	6.0	0.7
Blackwater	112	16.4	23.5	7.1	33.9	14.1	4.6	0.4
Duaringa rural	158	12.4	18.7	9.8	28.8	17.0	10.7	2.8
Moranbah	106	20.0	22.8	9.6	37.4	11.6	2.8	0.4
Belyando rural	184	9.1	13.5	8.1	33.6	17.9	12.1	5.7
Dysart	123	16.5	18.4	4.4	43.8	11.8	4.8	0.3
Broadsound rural	147	12.1	20.3	9.8	24.6	18.9	11.0	3.3

* Males per 100 females

Source: Pilbara, Arnhem and Eastern Goldfields, from P.W. Newton and R. Sharpe, "Regional Impacts of the Mining Industry in North-Western Australia", Paper delivered to the Regional Science Association (Australia-New Zealand Section) Annual Conference, Adelaide (December 1980); Bowen Basin, from *Population and Housing Census, 1976.*

established townships tend to increase their female representation over time. The figures for Arnhem reflect the side-by-side existence of two communities: the first is the mining community of Gove with its typical male dominance; the second, labelled "Balance Rural", is the Aboriginal population of the area with its evenly balanced sex ratio.

The sex ratios for the Bowen Basin indicate that it is the nature of the industry, rather than remoteness, which influences sex ratios. Thus upon casual inspection the ratios for Blackwater and Moranbah do not appear to be unduly high. However, this is due to the fact that the single men's quarters for these two towns are located near minesites and hence counted in the shire population. These figures indicate generally, and for the Bowen Basin in particular, that it is possible to have mining communities with balanced sex ratios if mining companies do not have a significant proportion of their workforce made up of single men. The prevailing general policy of having a single men component of around 25 per cent of the workforce means that a sex ratio of around 133 males for every 100 females would be the "normal" situation.

Age distribution

Frontier communities are essentially for the young: by their very nature they cannot be expected to be demographically "balanced" communities. Their male dominance is a direct consequence of personnel recruitment policies, and similar policies influence the age structure of these communities, as most mining companies have "preferred" age groups. In general this means males between twenty-five and thirty-five years of age, although the range is sometimes raised up to forty-five. Table 26 also presents the age structure of selected mining towns of Australia, and shows that the twenty to thirty-four age group generally accounts for at least a third and at times almost half of the total population. Older established areas or older mining towns generally tend to have less than a quarter of their population in this age group.

Brealey and Newton have noted that the uneven distribution of age groups throws increased strain on the towns' services and facilities, for example adolescents, the middle-aged and the elderly lack the numerical base to have their specialized demands met in a situation where there is normally an undersupply of facilities even for those in the majority (the young children and young adults).[20] However, while this observation is true, it is perhaps unrealistic to expect otherwise as, given the nature of the industry, there is little scope for the employment of the older age groups. Furthermore, with the possible exception of some of the West Australian port/coastal locations, the new mining towns are hardly suitable communities for retirement purposes. Thus, however much we decry these "towns without grandparents", this is probably inevitable.

The age structure of these towns also has a built-in element of instability, stemming from the fact that the dominant age category (twenty-five to thirty-five year olds) is made up of individuals who, both for career and family reasons, are likely to be relatively mobile people. Available evidence suggests that there is little scope for occupational mobility within a particular mining town or within a particular mining company. This means that any upward movement in occupational position involves movement out of town in the great majority of cases. The family related reasons for moving are somewhat more complex, but tend to focus around children's education and employment.

Generally speaking, families with household heads around twenty-five years of age are likely to have toddlers or children of early primary school age. Families with household heads around age thirty-five are likely to have children who are about to enter high school. These "mixes" of family composition tend to create differing education and employment demands. Thus, while it is relatively easy for the authorities to meet the educational needs of younger families by providing kindergartens and primary schools, the provision of next-level facilities, such as high schools, proves more difficult. This means that as children approach high school age, mobility out of town is often forced upon parents.

Migrants

Historically, mining in Australia has been associated with the presence of foreign-born people, the Chinese, for example. While perhaps not as colourful as the earlier association with foreigners discussed in previous chapters, the present wave of mining is also closely associated with the presence of foreign-born people ("overseas-born" is the more frequently used term in census and other official statistical data).

Table 27. Proportion of overseas-born population in selected mining towns

Towns	Percentage of Overseas-Born
Pilbara	
Dampier	28.5
Wickham	36.5
Karratha	22.7
Newman (1971-72)	59.3
Hedland	46.2
Northern Australia	
Gove	31.3
Nhulunbuy (1974-75)	40.3
Bowen Basin	
Moranbah	9.6
Blackwater	11.6

Source: For Newman and Nhulunbuy, P.W. Newton and R. Sharpe, "Regional Impacts of the Mining Industry in North-Western Australia", Paper delivered to the Regional Science Association (Australia-New Zealand Section) Annual Conference, Adelaide (December 1980); all other figures are from *Population and Housing Census 1976.*

Some idea of the importance of the foreign-born is indicated by the figures in table 27, which lists the proportion of overseas-born population in selected mining towns. It can be seen that in most mining towns the proportion of the overseas-born varies between 25 and 40 per cent. However, the difference in the Bowen Basin is striking; for the Bowen Basin as a whole the proportion of overseas-born is a low 7.7 per cent. This proportion appears to be evenly distributed as the mining towns of Moranbah and Blackwater return similar figures. The range of figures presented in the table serves as a warning that the importance of the overseas-born can be overemphasized, and indicates the need to place the role of overseas-born people within the broader context of the recent mining boom in Australia.

The current wave of mining has to be seen both in a spatial and a temporal context. Temporally the boom started in the mid sixties, turned sluggish in the early seventies, then returned to boom conditions once again. Spatially, the boom was associated with developments in the mid sixties in Western Australia, but shifted in the ensuing decade to northern Australia with developments associated with bauxite at Weipa and Gove. This period also saw the commencement of projects associated with black coal in the Bowen Basin and uranium in the Northern Territory. This brief outline is necessary as the spatial and temporal contexts of mining development in Australia are closely associated with the proportion of overseas-born in mining communities.

The importance of the temporal context stems from the fact that it has influenced flows within the Australian labour market. Thus during boom periods it has been extremely difficult for the mining companies, despite generous incentives, to attract labour to the mining projects. This has meant that the mining companies have had to cast their nets wide and attract personnel from overseas. (This may also have a direct bearing on the workforce turnover in these projects.) Similarly, the spatial context is also important. Most of the Western Australian and northern Australian developments have taken place in the context of a (virtually) non-existent regional labour market. In contrast developments in the Bowen Basin have taken place in the context of a labour market facing structural adjustment of the regional economy. Consequently, there was little need to cast the recruitment net too far, as the labour force which was being displaced in the rural sector (in particular, the sugar industry) was absorbed in the mining projects.

Single employer workforce

One of the thrusts of research into the new mining towns has come from a long interest in the social sciences in various aspects of "company towns". These aspects have usually been examined in terms of the company control of workers during work and non-work situations. A new dimension is added when these company towns are situated in remote areas, which tends to place the interaction between the company and its employees in sharper focus. Most of the research in this area has tended to emphasize the dominance of the company. What is quite clear, especially in remote environments, is that, while the company has considerable control, it also becomes the focus for any grievance that the town's inhabitants may have. Thus the end product of company dominance may be quite counterproductive, resulting in poor labour relations and frequent industrial strife.

It is too easy to ascribe blame to the mining companies, for such a situation can evolve "naturally" when certain policy elements are present, some of which are beyond the control of the mining companies. The workforce of most new mining towns is closely tied to a single employer, the mining company. Employment is often linked with housing, thus further reinforcing the link with the employer. As employment opportunities and mobility within the mining companies are generally limited, a job change (excluding those within the company) would more or less automatically mean a departure from the town.

This situation almost always applies, regardless of whether the town is closed or open. It can be argued that without a choice in housing and employment the distinction between open and closed towns is a meaningless one, as the real distinction is between towns with one dominant employer and towns without. Critics of present *laissez-faire* development of mining towns have argued that centralization would lead to a decreasing dependence on a single employer, especially if the resources in the region permitted development of multi-company towns. While the argument is logical enough there is no evidence from a handful of multi-company towns to support (or reject) it. However, the available evidence indicates that the single employer aspect is an important variable in the population turnover experienced by the new mining towns.[21]

Population turnover

The data in table 28 reveal one of the outstanding features of the new mining towns, namely, high population turnover. Two generalizations can be made. Firstly, it is clear that while turnover rates are generally high there is considerable variation in the rates from year to year and from place to place. The figures for the old established mining town of Mount Isa are useful for purposes of comparison. The figures also indicate a relatively stable labour force in the Bowen Basin. The second generalization is that the turnover of personnel in the "staff" classification tends to be lower than in the "wages" classification. The explanation for this difference lies in the characteristics and the motivation of the different staff classifications.

The figures indicate a general situation of high population turnover, with rates exceeding 100 per cent in some cases. This should not be surprising as the high turnover rates are related to the characteristics of individuals who are attracted to the mining towns. The age-sex structure of these towns is one of the important variables which would lead to high population turnover.

Table 28. Workforce turnover in selected mining towns (percentage turnover)

SELECTED WESTERN AND NORTHERN AUSTRALIA PROJECTS								
Town		1970	1971	1972	1973	1974	1975	1976
Hamersley Iron Pty Ltd								
Dampier	staff	25	32	24	19	26	21	
	wages	100	67	50	64	32	65	
Tom Price	staff	27	26	18	27	19	16	
	wages	61	44	40	49	61	68	
Paraburdoo	staff		20	38	18	24	18	
	wages		59	25	38	61	57	
Nabalco Pty Ltd								
Nhulunbuy	staff			21	46	33	25	27
	wages			36	62	50	44	26
Goldsworthy Mining Co.								
Goldsworthy	staff							43
	wages							102
Finucane Is.	staff							41
	wages							42
Shay Gap	staff							43
	wages							103
All Towns	staff				46	47	48	
	wages				108	115	109	
Mt Newman Mining Co. Pty Ltd								
Newman	staff					29	26	
	wages					69	102	
Port Hedland	staff					35	32	
(company sector)	wages					83	74	
Mt Isa Mines Ltd								
Mt Isa	staff	8	9	9	8	11	9	9
	wages	33	43	27	25	32	31	27

SELECTED BOWEN BASIN PROJECTS				
Town		1978	1979	1980
Central Qld Coal Assoc.				
Saraji	staff	5.2	6.5	11.3
	wages	9.2	6.5	8.2
Goonyella	staff	2.8	15.5	15.5
	wages	7.7	7.1	8.6
Norwich Park	staff	1.9	1.7	1.8
	wages	5.6	8.1	7.1
Peak Downs	staff	7.4	14.9	15.1
	wages	5.5	7.0	8.6

Source: For western and northern Australia projects, T.B. Brealey and P.W. Newton, "Migration and Mining Towns", in *Mobility and Community Change in Australia,* ed. I.H. Burnley, D.T. Rowlands and R.J. Pryor (St Lucia: University of Queensland Press, 1980); For Bowen Basin projects, unpublished data from Personnel Section, Utah Development Company, Brisbane.

In their surveys Brealey and Newton note that, although mobility rates tend to decline once the towns get established, the very nature of people coming into the mining towns creates a built-in element which encourages or results in high turnover.[22] They note that there are two basic types of workers who are attracted to mining towns:

- *"target" workers*: this group of workers makes up the majority and is interested in accumulating as much capital as possible in the minimum amount of time;
- *"transfers"*: government as well as company staff who move for promotional purposes.

It is clear that the majority of people who move to these new mining towns do not plan to stay there for a long time: four to five years is usual. Brealey and Newton found that the following percentages of samples in the towns listed indicated a stay of four years or less: Port Hedland 62.9, Dampier 68.5 and Nhulunbuy 75.2.[23] Similarly, 52.5 per cent of the Moranbah sample indicated that they planned to stay in the town for less than five years. Questions relating to future mobility intentions produced similar answers.

This leads to an examination of some of the broader issues. The popular notion in Australia is that most people would not work in remote communities (supporting mining or non-mining activities) unless the monetary rewards were high enough. The fact that the mining companies have been forced to recruit personnel overseas is an indication that they have not been very successful (except during periods of high unemployment) in recruiting personnel in Australia, despite generous financial inducements. There is also a popular conception of another category of individuals (usually employed by governmental instrumentalities of one sort or another) who have to "do their stint in Siberia". These popular notions are supported by survey data, which indicate that in most cases at least 70 per cent of respondents come into these towns primarily for highly paid jobs and have a short period of stay in mind. In a sense these "target" workers were involuntary migrants into mining towns, as they would rather have been elsewhere if circumstances had permitted. Thus for "target" workers the planned stay is dictated by money saved, while for "transfers" the period of stay is dictated by the employers.

Seen in terms of a commonly used classification in migration research, the migrants to mining towns are clearly "movers" rather than "stayers". However, in spite of the built-in factors which would favour high population turnover, evidence indicates a general decline of rates over time, suggesting that some of the movers change their minds and become stayers.

Survey evidence indicates quite clearly that very few migrants to the mining towns regard their move as a permanent one. However, after a certain period of residence in a township, a proportion of households revise their planned departure date. In the Moranbah survey it was found that 50 per cent of the sample had extended their original planned period of stay.

In their analysis of individuals who had extended their stays, Brealey and Newton found no striking patterns in the survey data. However, they did note that some variables were associated with a household's decision to extend or not to extend its length of stay in a new mining town. The noted associations can be summarized as follows:

- *marital status*: single persons were extremely unlikely to extend their period of stay;
- *age of children*: households with children in the ten to fourteen or over fifteen years age group were most likely to move to avoid difficulties associated with children's education and employment in these towns;
- *duration of residence in township*: the longer an individual or a family remained in an area the longer it was likely to continue to stay;
- *education of head*: the higher educated were less likely to stay.

Finally, Brealey and Newton did make the interesting observation that while high income was an attraction factor it was not a retention factor.[24] Higher income levels were not, on their own, sufficient inducement to warrant a household extending in length its residence in a township.

These observations conclude the examination of the physical and the social aspects of new mining towns. In a sense these are internal aspects. The next part examines the external aspects or the regional setting of the new mining towns. This examination accepts the earlier statement that despite the obvious growth of mining towns, the demographic impacts of the "resources boom" are unlikely to alter dramatically the settlement geography and the overall population distribution of Australia. Most of the current and proposed mining projects are largely extractive in operation with little, if any, processing of the extracted ore (the bauxite/aluminium industry being the only major exception). These projects are highly mechanical and capital intensive in nature, consequently the demographic impact at the national level is likely to be very slight. However, this does not preclude a substantial local/regional impact, especially on sparsely populated or demographically stagnant areas.

REGIONAL IMPACTS OF MINING: A CASE STUDY

The aim of this part is to examine briefly the impact of the new mining towns on their surrounding regions. Although the Bowen Basin of Central Queensland was chosen for case study, every effort will be made to

. 11. Shires and towns of the Bowen Basin

link the analysis to other such developments in Australia. Attention will be directed mainly to the socio-demographic aspects, as other aspects have been examined elsewhere in this book (though usually not from a regional point of view).

The Bowen Basin is a long settled agricultural region of Central Queensland. It has risen in national economic prominence with the development of the black coal export trade to Japan, and accounts for nearly 90 per cent of the state's coal production. As defined here the Bowen Basin consists of eleven shires (see figure 11). This definition has produced a somewhat overbounded region, but it has the advantage of including mining and non-mining shires for comparison purposes. Our analysis is restricted to the data from the Population and Housing Censuses of 1976, 1971 and 1966, as these cover the main period of recent mining development. Our indicators are designed to measure both quantitive and qualitative changes in the Bowen Basin.

Demographic Impacts

The Bowen Basin population is increasing at a rate higher than the national average. Between 1966 and 1971 it increased by 18.8 per cent and by a similar amount in the 1971–76 intercensal period. Over the 1966–76 intercensal decade the population of the region increased from 43,008 to 60,285.

Table 29. Population Trends in the Bowen Basin, 1966–76

				Percentage Change	
	1966	1971	1976	1966–71	1971–76
Banana shire	12,493	13,433	14,169	3.4	5.5
Moura	–	1,902	2,694	–	41.6
Bauhinia shire	2,094	2,319	2,372	9.9	2.2
Duaringa shire	2,060	4,910	7,693	97.2	56.7
Blackwater	–	1,984	4,638	–	133.8
Emerald shire	3,504	5,639	6,024	60.5	6.8
Emerald	2,197	2,923	3,161	33.1	8.1
Jericho shire	1,501	1,420	1,220	-5.6	-14.1
Peak Downs shire	1,054	1,177	1,239	6.6	5.3
Belyando shire	2,997	4,834	7,210	57.4	49.2
Moranbah	–	1,050	4,053	–	286.0
Broadsound shire	1,625	1,589	3,379	-3.2	112.7
Dysart	–	–	1,585	–	–
Mirani shire	5,379	4,772	4,889	-11.3	2.5
Nebo shire	479	777	800	62.2	3.0
Bowen shire	9,342	10,231	11,292	9.1	10.4
Bowen	5,159	5,880	6,707	14.0	14.1
Collinsville	1,909	2,147	2,403	12.5	11.9
Bowen Basin	43,008	51,082	60,285	18.8	18.0

Source: *Population and Housing Census, 1976*

Analysis of the data presented in table 29 indicates that the regional average is made up of conflicting population patterns. However, in general terms it can be said that the rural areas of the region are experiencing loss both in absolute and relative terms while the mining areas are experiencing population gain. Thus the three shires which have experienced major population gains — Duaringa, Belyando and Broadsound — have all done so through the development of the new mining towns. Overall, just over half of the regional increase in population can be attributed directly to the new mining projects.

Table 30. Bowen Basin: Age composition, 1966–76

Age Group*	1966		1971		1976	
	Male	Female	Male	Female	Male	Female
0- 4	6.57	6.17	6.07	5.51	6.07	5.96
5- 9	6.22	6.12	6.44	5.48	5.80	5.42
10-14	4.97	4.51	5.04	4.69	5.03	4.54
15-19	4.39	3.40	4.59	3.30	4.68	3.86
20-24	4.83	3.41	5.43	4.12	4.96	4.09
25-29	4.45	3.32	4.85	3.72	5.13	4.41
30-34	3.82	3.03	4.10	3.10	4.31	3.40
35-39	3.94	2.76	3.64	2.77	3.64	2.85
40-44	3.50	2.43	3.43	2.27	3.05	2.35
45-49	2.92	2.26	3.16	2.18	3.02	2.00
50-54	2.69	2.32	2.53	1.90	2.35	1.71
55-59	1.79	1.74	2.22	1.65	2.00	1.49
60-64	1.71	1.14	1.79	1.26	1.74	1.42
65 +	2.90	2.69	2.67	2.14	2.56	2.16

* Each age group expressed as a percentage of total population
Source: *Population and Housing Census 1966, 1971, 1976*

The change in numbers has had varying effects on other aspects of the demographic structure of the region; for example, it has had little effect on the age sex structure and birthplace composition of the population of the region. However, as table 30 indicates, while the overall regional effect has been minor, the local impact on the age sex structure is often marked. This effect can be illustrated by the data present in table 31. Here the areal units being considered are:

- *Moranbah:* a mining town in Belyando Shire
- *Rural Belyando:* Belyando Shire less population in townships
- *Belyando Shire:* total population.
- *Emerald Shire:* total population in a predominantly rural/agricultural shire
- *Bowen Basin:* total population.

Table 31. Age structures of selected Bowen Basin Areal Units

Age Category	Moranbah	Rural Belyando	Belyando Shire	Emerald Shire	Bowen Basin
0- 9	17.7	8.7	13.5	10.0	11.9
10-19	6.6	8.3	8.2	10.5	9.7
20-39	21.7	28.3	20.9	17.5	18.0
40-49	3.5	7.4	5.0	6.6	6.1
50+	1.8	12.1	6.3	10.9	8.8

Source: *Population and Housing Census, 1976*

The figures in the table refer to males expressed as a percentage of the total population of the specified areal unit.

If Belyando Shire is routinely compared with Emerald Shire, there are no striking differences apart from the age category fifty years old and over. However, if the Belyando Shire population is looked at in terms of Moranbah versus the remainder of the shire population, larger differences become apparent. Thus Moranbah has a high proportion of its population under ten years old and a very small proportion of its population over fifty years old. Belyando rural has a relatively large proportion of its population in the twenty to thirty-nine years age category and has a sizeable proportion of its population over fifty years old. The Moranbah age structure is a reflection of the mining company recruitment and housing policies: the mining company's preference for employees in the twenty to thirty-five age group is reflected in the age structure. Similarly the mining companies maintain a further distinction between married personnel who are housed in Moranbah and single men who are housed at the minesite in the rural part of Belyando Shire. As rural Belyando has experienced considerable outmigration it has a greater representation of population over fifty years old, while the location of the single men's quarters at the mine site increases its share in the twenty to thirty-nine years age category.

The above example from Belyando shire illustrates quite clearly how the demographic impact of mining development can become "lost" when strong counter currents are in operation. Thus, although the mining workforce is male dominant and concentrated in the twenty to thirty-five years age group, the age-sex structure of the region has changed little over the last decade. The data suggests that while mining developments have attracted population in the region, the outmigration of males has continued at a pace to reduce noticeably the expected impact on the age-sex composition of the Bowen Basin.

Another demographic characteristic which does not reflect the expected change is the proportion of overseas-born in the population. The change in the proportion of overseas-born during 1966–76 has been extremely small, with the 1976 figure of 7.7 per cent representing an increase of less than 1.00 per cent over a decade. More detailed analysis indicates that the regional average is also representative of the mining towns such as Moranbah and Blackwater. These figures for overseas-born are in stark contrast to other mining areas, where the proportion of overseas-born is usually higher than a third and sometimes approaches almost half of the population. The lower figure for overseas-born is an indicator of the degree of success the Bowen Basin mining companies have had in recruiting labour locally and within the state.

Employment Impacts

It is clear from the figures available that the major impact of the mining developments has been on the *workforce composition* of the region, and this is illustrated in table 32. The regional workforce has increased from 18,433 in 1966 to 26,532 in 1976. This rate of increase is slightly faster than the population growth in the region. Analysis of individual shires in the region indicates most of this workforce growth was concentrated in three coal mining shires — Duaringa, Belyando and Broadsound — which together accounted for almost 60 per cent of the regional growth of the labour force. The only "non-coal-mining" shire with a significant growth in the workforce was Emerald shire. The increase here was associated with the Fairbairn Dam project.

The above analysis provides an overall picture of the growth of the workforce of the region. If we disaggregate the workforce into its major industry components we find that important structural changes are taking place in the regional economy (see table 32). Two major industry

Table 32. Employment trends in selected industries in Bowen Basin shires, 1966–76

Shire	Agriculture			Coal Mining			Manufacturing			Construction		
	1966	1971	1976	1966	1971	1976	1966	1971	1976	1966	1971	1976
Banana	37.1	32.3	28.1	9.5	14.2	16.3	7.7	6.8	6.0	14.8	8.5	8.3
Bauhinia	57.5	52.0	48.9	0.6	11.3	8.4	2.4	1.2	0.5	14.5	8.0	5.4
Dauringa	50.2	25.1	15.0	1.6	31.5	33.2	3.2	2.2	2.1	21.0	10.0	12.1
Emerald	23.7	14.6	15.1	–	–	–	8.8	10.3	14.4	16.3	24.8	10.5
Jericho	42.0	38.7	35.9	–	–	–	4.4	4.0	0.1	18.6	16.9	10.3
Peak Downs	65.3	54.7	57.4	–	11.3	1.6	2.3	0.1	–	11.8	6.2	11.0
Belyando	42.4	21.4	17.7	2.5	13.3	34.4	5.1	2.2	2.5	13.3	30.2	8.1
Broadsound	56.5	54.6	31.7	–	0.1	29.4	1.7	1.8	–	19.3	20.1	14.2
Mirani	54.1	50.7	52.5	–	0.3	–	19.0	20.8	17.1	9.9	6.2	7.0
Nebo	71.7	49.5	50.0	–	5.7	–	0.7	0.1	2.7	17.1	30.7	19.5
Bowen	20.3	20.2	18.6	0.3	6.6	8.1	15.5	15.6	13.4	12.3	9.9	6.6
Bowen Basin	39.1	30.5	26.5	4.8	11.0	16.5	9.1	7.5	6.3	14.3	13.6	9.0

Source: *Population and Housing Census 1966, 1971, 1976*

groups are of minor significance for our analysis: the manufacturing industry has never accounted for a major sector of the regional economy, and the construction industry has tended to fluctuate widely and has varied in importance between shires.

The main element of the structural change is the steady decline of the agricultural sector and the rapid rise of coal mining. In both instances the changes have been both relative and absolute. In 1966 agriculture employed 7,200 workers and represented 39 per cent of the regional workforce. By 1976 there was an absolute decline of 160 workers with the share of the agricultural sector falling to 27 per cent of the total workforce. In the same period the workforce in coal mining has increased fivefold.

A more detailed analysis reveals that the changes at the shire level are not inconsistent with the above trend. Thus there has been a decline in the importance of the agricultural sector in every case, the only difference being in the magnitude of the decline. In contrast the growth of coal mining is highly concentrated in Duaringa, Belyando and Broadsound shires.

This analysis of the population and labour force changes in the region has thus indicated that the major gains have been made in coal mining shires, the other shires losing population either in relative or in absolute terms. It is clear that there has been a major decline in importance of agriculture in the region; however, this decline in absolute terms has been very small (160 persons in the industry over a decade). The important question that remains to be answered is whether the increasing workforce in coal mining is "displaced" local labour or has been attracted from outside the region. To answer this a mobility analysis of the population was undertaken. The data is from the 1971 and 1976 censuses and from a mobility survey carried out by the author in 1980.

The main measure of residential mobility in the census is the question relating to the respondents' residential location at the previous census. Table 33 presents the analysis for the Bowen Basin. The "area of origin" categories indicate whether the respondent was (1) in the same dwelling at the previous census; (2) elsewhere in Queensland; (3) interstate; or (4) overseas. Clearly, this is a crude measure of mobility as it identifies only the addresses at the last census and ignores any mobility during the intervening period. Furthermore, its broad aggregation level prevents us from identifying the origin area within the state. Thus respondents identified in this category could have moved from an adjacent shire or from another shire thousands of kilometres away.

The residential mobility data presented in table 33 indicates considerable residential mobility in the region, as almost half the population in each intercensal period is recorded as having made a move. It is interesting to note that the broad area of origin categories have been relatively stable during both intercensal periods. Even more interesting,

Table 33. Residential mobility in the Bowen Basin, 1966–76

Shire	Percentage Same Dwelling in		Percentage Other Dwelling in Queensland in		Percentage Other State in		Percentage Overseas in	
	1966	1971	1966	1971	1966	1971	1966[a]	1971
Banana	50.0	49.7	42.1	37.8	3.8	4.2	–	1.1
Bauhinia	48.4	43.7	39.8	31.3	6.1	6.4	–	1.4
Duaringa	27.3	30.0	54.2	45.6	10.4	10.5	–	2.4
Emerald	32.3	35.3	52.7	37.5	9.1	7.3	–	2.9
Jericho	48.6	58.4	43.7	31.9	2.4	0.7	–	0.6
Peak Downs	50.1	46.6	37.4	30.7	7.5	7.2	–	1.3
Belyando	74.4	26.2	21.3	50.2	3.0	8.3	–	2.3
Broadsound	50.5	24.4	38.5	49.4	4.0	11.7	–	2.0
Mirani	67.2	63.1	26.9	24.5	2.9	5.2	–	1.1
Nebo	42.1	34.4	46.9	53.2	8.6	2.8	–	–
Bowen	52.6	47.6	38.2	31.6	4.9	6.5	–	1.8
Bowen Basin	46.8	41.5	39.2	38.2	5.8	6.8	–	1.8

(a) Data not available
Source: *Population and Housing Census, 1976*

perhaps, is the fact that the interstate and overseas categories have not featured to any great extent as source areas. Disaggregation of the regional data to shire level produces somewhat different patterns.

Disaggregated data shows, not unexpectedly, that there is a relation between the rapidly growing shires and residential mobility. It can be seen in table 33 that the proportion of interstate and overseas migrants rises considerably above the regional average in the three coal mining shires of Duaringa, Belyando and Broadsound. These three shires alone account for 42 per cent of all interstate and overseas migrants into the region. The two other shires which record significant values in these two categories are Banana shire and Bowen shire (in the former the Moura mine, and in the latter, Collinsville power station complex, may be the source of attraction).

The origins of respondents in a sample survey of Moranbah conform to the broad patterns outlined above: interstate and overseas origins accounted for under 25 per cent, while the remaining origins were from within Queensland. Analysis of this data indicated a dominance of central Queensland origins, with the regional centres of Mackay and Rockhampton featuring as the two most important centres. Analysis of distances shows that approximately 60 per cent of the sample had moved distances less than five hundred kilometres.

The Bowen Basin and Other Mining Developments in Australia

While the choice of the Bowen Basin for a closer examination of impacts of mining is justified, there is some need to place the analysis into the broader context of similar developments elsewhere in Australia. Other such development complexes include the iron ore based developments in the Pilbara and the lesser complexes of Arnhem-Gove (bauxite) and the Eastern Goldfields (nickel).

The Bowen Basin development differs in a marked degree in that mining development has occurred within the context of an established (predominantly agricultural) regional economy. Agricultural development in the other three complexes has generally been minimal due to their climate being marginal for anything other than extensive stockgrazing. This has meant that the European population in these areas is generally negligible. The numbers of Aborigines have varied, with the Aboriginal population reaching significant levels in areas of bauxite development, and being too small in numbers to be of much significance in the Bowen Basin.

The location of the region is also of significance when considering regional impacts of mining. While the Bowen Basin falls within the tropics it has a relatively mild climate when compared to the Pilbara or the Eastern Goldfields, or the bauxite mining areas of northern Australia. The region is not as remote as the other three areas, being within easy driving distance of coastal regional centres such as Mackay and Rockhampton. As many inhabitants of the Bowen Basin mining towns plan to retire in these coastal centres, they are able to arrange the final move over a period of time, and this flexibility is a factor in producing a relatively stable labour force in the Bowen Basin.

It is important not to overstate the differences between the Bowen Basin and other mining areas. While the general contexts have important differences there are many similarities between these outback suburbias: they are generally planned with similar objectives and town planning principles in mind. These, together with generally similar recruiting policies, tend to produce a high degree of congruence in the overall profile of the new mining towns in the major mining complexes today.

CONCLUSION

While mining towns have existed for a long time in Australia, and although writers of fiction and non-fiction have produced vivid images of wild rip-roaring places, the new mining towns are markedly different entities. While a certain degree of chaos and disorder were the hallmarks of the old mining towns, the new towns are noticeable for their tranquillity and order. To some extent both types of town are a creation of their times, reflecting the prevailing town planning principles.

This chapter has looked at past and present aspects of mining towns; the future is unpredictable. If more mining towns are begun, it is very unlikely that any radically new type of planning will be attempted: the mining town of the future will probably be an enhanced version of the present ones. Should the mining industry take a major downswing, it is likely that many of the existing towns will simply be mothballed – a modern day version of the ghost towns of the past.

Notes

1. This term was suggested by Reba Gostand, author of chapter 10.
2. Experience shows that the mining industry generally operates in an enclave economy, with the immediate setting of the mining activity deriving the least benefits. In the Third World situation the main benefit accrues to the advanced industrial nations, whose capital and technical expertise is largely responsible for the mining activity. In Australia a similar situation has operated where the mining activity has most pronounced benefits for the capital cities, while benefits to other centres have been minimal. Thus the mining boom of the 1960s in Western Australia resulted in tremendous growth for Perth but, apart from the new mining towns, little growth elsewhere. In a similar manner the forecast "resources boom" of the 1980s and 1990s is likely to benefit the established major centres mainly.
3. For a statement on the CSIRO project see T.B. Brealey, *Living in Remote Communities in Tropical Australia* (Melbourne: CSIRO, 1972). Much of my work is yet unpublished; but see M. Guilfoyle and P.C. Sharma, "The Regional Impact of Coal Mining in The Bowen Basin of Queensland", Paper delivered to the Regional Science Association (Australia–New Zealand Section) Annual Conference, Adelaide (December 1980).
4. T.B. Brealey and P.W. Newton, "Migration and New Mining Towns" in *Mobility and Community Change in Australia,* ed. I.H. Burnley, D.T. Rowlands and R.J. Pryor (St Lucia: University of Queensland Press, 1980).
5. Most of the CSIRO studies fall into this category; see also J.C.B. Mercer, *The Australian Mining Town*, unpublished thesis for the Master of Urban and Regional Planning Degree, University of Queensland, 1979.
6. *Architecture in Australia* 63, 3 (1974); UNESCO, *New Towns in Isolated Settings* (Canberra: AGPS, 1976).
7. T.B. Brealey and P.W. Newton, "Mining Towns: the case for centralisation", *Mining Review* (July 1977): 7-9.
8. P.W. Newton, "Present and Future Settlement Systems in Sparsely Populated Regions", United States/Australian Seminar on Sparsely Populated Regions, Flinders University, Adelaide (July 1978), p. 13.
9. Brealey and Newton, "Mining Towns", p. 8.
10. T.B. Brealey and P.W. Newton, "Mining and New Towns", in *Rural Australia: Problems and Prospects,* ed. J. Holmes and R. Lonsdale (New York: Scripta/Wiley, 1982).
11. T.B. Brealey and P.W. Newton, *Living in Remote Communities in Tropical Australia: The Hedland Study* (CSIRO Division of Building Research, Special Report, 1978).
12. Mercer, *The Australian Mining Town,* pp. 3-11.
13. Closed towns are those towns which are entirely "owned and operated" by mining companies. In such towns the mining company is in effect the local government authority for the town residents. Open towns include all other towns which are not directly under complete mining company control. In theory individuals are free to move in and out of these towns as they wish. However, even in the open mining towns, company dominance is inevitable.
14. Newton, "Present and Future Settlement Systems", p. 18.

15. Population and Housing Census of Australia, 1976.
16. Brealey and Newton, *Living in Remote Communities*, p. 77.
17. Ibid.
18. Brealey and Newton, "Mining and New Towns", p. 14.
19. P.W. Newton and T.B. Brealey, "Remote Communities in Tropical and Arid Australia" in *Desert Planning: The International Experience*, ed. G. Golany (New York: Architectural Press, 1981).
20. Brealey and Newton, "Mining Towns", p. 8.
21. Ibid.; Newton, "Present and Future Settlement Systems", p. 23.
22. Brealey and Newton, "Migration and New Mining Towns", p. 59; the Moranbah data is from a survey conducted by the author in 1980.
23. Brealey and Newton, "Migration and New Mining Towns", p. 63.
24. Ibid.

The Economy

W.H. Richmond

Few Australians can be unaware that there has been a mineral boom which took off in the mid 1960s, and that the 1980s could see an even greater surge in activity based on the mining and processing of Australia's mineral wealth. However, there is much uncertainty about the implications of the boom for the economic welfare of Australians and the process of Australian economic development. The industry itself goes to some trouble and expense to tell everyone that it is now the "backbone of the country", and politicians speak proudly of the "development" so obviously represented by huge mining and mineral processing projects and new towns carved out of the wilderness. Clearly, much additional income is generated directly and indirectly by the growth in mineral production. Yet it seems that the majority of Australians feel strangely unaffected by all the dramatic and well-publicized developments, and are unconvinced that the boom has brought or will bring them much benefit; many see it as the cause of a change in the whole structure of the economy which is harmful to them.

The question is a complex one. The purpose of this chapter is to present a framework in which the various implications of the growth in mining activity can be analyzed and to comment on some of the available data concerning the relationships involved. The chapter focuses on the period since the early 1960s — the period of the mineral boom — but begins with a brief review of the place and role of mining in the Australian economy since the early nineteenth century, to place the boom in historical perspective.

The simplest measure of the mining sector's place in the economy is its contribution to aggregate production. Aggregate production is equal to the total of value added or income created by economic activity. Table 34 indicates the percentage of gross domestic product (GDP) at factor cost attributable to the mining sector for selected years from 1861 to 1979–80.[1] At first glance these figures suggest simply a long-term decline in the significance of the mining sector of the economy from an important place in 1861 (the first year for which reliable data is available), when mining activity accounted for nearly 16 per cent of GDP, to the

Table 34. Mining industry contribution to gross domestic product at factor cost, selected years, 1861 to 1979–80

Year	Percentage	Year	Percentage	Year	Percentage
1861	15.7	1900–01	10.3	1948–49	2.5
1866	11.7	1905–06	9.4	1953–54	2.2
1871	11.1	1910–11	5.7	1958–59	1.7
1876	6.7	1915–16	4.8	1963–64	1.8
1881	4.9	1920–21	2.5	1968–69	2.4
1886	4.1	1925–26	2.3	1973–74	3.8
1891	5.6	1930–31	1.9	1978–79	4.5
1896	7.1	1935–36	2.8		
		1938–39	3.3	1979–80	5.1

Source: 1861 to 1938-39, N.G. Butlin, *Australian Domestic Product, Investment and Foreign Borrowing, 1861-1938/39* (Cambridge: Cambridge University Press, 1962), pp. 10-11; 1948–49 to 1979–80, Australian Bureau of Statistics, *Australian National Accounts* (various issues).

late 1950s and early 1960s, by which time that contribution had fallen to less than 2 per cent of GDP. The figures also suggest that this decline was not uniform but punctuated by revivals: in the 1890s and 1900s, to a lesser extent in the late 1930s, and again since the mid 1960s. But none of these revivals returned mining to the place it held in the economy in the middle of the nineteenth century.

This place was due to the discovery and exploitation in the 1850s of gold.[2] Prior to this time (or at least since the 1820s when economic growth in the conventional sense can be said to have started in Australia), it was the pastoral industry which had underpinned the development process. A number of minerals – coal, iron ore, copper, silver and lead – were mined during the first half of the century, but mining played an insignificant part in the economy as a whole. During the 1850s, however, the mining sector became the most significant in the economy and gold replaced wool as the major export commodity. This supremacy was relatively short-lived. Gold production peaked just after the mid 1850s and declined steadily thereafter; by the late 1860s pastoral production had overtaken that of the mining sector. The discovery and exploitation of other minerals, notably copper, lead, zinc and silver in the 1870s and 1880s, though significant in their own right, were insufficient to offset the relative, and at times absolute, decline of the sector as a whole. By the 1880s the mining industry accounted for less than 5 per cent of GDP.

But the mining boom of the 1850s had fundamentally altered both the scale and structure of the Australian economy. The rush of people in the quest for gold led to a trebling of the population in a decade. In order to service their needs a major structural adjustment in the economy was required. Until the level of mining activity subsided the development of other sectors of the economy was constrained by the shortage and high cost of labour. But during the 1860s and subsequent decades the manufacturing and construction sectors of the economy in particular expanded at a relatively more rapid rate than the primary sector. Land-intensive industries, especially the pastoral industry, continued to be the main source of growth (defined as the increased in GDP per capita), but aggregate economic expansion (the increase in GDP itself) rested much more on the development of manufacturing and construction activities. In the process Australia became a highly urbanized nation.[3]

Thus while mining was the most significant sector of the economy in terms of share of GDP for only a short time, and declined in significance in subsequent decades, this interlude — or more specifically the large population increase induced by the gold discoveries — had a fundamental effect on the structure of the Australian economy, and the subsequent process of Australian economic development.[4]

A revival of gold mining in the late 1890s and 1900s boosted the contribution of mining to GDP once again, and for several years around the turn of the century the sector was responsible for around 10 per cent of GDP. The revival had significant regional effects but induced only a small amount of increase in population through immigration. There was thus little effect on the long-term course of economic development. Expansion of other sectors of the economy in the first decade of the twentieth century steadily diminished the significance of the stable mining sector. After the first world war the industry as a whole stagnated then declined in absolute terms, such that by 1930–31 its contribution to GDP had fallen to below 2 per cent.[5] During the 1930s there was a significant increase in the production of gold and some other minerals such that, by the end of the decade, the share of the sector had increased to 3.3 per cent of GDP. In the decades following the second world war mining activities diversified and production of a new range of minerals steadily expanded.[6] But other sectors of the economy expanded more rapidly, and the contribution of the mining sector to GDP gradually fell to a low point of 1.7 per cent of GDP in the late 1950s and early 1960s.

It is thus apparent that Australia's first mining boom — centred on gold — fundamentally altered the pattern of national economic development (even though the boom itself was relatively short-lived). But for a century after the 1860s mining was of gradually decreasing significance in the Australian economy. The industry's contribution to GDP is not the only measure of significance (a point that will be made in some detail with respect to the current mining boom); mineral exports have always made a relatively greater contribution to total Australian exports, and the possession of certain minerals, notably coal and the major base metals, has enabled and encouraged the development of manufacturing industry in Australia. In general, however, from the late nineteenth century until the mid 1960s, mining was not a significant determinant of the process of Australian economic development.

Since the mid 1960s the share of the mining sector in GDP has steadily increased. A new mining boom has been based particularly, though not solely, on greatly expanded production of iron ore, coal and bauxite. The expansion resulted from the coincidence of three major factors at this time: large Australian reserves of minerals which were augmented by new discoveries; strong growth of Japanese demand for minerals, particularly by the rapidly growing steel industry; and major developments in materials handling techniques. Table 35 reproduces the index of mineral output at constant prices presented in the *Australian Mineral Industry Annual Review*. Table 36 contrasts the increased production in the mining sector with the growth in GDP at current prices. These figures indicate the rapid growth in mineral production that has taken place since the mid 1960s and the rate of growth relative to GDP. They also indicate that growth was particularly rapid in the late 1960s and early 1970s. Over the decade to 1973–74 the gross product of the mining sector grew, in *real* terms, nearly three times as rapidly as GDP for the economy as a whole.[7] For the whole period since the new boom started, mining production has grown at more than twice the rate of GDP in total. It follows that the *increments* in aggregate production and income over the period of the late 1960s and the 1970s are due to the expansion of the mining sector to a larger extent than is suggested by the sector's average contribution to GDP. Yet the fact remains that the contribution of the mining sector to GDP, while growing steadily from its low point of 1.7 per cent in the early 1960s, was still only 5.1 per cent by 1979–80.

Table 35. Index of mineral output at constant prices (1969 = 100)

Year	Index
1961	40.9
1962	43.9
1963	46.5
1964	48.0
1965	51.4
1966	58.6
1967	67.2
1968	82.2
1969	100.0
1970	128.0
1971	150.4
1972	162.2
1973	188.0
1974	200.4
1975	205.3
1976	211.0
1977	218.7
1978	212.7
1979	222.3

Source: Bureau of Mineral Resources, *Australian Mineral Industry Annual Review 1979* (Canberra: AGPS, 1981).

Table 36. Mining industry contribution to gross domestic product at factor cost, 1964–65 to 1979–80

	GDP Mining Industry $m	GDP Total $m	GDP Mining as percentage of Total
1964–65	325	17,640	1.8
1965–66	367	18,403	2.0
1966–67	440	20,301	2.2
1967–68	488	21,584	2.3
1968–69	577	24,668	2.4
1969–70	925	27,522	3.3
1970–71	1,051	30,450	3.5
1971–72	1,289	33,982	3.8
1972–73	1,437	38,661	3.7
1973–74	1,746	46,045	3.8
1974–75	2,301	55,134	4.2
1975–76	2,626	64,301	4.1
1976–77	3,183	73,396	4.3
1977–78	3,563	79,991	4.5
1978–79	4,046	89,982	4.5
1979–80	5,108	100,395	5.1

Source: Australian Bureau of Statistics, *Australian National Accounts* (various issues).

While the growth of the mining sector tapered off in the mid to late 1970s there are indications of a renewed boom in the 1980s and there is little doubt that the mining sector will continue to grow in both absolute and relative terms. The rate of growth will depend on many factors — the future of the world economy and thus of existing and potential markets for Australian resources; the possible emergence of new sources of supply and hence competitors; the extent to which new discoveries of minerals are made; and Australian government policies which may influence the rate at which Australian resources are mined and exported. The growth of the mining sector is thus difficult to predict. One set of predictions suggests that by the year 2000 mining GDP will constitute as much as 7.3 per cent of GDP, though this figure is based on assumptions which are unlikely to be realized and may be somewhat optimistic.[8]

The aggregate amount of employment provided by the industry is another indication of the significance of the mining sector in the economy as a whole. Reliable figures are not available before 1891; however, census data since then indicate that mining has provided employment for only a small percentage of the workforce (a proportion smaller than the sector's contribution to GDP) and that this proportion steadily declined, at least until 1971 when a slight increase was recorded. Annual estimates of the total numbers of persons employed in mining show an absolute decline in employment in the industry over the period up until the mid 1960s, despite some short term fluctuations.[9] Since then employment in the industry has increased by about 60 per cent, but the percentage of the total workforce employed in mining has increased only slightly and, in 1980, was still less than 2 per cent.[10] Any new mining projects in the 1980s are likely to be even more capital intensive than those of the first phase of the new boom, and the aggregate increase in direct employment is likely to be modest.

In absolute terms an industry which accounts for over $5,000 million worth of production and employs around eighty thousand people is obviously important. But relative to aggregate output and employment it hardly seems to be of very great significance for the economy as a whole. Could one be forgiven for wondering what all the excitement is about? Or has the growth in mining activity had larger implications than these figures would suggest?

No sector of the economy, unless it operates as a virtual "enclave", can fail to have some impact on other sectors of the economy. It is possible of course that mining operations in an economy could be undertaken by a foreign company, using largely imported labour and capital and even obtaining many of the current inputs required for operations from overseas. If this were the case, and all the minerals were exported in their raw state, then such activity would have very little effect on the host economy. Virtually all the income directly created by mining activity (wages/salaries and profits) would accrue to overseas residents (with the exception only of wages and salaries paid to local labour and whatever part of profits the host government was able to capture in the form of taxes); and little additional production and income would be generated elsewhere in the economy except perhaps as a result of the demand for some current or capital inputs obtained locally. Such a situation has been approximated in some colonial economies; it is obviously not the case in Australia where mining activity is "linked" to other parts of the economy in a variety of ways. The nature and magnitude of these linkages determine the larger impact of the expansion of the mining sector on the economy as a whole and on economic welfare.

The growth of the mining industry has led to the development of a range of industries undertaking the processing of minerals. This is often described as a "forward" linkage effect of the growth of an industry whereby the existence of the industry and the product it produces leads to the development of other industries; these generate income and employment which would not otherwise be possible (unless the inputs for the "downstream" activity were imported). Now it is true that many minerals are exported from Australia in unprocessed form: the mining of a mineral is not a sufficient condition for the processing of that mineral to be undertaken in Australia. A decision to undertake such processing depends on many factors, including the level of technology associated with processing, transport costs, and the availability and cost of power.

Nevertheless, processing of minerals is likely to follow mining activities in many cases. Putting it another way, it is highly unlikely that many of Australia's mineral processing industries would exist had not the mineral been mined here. This is true of the refining of bauxite into alumina and the smelting of other base metals. It is possibly true also of aluminium smelting — the further stage of bauxite processing — though this has depended much more on the availability and cost of electricity. The production of basic iron and steel products (one of the major categories of mineral processing) should perhaps more appropriately be regarded as a backward linkage to manufacturing industry. However it is the convention in Australia to regard mineral processing as part of the *mineral sector* of the economy in the more general sense because of the linkage with primary mining activities. Subsequent discussion will thus be couched in terms of the mineral sector in the wider sense, though it is recognized that in some cases (particularly that of the iron and steel industry) the links may not be very direct.

For statistical purposes the processing of minerals, other than basic treatment undertaken at the mine site, is regarded as part of manufacturing industry. Data with respect to processing is available in the Australian Manufacturing Censuses, which indicate that since the mid 1960s value added in the mineral processing sector of the economy has increased along with the increase in the mining sector, though somewhat erratically and less rapidly overall. A comparison of the data in the Manufacturing and Mining Censuses also indicates the relative significance of mineral processing and mining activities. In the 1970s value added by mineral processing activities was approximately 40 per cent of that in the mining sector, so in round figures mineral processing accounts for an additional 2 per cent of GDP, and the mineral sector (encompassing mining and mineral processing) for something over 6 per cent of GDP. Approximately the same number of workers are employed in mineral processing industries as in mining. Growth in mineral processing activities, particularly in the processing of base metals and bauxite and alumina, is likely to occur during the 1980s.

Mining, thus broadened to encompass mineral processing activities, also stimulates production in other sectors of the economy and generates additional income and employment as a result of what are often termed "backward" linkages. These occur when firms in the mineral sector demand inputs — either current (fuel, services) or capital (plant, machinery, vehicles) — from firms in other sectors of the economy. In analyzing backward linkages it is useful to distinguish between the *construction* and *operating* phases of a mine or mineral processing plant. The establishment of a new mine or plant initially involves large capital expenditures often compressed into a short period of time. Depending on the location of operations, such expenditure may also include large scale infrastructure expenditure — on public utility, transport and communication facilities (roads, railways, ports) and even on whole new towns. (This expenditure may be undertaken by private companies or by government authorities.) Once projects are operational, inputs are then required to undertake the production process.

A significant characteristic of the mineral boom has been the very large *capital* expenditures associated with the establishment of mining and mineral processing projects. Since the mid 1960s the Australian public has become accustomed to regular announcements of projects which individually involve investment expenditure of hundreds of millions of dollars. Many of the mining projects, particularly in the case of iron ore, bauxite and to some extent coal mining, have been located in remote areas so that large infrastructure expenditure has been necessary as well. It is as a result of these large capital expenditures that the major backward linkages have occurred.

Table 37, which relates only to *mining,* indicates the marked, though fluctuating, increase in capital expenditure undertaken since the mid 1960s, and its significance in relation to total private capital expenditure and to GDP. Note the particularly high levels of expenditure in the late 1960s and early 1970s.

Table 37. Capital expenditure in mining

	Gross Fixed Capital Expenditure – Mining $m	Percentage of Total Private Gross Fixed Capital Expenditure	Percentage of Gross Domestic Product
1960–61	46	1.9	0.3
1961–62	48	2.1	0.3
1962–63	70	2.7	0.4
1963–64	68	2.3	0.4
1964–65	107	3.1	0.5
1965–66	217	5.9	1.0
1966–67	243	6.4	1.1
1967–68	325	7.8	1.3
1968–69	435	9.2	1.6
1969–70	533	10.3	1.8
1970–71	781	13.4	2.3
1971–72	874	13.8	2.3
1972–73	455	6.8	1.1
1973–74	525	6.7	1.0
1974–75	707	8.3	1.1
1975–76	656	6.4	0.9
1976–77	490	4.1	0.6
1977–78	804	6.3	0.9
1978–79	1139	7.8	1.1
1979–80	1081	6.7	0.9

Source: Australian Bureau of Statistics, *New Capital Expenditure by Private Enterprises in Selected Industries* and *Australian National Accounts* (various issues).

Investment in mineral processing activities has also varied from year to year but on average has equalled about one half of capital expenditure in mining.[11] A large part of this can be regarded as directly related to the increase in mining activities.

While there was a tapering off in mineral-related capital expenditure in the second half of the 1970s it is anticipated that there will be a considerable increase in the 1980s. Expectations were fuelled in the late 1970s and particularly in 1980 by the publication of a survey by the Department of Industry and Commerce (DIC) of major planned investment projects. The June 1980 report in particular showed a dramatic increase in projects at the "committed" or "final feasibility" stage, a very large proportion of which were in the mining sector or were associated with the processing or transportation of minerals or petroleum.[12] Projections of capital expenditure suggested that the "construction phase" of the 1980s developments would be concentrated in the early years of the decade with expenditures in excess of $5 billion per annum from 1982 to 1985.[13]

The DIC survey has been subjected to a considerable amount of criticism. It has been argued that survey estimates have contained an element of escalated costs, that some projects included in the survey were mutually exclusive in that the go-ahead of one might effectively mean the deferral or cancellation of others, and that some projects have been defined as "committed" where contracts were still to be obtained. Moreover, when the projects are looked at in aggregate terms there appear to be serious aggregate supply constraints — relating particularly to the supply of skilled labour, perhaps to the availability of funds for investment and also to the availability of necessary infrastructure (particularly with respect to transport) — which mean that it would be virtually impossible for all the projects listed to go ahead.[14]

In response to these criticisms some modifications were made to subsequent survey reports. More significantly, the decline in market prospects and prices of a wide range of minerals from the end of 1981, and the consequent deferral or cancellation of many projects, have led also to downward revisions being made in capital expenditure estimates for the 1980s. Nevertheless, the decade will still see a significant increase in capital expenditure related to resource-based developments.

Included in the increased capital expenditures that have been and are expected to be undertaken in the mineral sector is large expenditure on infrastructure. Mining companies in particular have had to spend large sums on basic social capital, including whole new towns, especially when operations have been undertaken in remote areas. However a considerable amount of expenditure on infrastructure related directly to developments in the mineral sector has been undertaken by governments, though this is difficult to quantify. Indeed the state governments have embarked on a massive overseas borrowing programme to fund infrastructure projects. A large part of this borrowing is for mineral sector related infrastructure expenditure, including major power stations in Queensland, New South Wales and Victoria. This public infrastructure programme will thus add considerably to the capital expenditure associated directly with the mineral boom of the 1980s.

The demand for *current* inputs has also increased during recent decades as the level of mineral production has increased; it will continue to increase if mineral production expands progressively during the 1980s with major new projects coming into production.

Both capital and current expenditures, to the extent that they are directed within the Australian economy, generate additional production, income and employment, and have significant stimulatory effects on other sectors of the Australian economy. Only fragmentary evidence has been available on the nature of these backward linkage effects.[15] The deficiency of information on the relationships involved led the Bureau of Industry Economics to instigate in the late 1970s a project to investigate the relationship between the operations of the mineral sector and the rest of economy.[16] This study attempted to quantify the impact on GDP of a resources development "scenario" with respect to open cut black coal, iron ore, uranium and oil and gas. The study indicated that both the development and production phases of projects in these areas during the 1980s would yield substantial benefits to other industries in the economy. However, it appears that these will be less than has often been supposed. The increase in GDP in the form of value added within the mining sector itself was predicted to constitute about 65 per cent of the total estimated impact of the projects on GDP. Thus the study concluded that, while the total benefits in the form of increased production and income would be substantial, "only a modest proportion will be distributed through the linkages associated with the development and operation of the mines".[17]

The mineral boom has thus led to increased levels of income generated not only within the sector itself but also in other sectors of the economy which have received a stimulus from mineral developments. However, additional production can only be undertaken with additional resources — labour and capital. The demands for these resources, created directly and indirectly by the rapid expansion of one sector of the economy, may be expected to increase inflationary pressures in the economy generally and cause difficulties for other sectors, as available resources are bid away from their existing uses and become more difficult to obtain and/or more costly. The problem is reduced to the extent that there are unemployed resources that can be called on to meet the increased demand, or that the existing supply of labour and capital can be augmented by immigration or foreign borrowing.

The mineral sector is a relatively small direct employer of labour, but its demands and those of other industries stimulated by the demands of the mineral sector have some special characteristics. The mineral boom, in all its overlapping phases — exploration, project construction and production — has placed considerable strain on the supply of certain types of labour — skilled tradesmen, engineers and geologists — and at times wages and salaries have been considerably bid up by the demand. Furthermore, many mining and mineral processing activities have been in remote and/or unattractive places and high wages and salaries have been necessary to attract people to work and live in such environments.[18]

Substantial problems were experienced during the 1960s when levels of employment were generally very high and competition for some categories of labour was acute. The pressure on wages and salaries eased during the 1970s as mining development slowed down and the economy generally experienced slower growth and increasing levels of unemployment. However, in the late 1970s, as growth of the mineral sector started to increase and new projects were initiated, shortages re-emerged, particularly of skilled tradesmen and engineers, and there were renewed pressures on wages and salaries. During 1980, increasing concern was expressed that the renewed strength of a mineral boom in the 1980s would place acute strains on the supply of labour, particularly that of skilled tradesmen, in the first half of the decade. This was expected to arise primarily from the demands of the construction phase.[19] Forecasts were widely made that the construction phase of a renewed boom would place great strains on the demand for skilled tradesmen and lead to labour shortages and/or wage increases which, given the centralized wage-fixing system in Australia, might be expected to flow on through the economy generally.[20] This could not only cause difficulties for firms in the mineral sector itself and related industries but also affect firms in other sectors of the economy which are large employers of labour.

The capital requirements of the expansion of the mineral sector through the demand for funds for investment have similar consequences in principle for other sectors of the economy. Capital expenditure directly associated with mineral development has been very large. The extent to which the demand for funds to undertake investment expenditure associated with mineral development competes with the demand of other sectors of the economy and limits the availability or increases the cost of funds to other sectors depends on the sources of the funds obtained by the mineral sector.

It has been estimated that for the period from about the mid 1960s to the mid 1970s, approximately 20 per cent of total capital raised was in the form of *internal* funds (undistributed profits, depreciation allowances and so on); 30 per cent was in the form of *share capital* (slightly more than half of which was raised domestically and the balance overseas); and approximately 50 per cent was in the form of *loan* raisings (55-70 per cent overseas, the remainder locally).[21] The limited funds available within Australia and the limited ability of Australian organizations to marshall funds competently for the development of mineral projects has led to heavy reliance on both debt and equity capital from overseas. Nevertheless it appears that, in the period under consideration, as much as 45 per cent of the total demand for new capital associated with mineral development was sourced in Australia, with Australian sources contributing slightly more to *share* capital than was raised overseas.[22]

It may seem therefore that these demands must inevitably have had a "crowding-out" effect on the availability of capital for other sectors of the economy such as manufacturing or housing. The relative fall in investment in manufacturing, accompanying the increased capital expenditure in the mineral sector discussed in the previous section, could be thought to lend some support to this. The reasons for this decline, however, have to be sought primarily in the range of factors adversely affecting profitability in the manufacturing sector, though obviously the growth of a more profitable sector of the economy might have attracted funds away from manufacturing.[23] The special characteristics of the mineral sector's demand for finance (related particularly to the scale of projects and the risks associated with them) mean that to some extent it is operating in a different market. It is only recently that there has been concern that the mineral producers' large demands for finance have had a significant influence on the cost and/or availability of finance to other sectors of the economy, but these doubts are increasing.[24]

One area that is rapidly becoming a major cause for concern relates to the large demand for funds by governments (mainly state governments) to finance the massive infrastructure programmes directly related to the mineral-based boom. To the extent that state governments (mainly through semi-government authorities) seek to borrow on the domestic market they are competing with the corporate sector. It was argued, following the Loan Council meeting in mid 1980 when the magnitude of state government programmes became apparent, that the current level of raisings by semi-government authorities was clearly starting to crowd out the private sector, and there was little scope for further public capital raisings without severe harm to the private sector.[25] It has also been argued that even if these funds were sought overseas, the commonwealth government would be forced to issue securities to absorb the extra liquidity in order to control inflation. This would then attract funds away from the corporate sector. So in *either* case the corporate sector faces a smaller financial market from which to draw and this contributes to increases in interest rates.[26] Concern has also been expressed that there will be a crowding out effect *within* the public sector. The Australian Institute of Urban Studies, for example, has argued that public funds normally available for deployment in the cities have been diverted by the large infrastructure commitments.[27]

Thus, while the competition for resources created by the growth of the mineral sector during the 1960s and 1970s was not great, it increased at the end of the seventies and the demands for both labour and capital arising directly or indirectly from any heightening of the boom during the 1980s might well serve to bid resources away from some other sectors of the economy, thus increasing costs and causing difficulties for firms in some existing industries. The pressure on resources could also serve to increase general inflationary pressures in the economy.

Further consequences of the growth of the mineral sector arise as a result of its implications for Australia's economic relations with the rest of the world. The growth of one sector of the economy — such as the mineral sector — which is heavily export-oriented will tend to generate surpluses in the balance of payments. This will arise primarily because of the increase in exports. The tendency may be reinforced if the expansion relies on overseas investment, in which case payments on capital account will add to the external surplus, at least during the expansion phases of the sector. The tendency to cause surpluses in the external account will be *offset* (and may even be balanced) by increases in overseas *payments* associated with the growth of the new sector. For example, there might be a rise in imports (due partly to the fact that a proportion of the increased income generated within the expanding sector will be spent on imports and partly to the increased demand, particularly for capital goods, of the sector). To the extent that the growth of the sector relies on overseas investment there could also be an increase in overseas payments of interest and dividends. In addition there could be an increase in capital exports if part of the increased income from the growth of the sector were used by its recipients to purchase foreign capital. However, it is likely that net surpluses will be generated.

In a free market situation such an imbalance in the external account would tend to be corrected by variations in the exchange rate of the country's currency, that is the price of the currency in terms of other currencies. In such a situation the exchange rate is determined like any other price: the higher the demand for the currency relative to supply the higher the price and vice versa. If a surplus in the balance of payments were created as a result of an increase in exports or capital inflow, there would be an excess supply of Australia's foreign currency holdings which would cause the price of the Australian dollar to *appreciate* in relation to the foreign currency. If this occurred the relative price of Australia's exports and imports would fall. Normally there would be a reduction in net receipts of foreign currency and equilibrium would be restored. The corollary of the process is that output and income would fall in the traditional export industries and in those industries producing goods which compete with imports. Furthermore both output and income earned within the new export sector would be reduced below the level which would have been the case under the initial exchange rate.

There is an alternative to the appreciation of the dollar, although the long-run result is the same. In Australia, as in many other countries, the nominal exchange rate — the rate at which the Australian dollar converts into other currencies — is not determined by free market forces but is set by the government. Any tendency to surplus can be corrected by an appreciation of the nominal exchange rate in line with that which market forces would have dictated. Failure to appreciate the exchange rate sufficiently will result in an increase in the base of the money supply because Australian receipts from abroad would be in excess of payments. In a situation of full employment of resources this will result in inflation: relative inflation serves as an alternative mechanism whereby changes in the *real* exchange rate occur if the nominal rate is not adequately adjusted. So a higher rate of domestic inflation will have the same effects on the competitiveness of traded goods industries, only the effects are felt through upward pressure on costs rather than downward pressure on demand. There are in fact an infinite number of combinations of nominal exchange rate changes and the relative rate of inflation between Australia and her trading partners which bring about the necessary adjustments. The consequences for traded goods industries in the economy are similar: inflation operates less rapidly but imposes wider costs on society.

The *necessity* for an appreciation of the real exchange rate in the face of a balance of payments surplus can be reduced by other measures. The most obvious of these measures is a reduction in the high levels of protection afforded some Australian industries through tariffs or quotas, which would lead to increased expenditure on imports and a reduction in the surplus in the balance of payments. In this case adverse effects on competitiveness and profitability would be confined to those sections of the manufacturing industry producing goods which compete with imports. These industries would be more severely affected than if adjustments had been made in the balance of payments through the mechanism of the exchange rate, but industries using imported inputs will benefit (principally those in the rural and mineral sectors but also some manufacturing industries). The necessity for exchange rate adjustment could also be avoided or reduced by measures to increase the export of capital (that is, the lending of funds overseas) either in the form of direct investment (through the branches or subsidiaries of Australian firms) or through institutional lending. If this were achieved the competitiveness and profitability of domestic traded goods industries would not be directly affected. The most obvious means of achieving this goal is the liberalization of restrictions on capital export, but it is not clear that this would have a very great effect in the short term.

To summarize then: the rapid growth of a new export sector of the economy is likely to cause surpluses in the balance of payments. If measures are not taken to accommodate this tendency to surplus — by appreciation of the nominal exchange rate or reduced protection for example — then inflationary pressures will be set in train. A higher rate of inflation relative to trading partners will in any case imply an appreciation of the real exchange rate. Whichever mechanism of adjustment is adopted, some other industries in the economy producing traded goods will suffer a decline in competitiveness and profitability and under certain circumstances (particularly if the mechanism of tariff cuts is adopted), some firms, and perhaps whole industries, may be forced out of business. Employment opportunities in these industries will diminish accordingly. The combination of mechanisms actually adopted will determine the relative effects on the different sectors; a reduction in protection, for example, will place all the burden of adjustment on import-competing industries whereas an appreciation of the exchange rate will affect *all* traded goods industries.

What has actually happened in Australia? In the decade after 1964 mineral exports grew (in real terms) at an average rate of about 25 per cent, and for several years in the late 1960s and early 1970s at well over 30 per cent. In addition, increased local production of oil led to a reduction in expenditure on oil imports.[28] To finance the very large capital expenditures associated with mineral developments, there was a large inflow of both debt and equity capital; this was particularly strong in the second half of the 1960s and the early 1970s. These increases were offset to some extent by payments of interest and dividends, which also increased steadily from the mid to late 1960s. However the net result was that large surpluses were recorded in the three years 1970–71, 1971–72 and 1972–73. By the end of this period Australia's international reserves had climbed to a figure many times that considered appropriate.

These surpluses cannot be attributed wholly to mineral developments: increases in other exports and speculative capital inflows played a part too. Nevertheless mineral exports and mineral-related capital inflows were significant factors. During these years, as foreign reserves were increasing, no action was taken to appreciate the exchange rate. In fact the Australian dollar was effectively *devalued* slightly in December 1971, resulting in a large increase in the domestic money supply and the initiation of a period of high inflation in the Australian economy. The dollar was eventually revalued (appreciated) in December 1972 and further revaluations took place in 1973.

In July 1973 a 25 per cent across-the-board tariff cut was made in an attempt to alleviate the pressure on resources in the domestic economy by increasing the level of imports.[29] Thus the inflation, the exchange rate appreciations and the tariff cuts which occurred in the early 1970s can all be attributed (in large part at least) to the rapid growth of the mineral sector. Each of these mechanisms was adopted, or occurred, in response to this growth. And each – in a different way – adversely affected either or both the export and import-competing sectors of the economy.

In an important article published in 1976, which spelt out the essential aspects of the analysis outlined above, R.G. Gregory attempted to indicate the magnitude of the effects of the increase in mineral exports during the late 1960s.[30] He did this by estimating the relative price shifts against domestically produced traded goods necessary to maintain the balance of payments equilibrium following the increase in exports of mineral primary products. These relative price shifts were expressed in terms of "tariff equivalents", which measure the extent to which an alteration in tariff levels, in the absence of the increase in mineral exports, would have had the same effect on traded goods industries as the increase in mineral exports actually had.

Gregory concluded that "the effect of the rapid growth of mineral exports on the rural exporting industries is approximately equal to the effect, in the absence of the mineral exports, of a doubling of the tariff level. For the import competing sector the effect of the mineral discoveries is estimated to be approximately equal to setting the average tariff at zero and introducing an import subsidy". On the basis of these estimates his conclusion that "a significant proportion of the difficulties now experienced by export and import-competing industries might be explained by the rapid growth of mineral exports over the last decade" seemed an understatement.[31]

Gregory's analysis and conclusions were criticized on several grounds.[32] The most significant deficiency in the analysis is the neglect of the effects of the expansion of the mineral sector on the creation of income within the domestic economy. Increased income is created both within the mineral sector and indirectly in other sectors as a result of the linkages discussed above. To the extent that this income is spent on goods produced within the Australian economy the adverse effects identified above are offset; and to the extent that it is spent on imports then the tendency for reserves to accumulate is reduced as well. Furthermore, interest and dividend payments becoming due as a result of foreign borrowing also reduce the tendency for reserves to build up. On the other hand, Gregory's model did not specifically take into account capital flows associated with mineral developments. The very large net inflows associated with the mineral boom (even though some of this was spent on imports) substantially increased the balance of payments surplus and increased the pressure for adjustment.

On balance it appears that Gregory considerably overstated the effect of the growth of the mineral sector on other sectors of the economy. The mineral boom of the late 1960s and early 1970s can nevertheless be regarded as a major cause of the changes in the real exchange rate, which caused the index of Australia's competitiveness to decline sharply from the end of 1972 until the end of 1974, and of the tariff cut of 1973 which further reduced the competitiveness of those industries for whom protection had been reduced. As a result many Australian traded goods industries, particularly those in the import-competing manufacturing sector, experienced decreased profitability and considerable economic difficulties during this period.

After 1973 the balance of payments moved into persistent deficit due largely to a lower rate of increase in mineral exports and mineral-related capital inflow, together with increases in imports (including increasingly expensive oil) and interest and dividend payments overseas. In response to the altered situation the Australian dollar was devalued in 1974 and 1976 and again in subsequent years by smaller amounts under a new system of altering the exchange rate. Protection for manufacturing industries against imports was increased in some areas after 1976. During this time the continued growth of mineral exports (and the continuing substantial levels of capital inflow related to mineral developments) merely helped to reduce the tendency towards deficits. At the end of the 1970s the mineral boom regathered strength and it seemed almost certain that the 1980s would see a renewed tendency towards external surpluses as a result. The boom euphoria faded rather suddenly at the end of 1981; nevertheless the capital inflow associated with the large mineral-related projects (in both the private and public sectors) which will proceed, and an increase in mineral exports during the 1980s, seem likely to more than offset increased expenditure on imports and increased interest and dividend payments overseas, and to result in a tendency to balance of payments surpluses. The same pressures as occurred in the early 1970s could thus recur in the early 1980s and bring with them the same implications for other traded goods sectors of the economy, though which sectors will be most affected will depend very much on the policies pursued by governments.

Some of these threads may now be drawn together. A “mineral boom” has a number of economic consequences which can be summarized in general terms. In the first place an increase in mineral production will directly generate increased income. In addition the growth of mining activities leads to an increase in production, and hence income, generated in other sectors of the economy: firstly in the mineral processing area, and secondly in those sectors of the economy supplying both the current and capital input requirements of both the mining and mineral processing sectors (that is, through backward linkages). However, these developments must impinge on other sectors of the economy. The expanding sectors of the economy demand resources — labour and capital — to undertake production, and bid these away from other sectors (an effect which can be modified to the extent that the supply of resources is expanded through immigration and the borrowing of capital from overseas). A further consequence arises from the fact that mineral production (at least in the Australian context) is heavily oriented to overseas demand. An increase in mineral exports will tend to create a surplus in the balance of payments; this tendency may be strengthened by capital inflows associated with the mineral developments. There will be some offsetting factors — increased imports, increased overseas payments of interest and dividends and possibly increased capital exports — but to the extent that the growth of the mineral sector results in the generation of net external surpluses, an appreciation of the exchange rate is necessary to restore equilibrium in the balance of payments. This may be brought about by a change in the nominal exchange rate or, in a less direct way, through a higher relative rate of inflation. Either mechanism will lower the relative price of traded goods and adversely affect the competitiveness and profitability of firms producing exports and goods which compete with imports; aggregate income and output in these sectors will be reduced; some firms will contract or go out of business. The need for changes in the exchange rate can be avoided or reduced if import tariffs or quotas, which exist to "protect" some industries from import competition, are reduced. If this is done imports will become relatively cheaper; the demand for them will increase, much of it being transferred from goods formerly produced domestically. The "burden of adjustment" in this case is felt relatively more (perhaps entirely) by producers of import-competing goods. An appreciation of the exchange rate could also be avoided if the surplus in the balance of payments were offset by capital exports. In this case domestic industries producing exports or import-competing goods would not be so affected. In the absence of an increase in capital exports, however, the growth of the mineral sector is a force for structural change in the economy.

All of these consequences are the result of the mineral boom which started in the 1960s. What are the overall implications for economic welfare? The pure economic logic, as presented here, is that labour and capital in the economy are transferred to relatively more productive uses, and so the change should imply an increase in aggregate income. However, this may not necessarily follow. If the growth of the mineral sector relies heavily on foreign direct investment, with a consequent high degree of foreign equity, a proportion of the income directly generated will accrue to overseas residents. Much will depend also on the amount of production and income indirectly generated by mineral sector growth, relative to the reductions in production and income caused by the process of structural adjustment induced by the expansion of the mineral sector. Some producers and employees in the traded goods sectors will be adversely affected and forced to make changes in their economic activities. Change usually involves costs. These may well be outweighed by the economic gains of transferring capital and labour to more productive and well-rewarded activities which have been made possible by the mineral developments, though this will depend on the strength of the relationships in each case. Of course new economic opportunities may not be equally available to people even if in aggregate terms they expand. Change for some people may be very difficult or expensive, or virtually impossible, and many may become significantly worse off. The mineral boom may, in other words, effect a redistribution in income which must also be taken into account when considering its impact on economic welfare.

A considerable amount of empirical work must be done before these various influences can be quantified precisely and the overall impact of the mineral boom on the economy and on the economic welfare of Australians be assessed.

Notes

1. Gross domestic product at factor cost is defined as the value added by the factors of production in the process of production. It is hereafter referred to simply as GDP.
2. See chapter 2.
3. See chapter 4.
4. For a fuller discussion of the effect of mining on the nineteenth century economy see W.A. Sinclair, *The Process of Economic Development in Australia* (Melbourne: Cheshire, 1976); and R.V. Jackson, *Australian Economic Development in the Nineteenth Century* (Canberra: ANU Press, 1977).
5. On mining in the post-war period see F.R.E. Mauldon, "The Decline of Mining", *Annals of the American Academy of Political and Social Science* 158 (November 1931), economic survey of Australia, pp. 66-76.
6. See H.G. Raggatt, *Mountains of Ore* (Melbourne: Lansdowne, 1968); and Geoffrey Blainey, *The Rush that Never Ended*, 3rd ed., (Melbourne: Melbourne University Press, 1978), chapter 29.
7. Industries Assistance Commission, *Structural Change in Australia* (Canberra: AGPS, 1977), p. 5.
8. Wolfgang Kasper, Richard Blandy, John Freebairn, Douglas Hocking and Robert O'Neill, *Australia at the Crossroads: Our Choices to the Year 2000* (Sydney: Harcourt Brace Jovanovich, 1980), pp. 189, 221.
9. N.G. Butlin and J.A. Dowie, "Estimates of Australian Workforce and Employment, 1861–1961", *Australian Economic History Review* IX, 2 (September 1969): 144.
10. Annual estimates of employment in mining are made by the Australian Bureau of Statistics from its regular population survey (published as *The Labour Force*) and from the annual *Mining Census*. The latter are reproduced in the *Australian Mineral Industry Annual Review*. There is considerable variation between these estimates and those derived from national census data. The highest estimate for mining employment in 1980 is 1.7 per cent of the workforce.
11. See Australian Bureau of Statistics, Mining and Manufacturing Industry censuses. Mineral processing as defined here includes the production of "basic iron and steel products" and the smelting and refining of alumina, aluminium and non-ferrous metals.
12. Department of Industry and Commerce, *Major Manufacturing and Mining Projects* (Canberra: AGPS, June 1980).
13. Australian Industries Development Association (AIDA), *Bulletin*, no. 321, August 1980, p. 10.

14. Such criticisms have been made by the Australian Industries Development Association, drawing the wrath of the Minister for Industry and Commerce in the process, and a number of industry economists. See *Australian Financial Review,* 9 July 1980 and 20 September 1980.
15. Some of these are summarized in Susan Bambrick, *Australian Minerals and Energy Policy* (Canberra: ANU Press, 1979), chapter 13 ("Relationships Again: Inter-sectoral Linkages and Multipliers"). See also T.D. Mandeville, "The Impact of the Weipa Bauxite Mine on the Queensland Economy" (Melbourne: Comalco Ltd, January 1980).
16. Bureau of Industry Economics, "Mining developments and Australian industry: input demands during the 1980s", Research Report no. 9 (Canberra: AGPS, 1981).
17. Ibid., p. 75.
18. See chapter 5.
19. The problem was first spelt out in detail in the report of a working party set up by the Department of Labour Advisory Council to analyze the expected demand for and supply of skilled labour in the early 1980s. See *Australian Financial Review,* 15 October 1980, p. 3.
20. See for example, Australian Industries Development Association, *Bulletin,* no. 321, August 1980, pp. 10-11. The policy problems and issues raised by the impending shortage are discussed in chapter 7.
21. B.L. Hamley, "Financing Australia's Mineral Developments", *National Bank Monthly Summary* (May 1977): 9. Figures are based on surveys conducted by the Industries Assistance Commission (IAC) and R.B. McKern. See IAC, *Report on Petroleum and Mining Industries* (Canberra: AGPS, 1976); R.B. McKern, *Multinational Enterprise and Natural Resources* (Sydney: McGraw-Hill, 1976).
22. Hamley, "Financing Australia's Mineral Developments", p. 9. That the contribution of Australian sources to share capital is not reflected in the level of Australian equity ownership is due to the fact that much of this equity has been raised by foreign owned enterprises after they established that a project was viable and a minority interest was floated off to the Australian public at a premium.
23. See *Policies for Development of Manufacturing Industry.* (Jackson Report), (Canberra: AGPS, 1975), pp. 76-78.
24. See for example, Barry Reece, "The Mining Boom, Capital Market Reform and Prospects for Housing in the 1980s", *Occasional Paper,* Planning Research Centre, University of Sydney, October 1980.
25. *Australian Financial Review* (editorial), 11 July 1980, p. 12.
26. Ibid., 10 July 1980, p. 6; 11 July 1980, p. 3.
27. Ibid., 29 September 1980, p. 15.
28. Australian Bureau of Statistics, *Overseas Trade* (various issues); and Bureau of Mineral Resources, *Australian Mineral Industry Annual Review* (various issues).
29. These policy decisions are discussed in Maximilian Walsh, *Poor Little Rich Country* (Harmondsworth: Penguin, 1979), chapter 3.
30. R.G. Gregory, "Some Implications of the Growth of the Mineral Sector", *Australian Journal of Agricultural Economics* 20, 2 (August 1976): 71-91.
31. Ibid., p. 72.
32. An early critique was given in N.R. Norman, *Mining and the Economy: an Appraisal of the Gregory Thesis,* a Report prepared for the Australian Mining Industry Council, September 1977.

7 Economic Issues and Policies

W.H. Richmond

For just on a century after the first great mineral boom of the 1850s had subsided, mining was, by and large, an uncontroversial activity. But the mineral boom which started in the 1960s has raised a great variety of issues and has become the subject of much public debate. The impact of mining on the environment and the particular issues associated with the mining of uranium are the subject of chapters 8 and 9. These issues have obvious economic implications, but the mineral boom has also raised a variety of more specifically economic issues. These relate essentially to the question of the size and distribution of the benefits (and costs) associated with the rapid increase in mineral production, and to the role of governments in influencing their determination. Several distinct though related issues have emerged.

Each of these issues poses policy problems for Australian governments which are of considerable economic — and political — significance. This chapter outlines some of the more important issues and discusses the policies adopted or proposed in response to them.

STRUCTURAL ADJUSTMENT AND MACRO-ECONOMIC POLICY[1]

The rapid growth of one sector of the economy — especially the demands this creates for labour and capital — may lead to problems of overall economic management, in particular with respect to inflation (arising from pressure on available resources and the bidding up of wages, prices and interest rates, and/or from increases in the money supply associated with exports and large scale capital inflow). Such growth may also create pressures for structural adjustment within the economy; the way in which these pressures operate, and the way in which they are influenced by government policy, will determine not only the extent of the macro-economic problems created but also the extent of structural adjustment necessary and its economic and social consequences.

Structural adjustment takes place through two main mechanisms. Firstly, firms in the new sector bid labour and capital away from firms in other sectors, causing difficulties especially for the less efficient users of those resources. Secondly, if the output of the new sector is exported, the increase in exports will tend to generate surpluses in the balance of payments which will in turn lead to an appreciation of the exchange rate; existing firms in other sectors, producing either goods which are exported or goods which compete with imported goods, will suffer a loss in competitiveness and profitability, and the least efficient may be forced out of business. It is this latter mechanism which is examined in detail here, for it is in this regard that the government has been, and is likely to be, forced to make some far-reaching policy decisions.

Surpluses in the balance of payments, as a result of the mineral boom, were first generated in the early 1970s. The government responded by revaluing the dollar (in 1972 and 1973) and by cutting tariffs (1973). But this action was only taken after the build up of foreign reserves had led to a marked increase in the money supply and added substantially to inflationary pressures already at work in the economy. The combined influence of inflation, revaluation and the tariff cut adversely affected firms in the traded goods sector of the economy (both exporters and, more particularly, manufacturers competing against imports), and a degree of adjustment did take place during these years, particularly in the manufacturing sector. This was virtually inevitable. What is interesting is the way in which it was achieved and the role of government policy (or lack thereof) in the adjustment. The delay in altering the exchange rate can be clearly traced to the effective pursuit by the Country Party of the interests of its constituents. In December 1971 the US dollar was devalued and, in the face of steadily mounting foreign reserves, the government was advised by the Treasury to maintain the existing parity with sterling, which would be in effect to revalue against the US dollar. The Country Party implacably opposed this move, threatening at one stage to terminate its coalition with the Liberal Party. Some pressure was also exerted from the mining sector itself which, as an exporter, would have been similarly disadvantaged by revaluation. Even when the Country Party nominally agreed to a substantial revaluation the outcome nonetheless meant a slight effective devaluation. Revaluation to the extent warranted by economic circumstances was a political impossibility during the remainder of the term of the government. It was left to the Labor government which took office in December 1972 (by which time foreign reserves had increased to over $4.5 billion) to take the unpopular but necessary step. The additional measure of cutting tariffs (via the 25 per cent across-the-board tariff reduction in July 1973) was a further reaction to the consequences of the balance of payments surpluses and the inflationary pressures they had created. The tariff cut came as a complete surprise to most people. Had there been any suspicion that such a move was to be made it would almost certainly have been pre-empted by political pressure both from producers and employees in the manufacturing sector. As it was, the policy was bitterly attacked by these groups after it was introduced and proved to be a major political disaster for the Labor government. The events of these years made it clear that both revaluation and tariff cutting was particularly difficult. Inflation, the policy of default, was the easy alternative until the government was forced to take other measures.[2]

After 1973, due largely to lower rates of increase in mineral exports and mineral-related capital inflow, the balance of payments showed a persistent tendency to be in deficit. However, the sort of problems which confronted governments in the early seventies may be expected to re-emerge in the 1980s if there is a resurgence of the mineral boom. Large capital inflows in the first half of the decade and progressively increasing levels of mineral exports as new projects come into production could generate net surpluses in the balance of payments. The events of the early 1970s, and the work of Gregory and others stimulated by these events, mean that the government (and, increasingly, the Australian people) would face a 1980s mineral boom with a much clearer idea both of the economic issues and the political difficulties involved. If there is a boom which generates net balance of payments surpluses, and if the inflationary pressures these imply are to be avoided, the government will be forced to revalue and/or to cut the levels of protection.[3]

This has been made clear by the Treasury whose views — which will be very influential, if not necessarily decisive — were spelt out in a speech by the Secretary, Mr John Stone, in November 1979. In his speech the Secretary argued that any increase in mineral exports in the 1980s must be accommodated by an increase in imports. The best way of achieving this, he considered, would be to reduce levels of protection. An across-the-board tariff cut, similar to that made in 1973, would be one means to this end, though it would be preferable to concentrate, particularly in the first instance, "on reducing the protection of the things we do worst", that is to give priority to areas where effective levels of protection are highest. The "only sensible alternative" to reducing protection would be exchange rate appreciation (that is if inflation, or monetary restriction to prevent inflation, are ruled out), but this the Secretary regarded as only "half-sensible" because it would adversely affect efficient as well as inefficient producers: "the difference between reducing import barriers and the other main alternative of exchange rate appreciation is that whereas the latter strikes at our efficient (exporting) industries as well as (indiscriminately) at our import-competing industries, the former is designed to let the readjustment occur to a greater degree at the expense of our inefficient (most highly protected) import-competing sectors."[4] This view has been reiterated (though not publicly) in other Treasury documents.[5] It was also supported in the report of a high-level inter-departmental committee on economic strategy tabled in parliament by the Treasurer in August 1980. The report emphasized the need for reduced levels of protection and possibly also for some revaluation, arguing that attempts to slow the rate of structural adjustment which a mineral boom implies can only be achieved at the expense of increased inflationary pressures, the retardation of some resources development and lower overall economic growth.[6] This position has also attracted considerable support from economists and financial commentators.[7]

Both of these policies would meet with stern opposition from those groups of producers and employees likely to be adversely affected by them. Predictably enough, producers of exports in both the rural and mineral sectors have indicated their opposition to revaluation and have joined the call for reduced protection. Manufacturers competing against imports are in a much more difficult position, facing reduced competitiveness and profitability from both revaluation and changes in protection policy, though more particularly from the latter. To argue for a retention of tariffs and quotas really implies some deliberate holding back of the mineral boom, a position some manufacturing groups have in fact adopted on the grounds that the benefits of the boom, accruing to a "closed club of beneficiaries, many of them not living in this country", are considerably outweighed by the costs borne by the producers and employees displaced as an inevitable result.[8] The possibility thus exists of quite severe conflict between producers (and employees) in different sectors of the economy, for the interests of each would be served, or harmed, by different policies.[9]

In the face of a strengthening balance of payments situation during 1980 and 1981 the Australian dollar was revalued and it appeared for a while as if the government would be faced with some difficult political decisions in the very near future. From the latter part of 1981 the rate of growth of the mineral sector has slowed, but this may only mean that the problem is less acute. It seems likely that large capital inflow to fund mineral developments will continue and mineral exports will still increase. Revaluation of the dollar could still be necessary over the longer term, and would be resisted by exporters. On the other hand manufacturing groups (and trade unions) would strongly resist changes to protection levels and indeed there is no evidence that the present coalition government is in any way inclined to countenance such changes. How strong this resistance would be and what policies the government would find *politically* possible would probably depend to a large degree on the benefits which accrue to Australians from the mineral boom. If the boom were to create new opportunities for producers and employees and if the community generally were seen to benefit, then the arguments against change and actual resistance to it would be weakened. Unless this were the case — and obviously so — then producers and employees in danger of being disadvantaged or displaced could hardly be expected to be much impressed by arguments relating to the efficient allocation of resources and aggregate economic growth. In this way the macro-economic strategy to be adopted in response to a heightening of the mineral boom might well depend to a large extent on the more specific policies adopted to influence the extent and distribution of the benefits the boom might bring to Australians. These will be discussed in the following sections.

FOREIGN OWNERSHIP AND CONTROL

Overseas investment has played a major role in the mineral boom. *Direct* investment, which occurs when overseas companies set up branches or wholly or partly owned subsidiaries, has been of particular significance and has raised a number of issues for governments because it has resulted in foreign ownership and, usually, foreign control of Australian resources.[10] The extent of foreign ownership and control increased rapidly during the period of rapid growth in the mineral sector. Australian Bureau of Statistics data for the *mining* sector suggest that foreign *ownership* of the industry as a whole increased from approximately 27 per cent in 1963 to 52 per cent in 1974–75 (the most recent year for which survey data are available) and foreign *control* from 37 per cent to 59 per cent over the same period.[11] There is of course considerable variation within and between the major industries in the mineral sector.[12]

Overseas investment in the mineral sector of the economy has brought many benefits. It has provided a considerable proportion of the financial resources necessary to permit mineral development; it has brought with it technical, managerial and commercial knowledge and skills otherwise scarce or unavailable in the Australian economy; and it has provided access to markets. It could even be argued that the very development of the mineral sector itself has been dependent on overseas capital. R.B. McKern, in his major study of foreign investment in the mineral industry in the 1960s noted that "there were very few Australian firms in existence at the beginning of the decade of the 1960s which could have served as vehicles for investment in the large mining projects which came to fruition during that period. The list of Australian-owned companies with the necessary financial, technical, entrepreneurial and marketing resources at that time is probably restricted to no more than three".[13] It is true that there were other large industrial companies and financial institutions but none had experience in the mineral industry and the financial institutions in particular regarded the risks involved as too great. McKern concluded that "without the multinational enterprise, it is clear that the minerals industry would not have developed at a rate remotely approaching that of the 1960s".[14]

Since the 1960s the ability and willingness of Australian companies and institutions to participate in mineral developments have increased dramatically, but the developments that have taken place have still relied heavily on foreign investment. The benefits this has continued to bring are substantial, but there have also been costs. The most significant of these is that a large proportion of the profits earned by the mining of Australian natural resources has accrued to overseas investors. To the extent that foreign investment has led to foreign *control* of our mineral industries it is possible also that policies have been pursued by individual companies, linked as they are to overseas parents, which may not have been in the best interests of Australia.

It is not easy to assess and weigh up the benefits and costs of foreign investment, particularly as the calculation will be influenced by other related policies pursued by governments. In this regard the ability and willingness of governments to ensure that foreign-owned companies do not earn "excessive" profits (but rather that these profits are captured in the form of taxation) and that companies are not able to avoid income tax through transfer pricing strategies is particularly important. (These matters are considered further below). But it is clear that a responsible government must assess the desirability of foreign investment and determine a general policy towards it. A "policy" can of course be directed at encouraging overseas investment, limiting it or placing conditions on the way it may be undertaken, or simply taking a neutral stance.[15]

In Australia both the commonwealth and state governments have powers which can be used to encourage or discourage foreign investment. In the case of minerals, state governments have sovereign power over onshore mineral deposits and thus the power to grant, or refuse, licences for exploration and development; they can impose *conditions* on development (relating to royalties, environmental impact, further processing, power or freight rates, the provision of infrastructure and so on). These powers relate to *all* investment in mineral development, but could be used in a way that would, for example, discourage foreign investment. In general they have not been so used, and the attitude of the states has been one of encouraging mineral investment regardless of its origins. The attitude of the Premier of Western Australia, for example, is summed up in his view that "if overseas investment and know how continue to be essential to our rapid growth then in the national interest, they should continue to be welcomed and not placed under a cloud of uncertainty".[16] More recently the Queensland Minister for Mines and Energy commented: "It is well-known that the Queensland government strongly supports the concept of free enterprise with a minimum of bureaucratic control. We accept the desirability of having substantial overseas equity in an export-oriented industry."[17] Greatest weight has been given to the goal of maximizing capital expenditure within the state, particularly if it can be argued to bring social benefits through the provision of infrastructure and of course income and jobs. There is also the more general political kudos associated with the achievement of "development".

The commonwealth government has more direct control over foreign investment and can directly prohibit or impose conditions upon the entry of foreign capital. In view of its more general responsibilities of national economic management it has tended to analyze the benefits and costs of foreign investment using rather broader criteria. During the 1960s and the early years of the seventies, the Liberal-Country Party policy towards foreign investment was in general very liberal; it was during these years, the time of the first stage of the mineral boom, that a high degree of foreign ownership in the expanding industries of the mineral sector was established. There was, it is true, a degree of ambivalence. The various costs of foreign investment, which were spelt out for example in the Vernon Report in 1965 and a comprehensive Treasury Paper in 1972, were known and accepted and some specific measures were taken by the Gorton and McMahon governments (including the passing of a Foreign Takeovers Act during the term of the latter); but in general the benefits were seen to outweigh the costs, in the mineral sector as in the rest of the economy.

The attitude of the Labor Party which came to power in 1972 was very different. The government and in particular the Minister for Minerals and Energy were highly critical of the degree of foreign ownership and control in the mineral sector (as well as being critical of a number of other matters relating to the sector). Early in his term the Minister made a commitment to reducing the degree of foreign ownership and control, but for some time no clear policies emerged and a state of uncertainty existed. In 1973 the Prime Minister made a formal statement in which he affirmed his belief in the benefits of overseas investment and the future need for overseas capital. He emphasized that "there is no general prohibition on foreign investment in Australia". But, he said, "my Government has the firm policy objective of promoting Australian control of Australian resources and industries. By the phrase 'the highest possible level of Australian ownership' we mean the highest Australian equity that can be achieved in negotiations, project by project, that are fair and reasonable to both parties and are within the capacity of our own savings to support". This policy was to be applied "in a pragmatic way".[18]

The statement did little more than create a climate of uncertainty, however, and a marked fall in foreign investment clearly indicated the need for a firmer policy. Such a policy was enunciated in September 1975. In introducing this policy the Prime Minister stated that

> the Government recognises that Australia will, for the foreseeable future, continue to require foreign capital, including equity capital, if we are to achieve our basic aim of ensuring that Australia's resources and industries are developed in such a way as to bring maximum benefits to the Australian people. Within this over-riding goal,

> however, the Government, while continuing to welcome beneficial foreign capital, has a major longterm objective of the promotion of Australian control and the maximum Australian ownership compatible with our longterm capital requirements and our need for access to markets, advanced technology and know-how.

The policy established a Foreign Investment Committee to examine all specific foreign investment proposed in terms of certain guidelines, though it was emphasized that these would be guidelines only and that proposals would continue to be examined on a case-by-case basis. The most important guideline in relation to mineral industries was that "the government will expect proposals for all new mineral development projects (other than those relating to uranium) to have no more than 50 per cent foreign ownership and with the foreign participants having no more than 50 per cent of the voting strength on the board of the development company". The development of uranium deposits was to be permitted only by companies with 100 per cent Australian ownership.[19]

The policy of the Liberal-Country Party relating to foreign investment in the mineral industries was released only weeks afterwards, in October 1975, and bore a remarkable similarity to that of the Labor Party. The "maximum control and ownership of our natural resources" was a clearly-stated goal, while the continuing need for foreign investment to promote economic growth and development was also recognized. The major difference in the policy was that the scrutiny of foreign investment proposed was to be taken over by an advisory board, the Foreign Investment Review Board (FIRB), instead of a departmental Committee. The "50 per cent rule" was to be applied (except in the case of uranium where 75 per cent Australian equity was to be required). However it was emphasized that the rule would be applied with some flexibility, particularly where Australian capital was unavailable.[20]

The Liberal-Country Party gained government shortly afterwards and it soon became apparent that, despite the fears voiced by some people, including the state premiers of Western Australia and Queensland, the guidelines were in fact applied flexibly and no major curtailment of mineral developments occurred. In June 1978 the guidelines were liberalized by a number of measures: the most significant of these allowed foreign-owned companies which had at least 25 per cent Australian equity and were committed to raising the level of Australian ownership to over 50 per cent to be granted "honorary" Australian status, thus permitting them to develop new projects.[21] In the following year the 75 per cent rule for uranium projects was modified and approval was given for development of two projects, both involving the Western Mining Corporation, where Australian equity was below 75 per cent. Such arrangements were now to be deemed acceptable if higher Australian equity could not be achieved, if the project would be of significant economic benefit to Australia and if Australian participants had the major policy-determining role in the project.[22]

Since early 1980, however, the guidelines have been applied more firmly and it appears that the government's policy has been hardening on the issue. Two decisions have been of particular interest. A proposal by Conzinc Riotinto Australia (CRA) to develop a steaming coal project at Blair Athol in Queensland which would have had only 25 per cent equity (deeming CRA to be an Australian company) was rejected, as was a proposal by Mount Isa Mines to develop the Oaky Creek coal deposit, even after it had increased its interest in the project to 78 per cent.[23] The tough enforcement of the guidelines has been criticized for inhibiting the development of large projects. The Queensland Premier referred to the rejection of the Blair Athol proposal as a "stupid decision" and the chairman of Mount Isa Mines pointed out the particular problems raised for companies which are only 50 per cent Australian owned.[24] On the other hand there are those who would argue that the necessary finance and expertise can be obtained to allow developments to proceed without yielding this degree of equity and control to foreign companies. However there appears to be a reasonably high degree of consensus that the present, somewhat firmer, policy represents an appropriate compromise between the goal of maximum Australian ownership and the need for foreign equity capital, both to fund and in other ways to facilitate further mineral developments. While it is desirable to have at least 50 per cent Australian equity in new projects to ensure that they are controlled by Australian companies, there is probably not a great deal of room for manoeuvre in this area without a substantial reduction in the rate of development in the mineral sector or large scale direct government participation.

However, there are some issues relating to foreign ownership in both existing and new mineral projects, other than the level of equity, where there is less agreement and, it could be argued, a need for government action. Two problems in particular may be mentioned: (1) foreign companies may earn "excessive" profits, which accrue to overseas residents; and (2) transfer pricing strategies may enable foreign companies to avoid Australian taxation of their profits.

The latter problem only arises in situations where foreign ownership is sufficient to imply foreign control. It may occur where overseas companies "adjust" their profits, and hence tax, by the way they price loan finance or payments for technical services by overseas parents. It may also arise in a situation where a multinational company undertakes different stages of production of a commodity in different countries. In these cases it is possible for the company to "price" the product of one stage which becomes an input for another stage (in a different country) such that profits are distributed in a way that minimizes the company's tax payment. (This is possible if the different countries in which the different stages of production are located pursue different taxation policies). In this way a host country, if it is a relatively high tax country, may lose out. In the case of Australia there are limited opportunities for such transfer pricing except in the case of the aluminium industry; even there it is difficult to say whether or to what extent it takes place, partly because the industry is so highly concentrated and vertically integrated on a world level. To the extent that this problem is in fact associated with the existing degree of foreign ownership, there is a case for keeping an eye on intracorporate purchases and sales, and particularly exports, to ensure as far as possible that they are made at "market" prices. The issue of controls over export prices is one with which the Australian government has concerned itself and is examined separately below; though as will be seen it is one issue which goes beyond that simply of transfer pricing.

The first problem exists if companies in the mineral sector do indeed earn more than what might be regarded as "reasonable" returns. The issue is of course a quite general one; the fact that there is a high degree of foreign ownership has only served to highlight it because part of the "excess" profits (if they exist) have been transferred overseas. The matter is really part of the more general issue of appropriate taxation policies.

TAXATION

The claims by government upon income earned by companies and individuals is invariably a contentious matter. The taxation of incomes earned within the mineral sector of the economy — and particularly by mining activities — has raised some particularly controversial issues and posed special economic policy problems.

Mining companies are subject to the same *company taxation* levied by the commonwealth government as other business organizations; though (as is the case with other categories of businesses) they are granted certain concessions. During the first stage of the mineral boom, up until 1973–74, these concessions fell into three major categories: (1) those which allowed mining companies to write off against income certain categories of expenditure not allowed as depreciable items at all for companies engaged in other activities; or to write off certain items of expenditure more quickly; (2) partial exemption from taxation of net income earned from mining certain minerals (in the case of gold all income was exempt); and (3) certain concessions to individual shareholders in mining companies.

These concessions were introduced in order to encourage mineral developments; no doubt they had that effect, though to what extent (compared to what would have occurred in their absence) is difficult to say. Their significance in terms of total taxation revenue received by the government received little consideration until the publication in 1974 of a report by T.M. Fitzgerald.[25] The report had been commissioned by the Minister for Minerals and Energy in the recently elected Labor Government; it was titled *The Contribution of the Mineral Industry to Australian Welfare,* though in fact it was concerned with the rather narrower issue of the direct fiscal relationship between the industry and government. Fitzgerald concluded, among other things, that over the six year period up to 1972–73 the tax paid by companies in the mineral sector (mainly as a result of the first category of concessions listed above) was very small in relation to profits earned. The report was subjected to considerable criticism.[26] In particular it was suggested that the period selected was exceptional, being one of rapid expansion involving large expenditures which, due to the accelerated write-off concession, could be offset against income in these years but would result in larger taxable income and hence larger tax payments in later years. Thus payment of tax was to a large extent only deferred.

Partly in response to the Fitzgerald Report, the Labor government in 1974 removed shareholder concessions, increased the minimum time period for writing off capital expenditure (while still permitting exploration expenditure to be immediately set against mining income) and removed the partial exemption from tax of income earned from minerals other than gold. The effect was to slow mineral developments as well as to further sour the general relationship between the mineral industry and the government. In 1976 the Liberal-Country Party restored some concessions and subsequently introduced others (for example in 1977 expenditure on petroleum exploration and development was made deductible against any income). In regard to company taxation the overall treatment of the mineral sector is now not markedly different to that of other sectors of the Australian economy, though some specific provisions (or lack thereof) remain the subject of controversy.[27]

But does company taxation, even if companies in the mineral sector are treated comparably to companies in other sectors, yield to the government, and thus to the people of Australia, the amount of revenue that should properly accrue as a result of the mining and sale of natural resources? The income earned by mining a natural resource includes a component which is termed *rent.* Rent is that part of income earned after all costs of factors of production, including "reasonable" or "normal" profits, are covered. "Normal" profits can be defined in terms of alternative opportunities for employment of the factors of production (in which can be included entrepreneurship) elsewhere in the economy. Rent is thus in effect the income that is earned by simple virtue of owning a natural resource, or being permitted by the owner of that resource to mine it.

In Australia natural resources are owned by governments (state governments where minerals occur in the states, the commonwealth government elsewhere). A government can in principle take all of the rent and a mining company will still be able to get a reasonable return on capital after paying costs of production, that is, make a "normal" profit from mining operations. If a government fails to take all the rent, which is the same thing as saying if it fails to "sell" the resource at its market value to the company which is going to mine it, then the mining company's profits include a component of "super-normal" profits which are in excess of those which would be necessary to induce the company to undertake the mining operation. Some or all of the income payable to the owner of the resource (the government) thus accrues, by default, to shareholders of the mining company.

In practice of course it is difficult for a government which owns a resource to know what the rent of that resource is; or to put it another way, what it should "charge" a prospective mining company to be allowed to mine the resource. The rent will be affected by the size of the deposit, which may not be known, and the market demand and hence price of the resource, which may fluctuate considerably over time. A possible solution is to auction mineral leases. Another possibility is for a government to permit mining on the condition that it takes that part of a project's returns which exceed a "threshold" rate of return comparable to that which could be earned elsewhere in the economy, allowing for risk. In this way mineral developments would be profitable for mining companies and rents would be captured by governments.

There has been a good deal of discussion, much of it fairly technical, about the principles of mineral rents and the manner in which rents can be captured by governments without discouraging the mining operations which yield them and provide other benefits to the community.[28] In practice the main device which has been used by governments to extract the economic rents earned by exploiting mineral deposits has been the imposition of *royalties*. This has been done principally by state governments in whom ownership of most onshore minerals is vested. Royalties have been imposed in an *ad hoc* manner and rates set, and changed, in a quite arbitrary fashion. Political rather than economic considerations have been paramount in many cases. The result is a complex system of royalty payments which vary from mineral to mineral and state to state. The resulting inconsistencies and anomalies have produced considerable dissatisfaction within the mining industry.[29] In some cases royalties have been related to profitability but more usually they have been set as a fixed amount per unit of output or as a proportion of the value of sales. The latter systems have been criticized on economic efficiency grounds, in particular because they tend to encourage "wasteful" resource exploitation. There is a widespread view that royalties are in general "too low". State governments and mining companies reply that higher rates would discourage further development, or simply assert that "enough" is

This illustration, taken from the cover of the March 1982 *Mining Review*, indicates mounting industry concern at the increasing share of the "mineral cake" taken by both state and federal governments, and the hardening of government attitudes towards the industry in general during 1981. (Cartoonist, Stuart Vaskess.)

paid already. More rational discussion of the matter has been inhibited by the inevitable arbitrariness of the royalty system.

Some other devices have also been used in effect to capture mineral rents. Several state governments, for example, charge higher rail freights on minerals. The commonwealth Labor government introduced a (per tonne) coal export levy in 1975 largely, it seems, in response to high profits being made, particularly by the foreign-owned Utah Development Co. The Liberal-Country Party government subsequently reduced the amount of the levy and committed itself eventually to phasing it out. On the other hand the same government committed itself to what is in effect a resources rent tax when it introduced, on grounds of economic efficiency, its policy of import parity pricing of oil. When the price of Australian oil was increased, domestic producers would have received a huge windfall gain had it not been for the oil levy, which diverted that part of the increased profit relating to "old" oil to the government. But this tax too will diminish rapidly after a few years and there is no suggestion of a tax on rents obtained from "new" oil.

In the context of these *ad hoc* and arbitrary policies of diverting mineral rents there has been a growing pressure to introduce a more comprehensive and rationally based resources rent tax. In 1977 the commonwealth government floated the idea of such a tax relating to petroleum and uranium production, but dropped the idea in the face of intense opposition from industry representatives. In announcing this formally in July 1978, the Minister for Trade and Resources and the Treasurer argued that such a tax would have "adverse effects on exploration and development decisions and on investor confidence".[30]

On the other hand the Labor Party, which first adopted the policy of a resources rent tax in 1977, has become increasingly attached to the policy and used it as a major issue in the 1980 election. To support his advocacy of the tax the Leader of the Opposition quoted from a confidential Treasury document which gave cautious support to the notion, mainly on the grounds that a resources rent tax would serve to prevent a large proportion of the benefits from resource developments accruing overseas: ". . . taxation and royalty arrangements, if excessively generous, can 'give away' much of the income benefit of resource development, particularly where it is foreign owned and largely export-oriented".[31]

The implementation of a resources rent tax would entail many problems both of principle and practice. Decisions would need to be made as to the threshold rate of return above which profits are to be deemed "excessive", and what *rate* of tax is to be applied to these. The treatment of fluctuating levels of profits (and on occasions losses) also needs to be considered. Errors of judgement on these matters could serve to discourage investment in mineral projects as well as to unfairly discriminate against mineral companies. The relationship of a tax on rents to the existing systems of company taxation, royalties and other arrangements which are accepted as *de facto* taxes on mineral rents must also be decided. In principle a comprehensive resources rent tax could replace royalties and other taxes, but there are many formidable problems associated with the imposition of such a tax within the existing Australian federal financial system. Finally of course considerable resistance may, not surprisingly, be expected from mining companies themselves.[32]

It has been pointed out that when the revenue obtained from the oil levy diminishes through the early 1980s the government will be looking for a new form of tax. The economic arguments for a resources rent tax, together with considerable public support and Treasury backing for the policy, suggest that it is only a matter of time before it is introduced. There is indeed some evidence to indicate that mining industry leaders, despite their protestations, are coming increasingly to accept the tax as inevitable. For the moment, however, the government remains quite explicitly opposed.[33]

EXPORT PRICES AND CONTROLS

Most of the output of Australian minerals is exported.[34] The manner in which export sales are negotiated depends on the structure of the market concerned; however, in most cases — including those of iron ore, coal and bauxite/aluminium (which together account for well over half of total mineral exports) — sales are made by way of long-term contracts.[35] These contracts are usually relatively flexible (within limits) both with respect to prices and quantities. However, they represent a basic commitment by both buyers and sellers to trade and indeed provided the essential security which allowed the first stage of the mineral boom to proceed.

The negotiation of the initial terms of contracts and of variations over time within the framework of these initial terms is essentially a process of bilateral bargaining. It has widely been thought that Australian mineral producers have "lost out" in the bargaining process because they have been relatively fragmented whereas overseas buyers have been much more organized and in some cases have virtually operated as a buying cartel. This may have been so particularly in the case of sales of iron ore and coal to the Japanese steel industry. Australian producers have tended to compete with each other rather than face Japanese buying consortia collectively, the result being that the gains from the bilateral trade involved have gone mostly to the Japanese. There were signs during 1980 that some Australian coal producers were acting in a more united way in negotiations with Japanese buyers particularly in the case of steaming coal for which demand is becoming increasingly strong. But Japanese buyers of both coking and steaming coal were also reported as co-ordinating their activities to achieve "more orderly" purchase of coal and Australian producers (particularly of coking coal) were still attracting criticism for the way they were undermining prices by competing with each other.[36]

The outcome of the bargaining process particularly with respect to price is, obviously, of significance to the company involved and to other competing firms. It is also a significant determinant of the contribution of the mineral industry to economic welfare in Australia: clearly the higher the profits made by Australian mineral companies the larger will be the benefits to shareholders (though a number of these are overseas residents), and the higher will be the contribution in the form of tax (the extent depending of course on the system of taxation of mineral companies). The question has been raised whether the Australian government, by intervening in the bargaining process, can and should assist (or force?) Australian companies to negotiate higher prices. (It is *able* to do this by virtue of its constitutional power over exports, which gives it indirect control over the price at which exports are sold.)[37] A number of suggestions have been made as to how governments can overcome this problem. One possibility is for the government itself to establish a selling agency, which compulsorily acquires all or a large part of production from producers. Less dramatically it could simply monitor trade dealings and indicate what prices it thought were "reasonable"; exports at prices below the level would not be permitted. Alternatively, a government could attempt to alter the framework in which bilateral bargaining took place by encouraging or perhaps even enforcing the creation and maintenance of private cartels; in this case it would not play any direct role in the bargaining process itself though it might advise producers on strategy and take an interest in their negotiations.

There are problems associated with each of these proposals. It is not clear that a government organization, whether acting as a selling agency or arbiter of contract terms, would be sufficiently well informed of the basic long-term economic and commercial realities which set the limits to bargaining. It is possible that it would take too optimistic a view of its bargaining strength and of the bargaining "margin" which it can exploit. Certainly the short-term bargaining strength of an Australian government controlling mineral sales (especially to Japan) is very considerable, but exercise of that strength may not be in the nation's long-term trading interests. It has been suggested by one economist that the issue is really only about who gets the icing on top of the cake, and it would be foolish to allow bargaining over the icing to be conducted in such a way as to prevent continued consumption of the cake. "Over-bargaining" — indeed any direct involvement by governments in the bargaining process — could also lead to a general politicization of trade which might not be in Australia's long-term interest. Thus it has been argued that "it is not obvious that governments should take any hand in the trade bargaining process itself. Such . . . intervention can only be justified by a conviction that governments (or government agencies) are in a better position to assess bargaining limits and bargaining strategy than are the representatives of private companies. Although such a conviction may come easily to politicians and bureaucrats, it may hold considerable dangers for the long-term development of the bilateral trade relationship".[38] It is also very difficult to try to cartelize producers and thereby alter the bargaining framework unless a certain amount of force is used and/or the government itself is prepared to take a formal role in the operation of the cartel — and there would be considerable opposition within the industry to that. The interests and objectives of individual producers are too diverse.

It thus appears that the choice is really between accepting the losses due to fragmentation of Australian producers and taking the risks associated with government intervention. If, despite attempts to get Australian producers to act in a more co-ordinated way, it appears that the relative bargaining strength of united buyers remains considerable, or if it appears that the interests and objectives of individual producers (particularly if those interests are linked to those of an overseas parent) are pursued in a way that clearly conflicts with the national interest, then the case for some form of direct government intervention in the bargaining process becomes stronger, even in the face of the difficulties and dangers associated with it. There is also the special problem of exports which are merely intracorporate "sales" where transfer pricing can occur. As noted in the discussion of foreign investment, this can arise where there is a high degree of foreign ownership (though in Australia the problem seems to be mainly confined to bauxite and alumina sales), and the only way of avoiding it is by direct surveillance of prices.

Australian governments did in fact take action during the 1970s to secure higher prices for exports of Australian minerals. The Labor Minister for Minerals and Energy applied the power of the commonwealth government to all exports in order that all contracts could be examined to see that "acceptable" prices were being obtained. (Hitherto the power had only been used on odd occasions.) The Minister attempted to get Australian producers to adopt a more united stance and to negotiate higher prices, and on one occasion he intervened personally in negotiations. He achieved some success in securing higher prices. The Fraser government after 1975 adopted a less interventionist position, while encouraging producers to exchange information and discuss their bargaining strategy with the relevant government department — the third of the three broad options outlined above. A coal exporters' committee was formed under a federal government directive. The Minister for Trade and Resources also made representations from time to time to the Japanese government in support of Australian producers. It appears, however, that he became increasingly unhappy with the terms of some contracts, particularly for iron ore exports. In the middle of 1978 the major iron ore producers were widely thought to have been adroitly played off against each other and forced to accept lower prices. In response the Minister announced in parliament in October, that "exporters who wish to enter into negotiations under new or existing contracts will be required to obtain specific approval before making any offers or entering into any commitments. . . . I will determine the parameters within which a company or companies will be authorised to negotiate. Such parameters will include, as circumstances require it, pricing, tonnage, duration or other provisions of commercial commodity contracts". He argued that "we are facing a situation where buyers are imposing settlements on individual sellers which are less than can reasonably be expected in the market situation".[39]

The export guidelines policy ran into immediate criticism from a number of government backbenchers and most particularly from the Premier of Western Australia (Sir Charles Court) who referred to the policy as "socialist". As a result iron ore was effectively removed from the list of commodities which the proposed guidelines policy covered. The reaction of mineral producers themselves to the sudden adoption of the more interventionist stance by a government philosophically opposed to such action was generally unfavourable, though many saw some advantage in being able to state in the process of negotiation that prices below a certain level would be unacceptable to their government — provided of course that the government's price was not such that the cake would be lost in the bid for the icing.

The way the policy has been administered with respect to the other commodities and what effect, if any, it has had on prices negotiated is unclear. Indeed the present state of the policy is itself somewhat uncertain. The government still officially has an export guidelines policy which requires the commonwealth's approval before newly negotiated export contracts come into effect, but it has been suggested that the policy is no longer effectively operative. A report in early 1980 that a new export guidelines package for steaming coal was being "shaped" in Canberra was immediately denied by the Acting Minister for Trade and Resources, who maintained that "the Government does not have a policy of interfering with the market mechanism".[40] On the other hand it has been reported that the government has been attempting to form a "council" of industry and government representatives to negotiate export conditions with Japanese coal buyers, and further that it has been promoting co-operation between Australia and Brazil on a number of aspects of iron ore (though not, directly at least, on the matter of pricing). How effective these arrangements will be is uncertain, particularly in the light of what appears to be a general lack of enthusiasm on the part of Australian producers.[41]

INFRASTRUCTURE

Many mining and mineral processing developments have been undertaken in remote areas or in areas where they have made a substantial regional impact. In addition to large scale expenditure on mine and plant development, considerable expenditure has also been necessary on infrastructure — transport and communications facilities (roads, aerodromes, ports, telecommunications); water and power supply; education, health and recreational facilities; indeed all the overhead capital necessary to service a mineral project and its employees. The question has arisen: who should provide and then operate and maintain these facilities, the company or consortium undertaking the development or the government? Where the infrastructure serves only a specific project (which is likely to be the case in a remote area) and particularly where that project is likely to have a limited life, then there is an argument for only very limited government expenditure on infrastructure. It would be reasonable for an isolated mining community to be provided with those infrastructure items (schools, health facilities and so on) which are ordinarily provided from government revenues without specific charge. However, there is no case for provision of facilities such as railways, power and water where users are normally required to pay for these services, except in cases where benefits from their provisions accrue to the community at large in addition to those people associated with the project. The proportion of infrastructure expenditure that is appropriately incurred by mineral companies on the one hand and governments on the other will vary from project to project. The determination of this proportion is, in both principle and practice, likely to be very difficult in many cases.

In the first phase of the mineral boom in the late 1960s most infrastructure expenditure was undertaken by mining companies. During this period it appeared that around 25 per cent of all mining investment consisted of infrastructure items; a survey by the Australian Mining Industry Council of eleven projects developed during the 1960s showed that infrastructure expenditure constituted in these cases nearly two-thirds of total capital expenditure.[42] In recent years, however, state governments have shown a greater willingness to undertake infrastructure expenditure related to resource-based projects and particularly to certain types of mineral processing. Since 1978 a number of state governments have embarked on what have been termed "grandiose" infrastructure programmes with the explicit purpose of attracting investment to their state. The major programmes include several power stations, coal-loading facilities, railway developments, and (in Western Australia) a natural gas pipeline and infrastructure for the Alwest bauxite project. To undertake these infrastructure projects the states have had to initiate large borrowing programmes which have been described by one financial commentator as "a splurge reminiscent of the period following the first World War".[43] Concern has been expressed at this vastly increased state expenditure on infrastructure related principally (though not solely) to resource projects for two main reasons.

Firstly, it seems in some cases that private companies should be bearing a greater proportion of the cost, although what is "appropriate" really has to be assessed on a case by case basis. Some projects, including several electricity generating projects, have been the subject of specific criticism, particularly in relation to the prices at which electricity is, and will be, sold to consumers such as the aluminium industry. There is considerable evidence that states are in effect competing with each other to attract "development" by cutting the price of electricity well below its real resource cost.[44] On a different level the huge borrowing programmes instituted by the states have also been criticized for their likely effects on the availability and cost of capital to the private sector and more generally for the economically irresponsible manner in which aspects of the programme are being undertaken. In mid 1980 the commonwealth Treasurer himself expressed concern at the pressures likely to be exerted on the capital market with consequences for interest rates and the rate of inflation. It was also pointed out that in some cases infrastructure borrowings are being made in part to cover interest costs (that is, interest costs are being capitalized) — a classical financial abuse. The Treasurer's submission on this matter has indeed been described as "a damning indictment of the lack of financial probity of the States".[45]

An interesting specific case which came to a head in late 1980 concerned Gladstone in Queensland, the site of several major mineral processing projects developed in the 1960s and 1970s and of several others proposed for the 1980s. These developments have already placed enormous pressure on housing, schools, water supply, sewerage, roads and other basic facilities in the town. To date no contributions have been made from developers; agreements have been negotiated for contributions from the developers of the Rundle shale deposits near Gladstone and are expected to be negotiated with developers of other major projects planned for the 1980s. In the short term however, there is acute pressure on basic facilities resulting from the failure to secure such contributions from developers in the 1960s and 1970s.[46]

MINERAL PROCESSING

A significant proportion of the minerals mined in Australia, particularly those which have been the basis of the new mining boom, have been exported in unprocessed or semi-processed form, although an increasing amount of processing is being undertaken in some industries, notably aluminium. The view is often expressed that it would be desirable if more processing were undertaken in Australia, thus creating more employment and income. A frequently quoted statement of former Prime Minister John Gorton made in an address to the Australian Mining Industry Council in 1970 expresses the view implicitly or explicitly held by many.

> One million tons of bauxite — from Weipa or from Gove — earns $5m from export. If it is converted to alumina, the equivalent of its earnings is some $30m. If that alumina were converted to aluminium, ingot aluminium, it would be worth $120m, and if, finally, that ingot aluminium is fabricated into aluminium products then it would be worth $600m. There really is a premium . . . on getting $600m or $120m rather than $5m, not only a premium in what can be earned overseas by it, but a premium in the factories that are provided, the smelters that are provided, the jobs that are provided, the decentralization that is provided, if we can more and more get into this field.

The West Australian Premier is another public figure who has consistently argued for "the maximum degree of mineral processing". The corollary usually drawn is that governments should "encourage" a greater degree of mineral processing.

However, such encouragement is not without cost. If a mineral is not processed in Australia it is usually because it is not profitable to do so. To encourage mineral processing in such a circumstance means that a government must directly or indirectly subsidize companies to establish processing plants. These subsidies can be in the form of tax concessions, royalty rebates or direct subsidies or bounties; alternatively governments can provide infrastructure or subsidized power or freight. Such subsidies are funded by taxpayers generally; the cost of them needs to be carefully weighed against the benefits gained by encouraging mineral processing industries in this way. Policies of this sort might simply encourage an economy's resources to be diverted from the production of goods at which it is more efficient. In a situation of unemployment this argument may not hold in the case of labour, though it should be considered whether the jobs created are the sort available to unemployed people or whether they serve only to attract labour away from and create difficulties for industries which are more efficient.

Remarkably little cognizance has been taken of these sorts of arguments. The promotion of mineral processing has been a virtual article of faith, at least of state governments, and policies have been implemented to encourage or coerce mineral producers accordingly (usually through appropriate financial incentives). Processing conditions have been written into lease agreements (though often these have only required mining companies to "investigate" the possibility of further processing) and two-tier royalty systems have also been used whereby lower royalties are payable on minerals processed within the state. More recently, states have been prepared to spend large amounts on infrastructure and to offer subsidized power in a bid to attract mineral processing industries.

The development of mineral processing industries clearly brings substantial benefits to Australians; but if it can only be achieved as a result of subsidies (be they direct or indirect) then the benefits must be weighed against these costs. The policies being pursued by state governments to attract mineral processing activities seem to be widely accepted. Whether or not they are always in the national economic interest is another matter.

CONCLUSION

The purpose of this survey has been to outline the major economic issues arising from the mineral boom, to indicate the problems they present for Australian governments and to discuss the policies adopted or proposed with respect to them.

All the issues and policies discussed are of considerable economic significance. There are also broader political implications. Reference was made to the potential conflicts that are already apparent between, on the one hand, producers and employees in export industries (including the mineral sector itself) and, on the other hand, groups whose economic interests are aligned with industries (principally in the manufacturing sector) competing with goods which are or which could be imported. Perhaps of more importance (and closely related to these conflicts) are the conflicts emerging on the broader level between the states rich in resources (notably Queensland and Western Australia) and those with established industrial bases, and between these states and the commonwealth government. A basic issue to be resolved is how the benefits deriving from mineral wealth should be distributed in a federation where the ownership of minerals themselves is vested in the states. This issue has not been discussed here but it is clear that, in the development of our mineral wealth, there are profound implications for Australia as a federal political entity.

Given the importance of these issues, it is surprising that a more coherent set of policies aimed at maximizing the national benefits of developing our mineral wealth has not emerged. Most of the issues discussed above remain unresolved. State governments who have considerable power in this area often seem to be more impressed by simple notions of "development" than with attempts to analyze seriously the benefits and costs (including opportunity costs) of development and policies related to it. In two cases where firm policies have been enunciated by the commonwealth government – with respect to structural adjustment (or non-adjustment) and taxation of rents — the policies (adopted in the context of an election campaign admittedly) seem to favour minority pressure groups rather than the nation as a whole.

In the early 1970s, after the first phase of the mineral boom, one economist described Australia as a "pleased but puzzled" owner of a vast stock of mineral wealth, and another noted that "there is no plan to take advantage of this wealth, as though the nation were confounded by riches".[47] These comments remain relevant in the 1980s.

Notes

1. The discussion is this section follows from the analysis in the previous chapter. The last three sections of that chapter in particular could usefully be read before this section.
2. These policies are discussed in Maximilian Walsh, *Poor Little Rich Country* (Harmondsworth: Penguin, 1979), chapters 3 and 4.
3. There is a body of opinion which holds that the need to do one or both of these things can be avoided if capital exports increase sufficiently to offset the balance of payments surpluses that may occur. It is argued that much of the income earned

directly from the sale of mineral resources is in the form of dividends and interest, with the result that recipients are essentially faced with portfolio decisions; these could well include the purchase of foreign capital. Increased capital export will require a change in attitude towards Australian investment overseas. Perhaps more importantly it requires a relaxation of the present government-imposed restrictions in this respect. Thus there is argued to be a policy option in addition to revaluation and reduction of protection. Whether such a liberalization of restrictions on Australian investment overseas would lead to a sufficient increase in capital exports to avoid the need for other policies is a moot point.

4. *Australian Financial Review* (*AFR*), 20 November 1979, p. 12.
5. For example the report of a document "leaked" during the 1980 election campaign in *AFR,* 14 October 1980, p. 3.
6. *AFR,* 28 August 1980, p. 1.
7. For example *AFR,* 22 October 1980, p. 14; 29 October 1980, p. 12.
8. For example the remarks of the President of the Australian Confederation of Apparel Manufacturers, *AFR,* 13 October 1980, p. 7.
9. The point is elaborated in Tony Corrighan, "The Political Economy of Minerals", *Journal of Australian Political Economy* 7 (1980): 28-40.
10. *Direct* investment is distinguished from *portfolio* investment (the purchase of Australian shares by overseas investors with the aim of earning income or capital gain but not control over the company) and "*institutional*" investment (loans raised overseas from financial institutions and other companies which do not entail any ownership or control over the borrowing company). However, the distinctions are not always clear-cut, e.g. investment may be made in a coal-producing company by overseas steel mills as part of a marketing arrangement: while not strictly direct investment as usually defined this can entail a considerable share of ownership and an element of control. Of course overseas investment of *all* types affects the balance of payments and raises issues of macro-economic management: these have been discussed in the previous section.
11. Australian Bureau of Statistics, *Overseas Participation in the Australian Mining Industry,* 1968; *Foreign Ownership and Control of the Mining Industry, 1973-74 and 1974-75* (Canberra: AGPS, 1976). Foreign control is deemed to exist if direct foreign ownership in an enterprise exceeds 50 per cent or if direct ownership by a foreign entity exceeds 25 per cent and there is no Australian enterprise or individual with a larger holding.
12. Data on foreign ownership within the more significant industries in the mineral sector are set out in chapter 3. They are calculated on a different basis but confirm the high degree of foreign ownership in the sector overall.
13. R.B. McKern, *Multinational Enterprise and Natural Resources* (Sydney: McGraw-Hill, 1976), p. 13.
14. Ibid., p. 189.
15. Overseas investment *in general* has been a major issue of Australian economic policy since the second world war. For a comprehensive discussion of the issue see the Treasury Economic Paper No. 1, *Overseas Investment in Australia* (Canberra: AGPS, 1972).
16. "The Role of Government", in *Minerals Investment and Australian Development* (Sydney: University of Sydney, 1971), p. 45.
17. Quoted in *AFR,* 8 April 1980, p. 54.
18. E.G. Whitlam, *Press Statement,* 30 October 1973.
19. Ibid., 24 September 1975.
20. *Press Statement,* 13 October 1975.
21. Foreign Investment Review Board, *Annual Report, 1978* (Canberra: AGPS, 1978), pp. 7-9, 41-45.
22. Foreign Investment Review Board, *Annual Report, 1979* (Canberra: AGPS, 1979), pp. 8-9, 33-35.

23. *AFR,* 16 April 1980, p. 4; 25 September 1980, p. 3; 30 October 1980, p. 1; 5 December 1980, p. 1.
24. *AFR,* 30 April 1980, p. 53; 19 November 1980, p. 55.
25. *The Contribution of the Mineral Industry to Australian Welfare* (Canberra: AGPS, 1974).
26. See in particular the response by the Australian Mining Industry Council, *Mining Taxation and Australian Welfare* (Canberra, June 1974); and Susan Bambrick, *The Changing Relationship: The Australian Government and the Mining Industry* (Melbourne: CEDA, 1975), esp. chapter 3.
27. The continued exemption from tax of profits made from gold mining is difficult to defend, given the high price of gold and the high profits being earned in the industry. On the other hand the inability of companies to offset expenditure on minerals exploration against income other than mining income (in contrast to the provisions for petroleum exploration and capital development expenditure) is said to discriminate against minerals exploration, at least by smaller companies.
28. For a comprehensive discussion with particular emphasis on the claims of a federal government to rent see Anthony Scott, *Central Government Claims to Mineral Revenues,* Centre for Research on Federal Financial Relations, Australian National University, Canberra, Occasional Paper no. 8, 1978.
29. *National Times (Business Review),* 12-18 October 1980, p. 3.
30. *AFR,* 3 July 1978, p. 1.
31. *AFR,* 14 October 1980, p. 3.
32. Scott, *Central Government Claims to Mineral Revenues;* and for example, MIM Holdings Ltd, *Annual Report 1980.*
33. This was made clear by the Prime Minister during the 1980 election campaign.
34. Some details are given in chapter 3.
35. However, in the case of bauxite and alumina somewhat less than one half of total export sales are made to independent purchasers; the rest are intracorporate sales. See chapter 3. For a detailed discussion of long-term contracts see Ben Smith, "Long-term Contracts in the Resource Goods Trade", in *Australia, Japan and Western Pacific Economic Relations*, Sir John Crawford and Dr Saburo Okita (a Report to the Governments of Australia and Japan) (Canberra: AGPS, 1976), pp. 299-325.
36. See further details in chapter 3.
37. The power to control exports has also been used for other purposes — to conserve supplies for domestic use (the decision to ban iron ore exports in the 1930s) and to stop production of a mineral (mineral sand mining on Fraser Island). Here we are only concerned with the use of the power to ensure "adequate" prices.
38. Ben Smith, "Bilateral Monopoly and Export Price Bargaining in the Resource Goods Trade", *Economic Record* 53, no. 141 (March 1977): 49.
39. Australia, House of Representatives, *Debates* 1978, no. 111, pp. 2186-89.
40. *AFR,* 25 March 1980, p. 17; 31 March 1980, p. 8.
41. *AFR,* 8 April 1980, p. 1; 3 September, 1979, p. 1.
42. Susan Bambrick, *Minerals and Energy Policy in Australia* (Canberra: ANU Press, 1979), pp. 48-49.
43. *AFR,* 11 July 1980, pp. 3, 12.
44. The matter is well known to be a source of concern to, and under investigation by, the commonwealth Treasury.
45. *AFR,* 11 July 1980, p. 12. See also chapter 6.
46. *AFR,* 19 December 1980, pp. 10-11.
47. Richard E. Caves, "Policies towards Australia's resource-based industries", in *Australia, Japan and the Western Pacific Economic Relations,* Sir John Crawford and Dr Saburo Okita (Canberra: AGPS, 1976), p. 172; A. Fitzgibbons, "Mining and the Future Structure of the Australian Economy", *Australian Quarterly* 45, 2 (June 1973): 86.

POSTSCRIPT TO CHAPTERS 6 AND 7

Since these chapters were first written, and particularly since the end of 1981, there has been a general decline in the profitability of companies in the mineral sector due to lower levels of economic activity in major world economies, including Japan, and (largely as a result of this) lower world prices of most minerals and metals. As a consequence the rate at which new projects have gone ahead has slowed markedly and several have been deferred at quite advanced stages (notably in the aluminium industry). Indeed it was widely suggested, from the second half of 1981, that the "boom" was, in the words of one industry leader, a "mirage". In the early stages of this downturn some over-reaction was evident and there were some obvious attempts to "talk down" the boom and to emphasize its fragility in order to counter some of the problems (such as the mounting pressure for increased taxation of mining companies) created for it by the "boom psychology" which had been building up (due in no small part, it must be said, to statements, by the government itself). In response, it was argued that the boom was "a very real phenomenon [which] has more the characteristics of an irresistible Juggernaut than a fragile blossom".[1] This view seemed to be held also by the Treasurer who, in an important speech in October 1981, echoed many of the views put forward by the secretary to the Treasury in November 1979.[2] But further significant falls in mining company profits were announced during 1982. Announcements of deferral of new projects became more frequent and the potential effects of the downturn on the rest of the economy were becoming increasingly apparent (for example the Australian Federation of Construction Contractors quarterly survey in March 1982 predicted a real *decline* in construction activity in 1982–83 of 9 per cent — this followed a prediction of a 1 per cent real increase in its December 1981 survey and a prediction of a 25 per cent increase three months earlier).[3] Within the space of less than 12 months a mood of euphoria turned to one of gloom.

It is therefore necessary to modify somewhat the conclusions in chapters 6 and 7 in so far as they were applied to the 1980s. To the extent that profits decline, that the level of exports grow less rapidly than expected, and that investment expenditure on new projects and hence the demands on labour and capital resources are lower than anticipated, the magnitude of the effects outlined in chapter 6 will be less pressing than would have been the case in the more full-blooded boom situation which, until late 1981, seemed likely to occur. In particular it may be noted that lower than anticipated exports and higher levels of imports since 1981 have altered the balance of payments situation such that an appreciation of the Australian dollar — or reduced level of protection as a means of minimizing adjustments to the exchange rate — which seemed inevitable in 1980 became unnecessary. Indeed by 1982 regular *de*valuations of the dollar were being made. A revival of the boom could lead to renewed pressure for revaluation and/or reduction in levels of protection though it is perhaps more likely that the growth of the mineral sector will be "stretched" in such a way that its macro-economic effects and their consequences will be somewhat less marked than suggested in chapters 6 and 7. In view of the comments made in chapter 6 and footnote 3 (chapter 7) it is also interesting to note that in July 1981 — while the boom was still proceeding apace — the government acted to relax restrictions on outward portfolio investment. In the context of high levels of capital inflow which were largely responsible for the need to allow the exchange rate to appreciate, and of other measures to offset their monetary impact, it was, in the government's view, "something of an anomalous situation that restrictions were maintained on portfolio investment overseas . . ."[4]

With respect to the issue of foreign ownership and control, the government continued to enforce its policy guidelines relating to new resource projects and to affirm its policy of achieving a "high level" of Australian equity in resource projects. A somewhat surprising expression of this policy was the Treasurer's "invitation" in August 1981 to the Shell Co. of Australia to "discuss with the Foreign Investment Review Board the possible introduction of 25 per cent equity in its Australian operations".[5] A quarter of Shell was estimated to be worth between $500 million and $700 million, and such a move would have placed enormous strains on the Australian domestic equity capital market, at a time when the government was itself arguing (with justification) that a high level of foreign investment was necessary because domestic capital resources were insufficient to pursue resource development at a desirable level; at the same time little would have been achieved in terms of increased Australian *control* of the company's operations. In the event (and probably fortunately) Shell declined the "invitation". Following a review of foreign investment policy the government announced in early 1982 that mineral processing activities, with respect to which there had been some ambiguity, were to be made subject to stronger Australian equity requirements, although this was to be applied flexibly and with regard to the circumstances in each case.

A hardening of policy with respect to taxation, on the part of both the state and commonwealth governments, was also evident — at least for a time. Higher royalty charges were announced by the Western Australian and Northern Territory governments and the Victorian government announced plans to impose a resource tax in the form of a $20 million "licence fee" on Esso–BHP onshore pipelines. Both the Queensland and New South Wales governments increased rail freights payable by mining companies — widely recognized as a *de facto* resources tax. The commonwealth government, in its 1981–82 budget, extended the duty on coal exports to steaming coal (having previously promised to abolish the tax altogether), eliminated some other concessions to the mining industry and asked the Industries Assistance Commission to investigate the various forms of budgetary assistance given to the mining and oil industry. However, as the profits made by most mining companies fell drastically after 1981 several of these policies were revised. In a package of general industry assistance measures announced in July 1982 improved capital expenditure write-off provisions were granted for mining investment; the levy on steaming coal (though not the higher levy on coking coal) was removed the following month; and the royalty increases announced by the Northern Territory government were reduced. The *ad hoc* and arbitrary nature of most of these changes reflect the fact that Australian governments (federal and state) have yet to formulate a coherent policy whereby an appropriate proportion of the profits of mining companies can be diverted to the Australian people in a manner that is seen — both by Australians in general and the mining companies themselves — to be rational and equitable.

Mining companies were also required to make larger contributions to infrastructure costs, particularly by the Queensland government, which decided in principle that large development projects should contribute not only to infrastructure directly related to their operation but to community facilities and to the upgrading and maintenance of relevant regional road networks; legislation was introduced to allow the government to step in and co-ordinate the requirements of the local shire and state governments in relation to infrastructure planning. Again, however, there appeared to be some backtracking on this policy in 1982 as economic conditions in the industry deteriorated.

Notes

1. *AFR*, 3 September 1981, p. 12.
2. "Manufacturing Industry and Resource Boom — a Balanced Approach", Address by the Treasurer to the Metal Trades Industry Association National Affairs Forum, 19 October 1981.
3. *AFR*, 24 May 1982, p. 1.
4. Treasurer, *Press Release* no. 137, 19 July 1981.
5. Treasurer, *Press Release* no. 151, 12 August 1981.

8 The Environment

Patricia Dale and Errol Stock

The extraction of minerals from the earth's crust involves an interaction between a particular human culture and the physical environment. A mineral is perceived to be a *resource* when a *use* for it is recognized and when technological and economic constraints can be met. Thus hunter-gatherers and "primitive" subsistence farmers are not involved in major mining ventures! Whether a specific mining operation takes place depends not only on the status of the resource but also on its location with respect to processing and shipment facilities, markets and labour supply. There is a need to consider the effects of mining on the natural environment and on future land use options.

This chapter has four aims. The first is to emphasize the need for and present status of information on the environmental effects of mining and to describe the related public attitudes and values. These are important in influencing political decisions about whether to mine and how much to control the impacts of mining. The second aim is to provide an overview of the mining process and its effects on the biophysical environment. Thirdly, specific mining industries are considered in detail: black coal in the Hunter Valley, iron ore in the Pilbara, bauxite at Weipa, mineral sands in eastern Australia and uranium at Ranger. Finally, there is an outline of legislative provisions for the control of mining and a discussion of their impacts in Australia.

IMPACT STUDY AND PUBLIC ATTITUDES

The various impacts of mining on the biophysical environment in Australia have not been studied systematically. Indeed there is no popular expression of the need for such a study either by the mining industry, state and federal politicians and public servants or the general public. Rather, most Australian people since the 1850s gold-rush days seem willing to accept whatever is necessary to win the minerals, including accompanying disturbances, without question.

Over the last twenty to thirty years in Australia, there has been a growing awareness of, and reluctance to accept, the long-term effects of mining. At least two different reasons for this can be identified: the expansion of residential areas adjacent to mining (quarrying) operations and the movement of public recreation around areas affected by mining (specifically coastal mineral sand mining). Over the years public expression of concern has encouraged government and mining companies to recognize, research and attempt to control the problem.

For example in 1951, several mining companies, the New South Wales Soil Conservation Service and the New South Wales Mines Department commenced field studies on the coast aimed at improving revegetation techniques after sand mining for heavy minerals.[1] But not until the mid-1960s did this activity accelerate. The intensification of effort was primarily a response by mining companies and government departments to considerable public criticism of the visual impact during mining and the apparent low priority and expenditure given to revegetation after mining.

Most studies and reports of mining impacts on the biophysical environment are either detailed and rather local in scope (for example, unpublished submissions to Fraser Island and Diamond Hill inquiries) or brief and very broad summaries such as in Heathcote.[2] There are many specific investigations conducted primarily to determine land use practices during and after mining, so that current and future users should not be affected, or be affected only to an "acceptable" level. These are contained in unpublished documents loosely called "Environmental Impact Studies", but they have received very little *post hoc* evaluation. The mining industry holds regular workshops for industry specialists to share practical skills in landscape reconstruction, but discussions of broader issues are not usually agenda items. Only recently have there been some attempts to stimulate the development of *general principles* for the planning and management of mined lands in Australia.[3]

Parbo quotes the Australian Mining Industry Council claim that less than two parts in ten thousand of Australia's land surface is affected by mining.[4] Conzinc Riotinto claim that "only a minute part of Australia's land surface (0.0035 per cent)" is affected.[5] It is not clear how these proportions are derived, but such statements tend to divert attention from important matters such as the location of mining (relative to, say, urban areas, erosion-prone land and natural drainage) and the degree and kind of disturbance for other land users. While it is true that farming, grazing, forestry and housing industries disturb very much more land than the mining industry, it does not follow that this should exempt the impacts of mining from closer examination. Nor is it necessarily a valid argument to state that the impacts of some mining operations may be excused because they are located in "remote" areas *currently* of little "use".

A mining company receives direct instructions of its legal responsibility towards land and other land users in leasehold tenure documents. Company officers also learn current values and attitudes towards land

and other land users through discussion with ministers and public servants from several departments. Few mining companies are likely to follow procedures beyond legal requirements or standards communicated by public authorities, as the economic and technological implications are too open-ended. This does not excuse any impacts which may accompany mining but points to a vital need for both company officers and public authorities to discharge their joint responsibility as stewards of the land. Some mining companies have commenced a pioneering role here but much more needs to be done.

In our opinion, mining impacts are under-estimated. They require much more scrutiny, not less. This is especially so if the 1980-90s "mining boom" forecast by mining industry spokesmen and politicians becomes a reality.

THE MINING PROCESS AND ITS IMPACTS

In the Australian Standard Industrial Classification, mining covers a broad range of extractive operations. This includes the extraction and processing by dressing or beneficiation of naturally occurring minerals in solid, liquid or gaseous form. Coal mining and petroleum and natural gas extraction would be included as "mining" under such a classification. We also include dredging and quarrying to extract, for example, limestone (coral), gravel, sand and crushed rock, as mining operations. Unless otherwise stated, the data reported here relate to this broad definition.

The effects of mining may be direct, such as the removal of surface material, or indirect, such as disturbances due to associated transport, processing and settlement. Both direct and indirect effects are interrelated. For example, the growth of Gladstone in Central Queensland is a consequence of coal mining expansion and alumina processing. If it can be argued persuasively how important coal and alumina industries are to local, regional, state and federal economies, then it is legitimate to consider the biophysical impacts within the same boundaries. However, this chapter accepts the artificial device of excluding the impacts of suburban growth of mining towns and mineral processing towns, as aspects of this are covered by other authors. The impacts of such structures as aluminium smelters are included within the ambit of mining impacts.

Our interest is in the direct, indirect and interrelated effects of mining on the Australian biophysical environment. For this purpose "biophysical environment" is restricted to the rocks, soils, vegetation, fauna, drainage systems, water bodies and atmosphere. The Australian assemblage of these components is unique on a world scale and contains some extremely fragile ecosystems.[6] As there is no opportunity here to attempt the detailed study the subject deserves, our intention is to suggest a simple framework in which mining activities and their impacts could be examined to illustrate the main points of our argument through selected examples.

Models of the Mining Process

For more detailed examination of the impacts of mining it is useful to refer to simple models of the mining process. Though the process varies in detail according to whether the mining activity is for, say, base metals, industrial rock or petroleum, a model based on base metal mining is appropriate for this discussion.

The mineral economist Mackenzie lists six stages in mining operations which are summarized in figure 12. It is interesting that Mackenzie does

Fig. 12. Stages in a mining operation

I EXPLORATION

Primary Exploration Stage – reconnaissance geological, geophysical and/or geochemical surveys over anomalies; exploratory drilling of targets or anomalies.

Delineation Stage – surface drilling with or without underground exploration (continued as long as the net value of additional information is thought to exceed the cost of obtaining it).

II DEVELOPMENT

Mine Development Stage – stripping overburden, shaft sinking and haulage way construction, stope preparation, installation of mining and material handling facilities and equipment; construction of workshops, offices, roads, housing townsite, power and water facilities; petroleum industry requires production platforms and pipelines.

Construction of Processing Facilities Stage – industrial and construction minerals require relatively simple facilities for upgrading products; base metals require milling plant and equipment to separate and concentrate.

III PRODUCTION

Mining Stage – stripping waste, stope preparation, mining ore (blasting, drilling, material handling, filling mined-over stopes), associated technical and planning services.

Processing Stage – crushing, grinding, differential floatation, drying, tailings disposal, loading concentrates for shipping.

IV RECONSTRUCTION

Reclamation Stage – demounting buildings and facilities, decommissioning roads and transmission lines, stockpiling and spreading topsoil, spreading overburden and contouring wastes.

Rehabilitation Stage – ecological surveys and monitoring for revegetation, habitat recreation and fauna management; maintaining nursery; seeding, planting and watering.

Source: Stages I-III from B.W. Mackenzie, *Mineral Economics*, vols. 1 and 2, Notes prepared for Workshop 121/79 (Adelaide: Australian Mineral Foundation, 1979), pp. 42-46.

not include a "reconstruction" stage in his model, which was presented at a workshop for mineral industry personnel. Two reconstruction stages have been added to figure 12 by the authors of this chapter.

This subdivision of stages in a mining operation reveals the different expertise required and different spatial aspects of a project, but cannot show that many stages and sub-stages overlap in time. For example, to save costs in double-handling, good managers would co-ordinate where possible, the stripping of overburden and wastes during "development" and "production" with spreading and contouring during "reconstruction".

Figure 13 shows Mackenzie's scheme of the distribution of cash flow associated with a mining project, with a notional size and timing of expenditure required for reconstruction added. As with exploration and development, some reconstruction expenditure would be considered as "lead-time" costs to be paid for initially from funds raised before the mine generated revenue. The remaining portion of reconstruction costs could be funded directly from the producing mine with, perhaps, the bulk of the expenditure occurring after production has ceased.

As presented in figure 13, reconstruction is like exploration and development — a necessary activity to produce the mineral but occurring mainly in a different time frame. Just as a mining company pays for exploration and development from loan funds or by charging against another operating mine, the costs of reconstruction at whatever stage necessary could be handled by some suitable accounting procedure. A

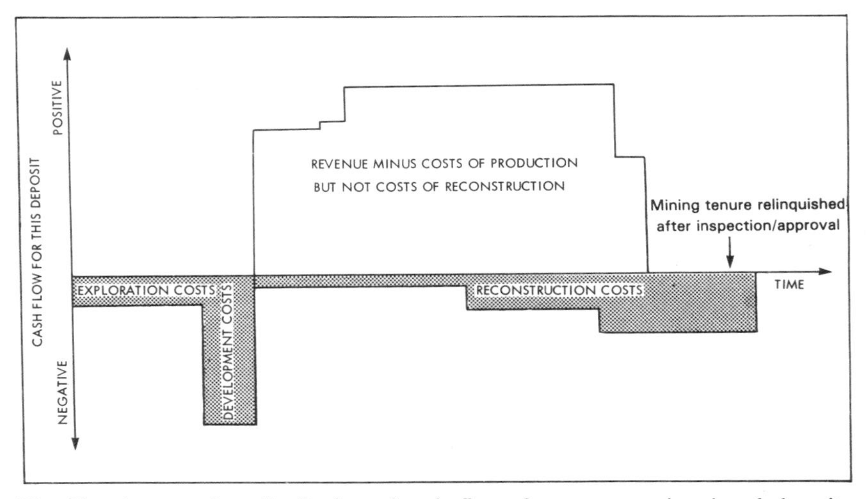

Fig. 13. Average time distribution of cash flows for an economic mineral deposit. Modified after B.W. Mackenzie, *Mineral Economics,* vols. 1 and 2, Notes prepared for Workshop 121/79 (Adelaide: Australian Mineral Foundation, 1979), pp. 42-46.

diagram such as figure 13 cannot show the scale or the kinds of mining impacts, but it is important to recognize that the reconstruction works required to attend to the impacts are spread across the life of an idealized mining project.

Most areas of land, and the sea and seabed, are never affected beyond the exploration stage because relatively few mineral deposits are located and even fewer are rated as "economic" by the different companies. Thus a large part of Australia has been exposed to this *essential* first stage of mining yet the impacts of exploration are virtually ignored by the industry and public authorities.

Exploration

Exploration or prospecting is the search for and appraisal of both new and known mineral occurrences. This usually involves geological, geophysical and geochemical research and field techniques which include drilling, but not the development of mines or sites (for example by removal of overburden). During the period 1971–77, which was characterized by a downturn in exploration activities in Australia, total expenditure (excluding petroleum) amounted to $735,971,000, of which $202,891,000 (27.6 per cent) was spent on drilling. This amount financed 15,359,000 metres of augering, percussion and diamond drilling.[7]

The impacts of the exploration stage of mining are many and varied. Perhaps the most significant is the result of extending a network of tracks for exploration and service roads to drilling sites. Vegetation is cleared from tracks and stacked in windrows for burning or left to decay. Some tracks may involve upgrading and maintenance, requiring borrow pits for gravel, and most require at least one pass of a dozer. Soil and subsoil are disturbed along the route, and on the sides of a track the irregular microtopography and surface texture provide sites for natural plant regeneration which are very different from the undisturbed situation. Very rarely are such tracks sited for long-term use; they are commonly placed to conform with the grid pattern a geologist has determined without specific regard to terrain. Such tracks tend to interrupt natural drainage lines and promote accelerated runoff and soil erosion. Road traffic affects compaction and soil permeability.

Recent airphotography throughout Australia reveals the surprising extent of exploration tracks which are an inheritance of prospecting activities primarily during the "boom" of the late 1960s and early 1970s. No mining/exploration company or local or state government is known to have a policy to co-ordinate and plan the location of such tracks, to organize the decommissioning and repair of most tracks and to retain and maintain tracks which may serve a function after exploration tenure is lifted. Rather, tracks are generally abandoned, though some may be upgraded by a grazier or reopened by an off-road vehicle user in search of "recreation". The significance of tracks for access and impact on flora and fauna is well documented.

Thus the mining industry unwittingly promotes effects which were never anticipated by either company employees or public servants associated with mining activity. Considering the singular importance of exploration tracks, their impact on the biophysical environment needs further elaboration than can be given here.

Similar disturbances are associated with survey lines cut to mark out boundaries or provide line-of-sight, though most of the impact is on vegetation and much less on soil and rock. Trenching and costeans which are excavated to reveal features in the near sub-surface are usually backfilled to reduce hazards to people, stock, wildlife and vehicles; the end result rarely conforms to the original landscape. Though each excavation is relatively narrow and of small extent, there are thousands of hectares so disturbed in Australia.

The role of fire in the Australian environment is still not well understood, though it is used extensively as a management tool by graziers. Those geologists who are known to burn off large areas to improve mapping and working conditions are no doubt following local example. Tracks, rubbish tips, abandoned equipment, oil and grease spills and cut vegetation are common around exploration and drillers' camps. Rarely are such sites of disturbance considered for reconstruction other than basic reclamation, such as burying rubbish and salvaging materials that may have some future use.

The duration of such impacts depends on the natural recovery potential of the local environment, though this is complicated by the possibility of unusual climatic conditions during and after disturbance. As a very general rule, semi-arid and interior environments will sustain greater amounts of damage and recover less rapidly than humid and coastal environments. The sand masses of coastal environments are an important exception to this.

Petroleum drilling and testing of exploratory wells may have greater impacts on the environment than prospecting for solid minerals because of the potential of spills and leakages. This is especially so in marine environments. In addition, it was common practice until relatively recently to abandon equipment on the sea floor because of recovery problems after exploration. As a result of prolonged lobbying by fishermen, who suffered net and equipment damage in operations in Bass Strait, this is no longer accepted practice in offshore drilling in Australia. The Australian Petroleum Exploration Association has compiled a brochure outlining its policy towards reducing environmental impacts.

Development, production and reconstruction

Only a small number of the mineral deposits discovered by exploration become producing mines. Economic, technological and political variables have to be favourable for mining company executives to commit resources to a project. It is during the development and production stages that the most rapid and pervasive impacts occur on the local biophysical environment.

Ideally it is at the project development stage that good managers would plan for reconstruction to allow some use after production ceases. However, two important factors constrain their ability and willingness to plan in detail so early.

Firstly, few state and local government authorities have conducted land-use planning programmes which would provide the framework for post-mining land-use after leasehold tenure is lifted. In the absence of forward public planning, mining personnel are left with little option but to plan according to lease conditions. Reconstruction specifications, usually prepared by negotiation, are commonly based on broad and minimal concepts such as "prevent further visual pollution and soil erosion" or "establish pasture for grazing". The ambiguities associated with instructions to "restore to 'natural' conditions" are and will continue to be sources of contention between company employees, public servants and the public.

Secondly, mining companies are institutions whose primary function is to produce minerals. Their expertise is commonly judged on their ability to supply the demand for mineral products efficiently and effectively. Few mining companies would regard their success in reconstruction after mining as a valid criterion for evaluating their performance as mineral producers. Thus, planning for reconstruction is very low in priority on the complex schedule of total planning for mining.

Reconstruction then usually consists of reclamation and revegetation only to a standard acceptable to the appropriate government officers and thus, it is inferred, acceptable to the public. "Acceptable" involves both the relatively objective activity of measuring the degree of reconstruction and the subjective activity of judging the level of performance. Until the criteria for reclamation are spelled out in detail and made public, the ability to discuss value judgements separately from reasonable performance standards will remain elusive.

Many mining companies have established expertise in some reconstruction techniques required to ameliorate the impacts of development and production. Perhaps the best and most well known example is the transformation of the 1930s desert around Broken Hill to the oasis of today. The nickel mining town of Kambalda in Western Australia was planned and established by giving attention to local biophysical conditions. A common feature of these and other cases, where innovative company employees gave the example to follow, is that the rewards were both tangible and essentially immediate. With most reconstruction after mining, the effort is for land owned by the Crown and for some unknown future public.

The impacts of the development and production vary widely according to the mining project and the appropriate reconstruction similarly is determined by the project and its location. In the following section the impacts of development and production and associated reconstruction are discussed for several major Australian mining operations.

EXAMPLES OF MINING OPERATIONS

The five mining operations discussed here were selected not only for consistency with the book as a whole, but also because they represent mining operations at different levels of national importance and with different projected growth rates. They largely represent surface operations which result in major alterations to the landscape. Other ventures which primarily use underground mining methods are not considered here.

Black Coal in the Hunter Valley, New South Wales

Just under 60 per cent of Australian black coal is produced in New South Wales and roughly the same proportion of New South Wales production comes from mines in the Hunter Valley. The Hunter Valley coal mining area is situated about 250 kilometres north of Sydney in eastern New South Wales. Newcastle is the main coastal port, and the valley extends inland for 150 kilometres to the Great Divide. In the lower valley (Newcastle-Cessnock area), coal has been extracted by underground methods for the last hundred years or so. In the upper valley (Singleton and Muswellbrook), surface mining began in the 1950s and by the late 1970s accounted for about 25 per cent of New South Wales production. In 1978 there were nine surface mines in the region with five more planned. This section is confined to the environmental effects of surface mining, since these are greater than in the case of underground mining.

The natural landscape is gently undulating to hilly, with local relief up to a hundred metres. The underlying rocks are mainly sandstones, siltstones and conglomerates on which a variety of soils have developed. The climate is humid with mild winters and hot summers.[8] Rainfall is unreliable and seasonally of high intensity which causes local flooding and severe soil erosion. The original vegetation consisted of open savannah woodland and much of the present area is used for grazing.

Effects of development and production

Strip mining involves the clearing and levelling of the strip and loosening by means of explosives. Loosened material is removed to the top of the coal seam. This opens a strip about seventy metres wide and spoil is dumped on to natural ground surfaces between the mine and haul road. Spoil from the subsequent strips is back-piled on to the previously mined strip.

The landscape during and immediately after mining consists locally of a series of longitudinal ridges and furrows whose alignment may cut across natural waterways and be inconsistent with the overall "lie of the land". During mining, natural waterways have to be protected and elaborate temporary diversions are constructed as required.

In the upper Hunter Valley, the bulk of the reworked material is porous and elevated above the surrounding land, and so suffers from rapid dessication. The surface "soil" structure is poor, lacks organic matter and seals on wetting. This results in rapid run-off and is a potential erosion hazard.

In the Cessnock-Maitland area of the lower Hunter Valley there are additional problems in the form of a high sulphur content, both in the coal seams and in the overlying strata. As a result of oxidation the soils on mined areas become acidic to the point of not being able to support vegetation. Additionally, water pollution problems ensue and vegetation may be severely damaged along water courses.

Reconstruction

Details of the reconstruction programme in the Hunter Valley have been described by Hannan.[9] The aim is to return the land to a state suitable for grazing. To this end, the research programme investigates the nature of the spoil material during the production stage, carries out field trials for grasses and trees, cultivation trials for top-soiling strategy, seed casting and ploughing.

The implementation of the programme follows a sequence:

1. Spoil dumps become available for rehabilitation.
2. Landscape is shaped, dam sites and stream channels are prepared.
3. Rocks greater than fifty centimetres in diameter are removed and a five centimetre layer of topsoil is spread and fertilized.
4. The land is contour-ploughed to twenty centimetres depth.
5. Grass and legume seeds are broadcast.
6. Further management includes the prohibition of grazing and traffic for two years, or until the pasture is well-established; top dressing is applied every six months and trees are planted later.

These two photographs indicate the extent of disturbance to the landscape caused by open cut coal mining (above) and the restoration after such mining (below). *(Photographs courtesy of Utah Development Co.)*

Reconstruction costs reported in 1979 were between $450 and $550 per hectare.

Where there are problems of water pollution and soil acidity, reconstruction involves replacing all washery waste in worked out excavations which are then covered with untainted overburden and soil, and then revegetated. Problems do exist in older sites where the sulphur-rich material has been placed on the surface. The usual method is to cover the material to a depth of at least thirty centimetres with overburden and treat with lime to reduce acidity.

Bauxite Mining at Weipa, Queensland[10]

Queensland produces nearly 40 per cent of the bauxite of Australia, most of which comes from the Cape York Peninsula. Specific reference is made to bauxite mining at Weipa. The Commonwealth Aluminium Corporation Ltd (Comalco) bauxite mine at Weipa is one of the largest open-cut bauxite mines in the world. There are three mining areas — Weipa which began production in 1963, Andoom beginning in 1972 and Para which is not yet producing.

The economic bauxite deposits of Cape York are located on the north-west coast of the peninsula. This is a relatively flat area with many estuaries and the land rises gently to the east. The bauxite of varied quality occurs as part of a deeply weathered profile and the highest grade ore body is a flat to gently dipping laterite varying in thickness from 1 to 4 metres. About 0.6 metres of topsoil covers the ore and it has poor physical and chemical characteristics. Much of the area is dominated by eucalypt open forest. The climate of Weipa is humid tropical with maximum rainfall in summer and a small annual temperature range.

Effects of development and production

Indirect effects include those associated with the construction of fixed plant railway, roads, seaport, airport, administration buildings and residences for a community of around four thousand. The direct effects of operation are produced by surface mining and the construction of tailings dams.

The sequence of mining operations and part of the reclamation is as follows:

1. Surface vegetation is pushed over, stacked in windrows and burnt. Usually, the destruction of original vegetation is timed to take advantage of moist surface conditions and to avoid the hard surface conditions of the dry season.

2. Topsoil (overburden), usually at depth 0.6 metres, is scraped off and spread on an adjacent excavated and recontoured mine floor. Stripping and placement of the topsoil is best timed to avoid wet conditions and attendant difficulties of creating a suitable surface for revegetation. To improve root penetration of the compacted mine floor and ironstone the area is ripped using a dozer and rear-mounted ripper.
3. Exposed bauxite, at two to six metres depth, is excavated by front-end loader, carried by haul trucks to a central loading station and railed to Weipa. As several mine faces may be operated to maintain quality control of the bauxite, this presents some complications in reconstruction sequences.
4. At Weipa, the bauxite is blended, graded and washed prior to shipment. The washed fines are pumped in a slurry to local tailings ponds.

Reconstruction

The objectives of the reconstruction programme are to minimize impact on the environment, to restore the area to an acceptable condition and to assess the potential of reclaimed land to support a secondary industry for Weipa. To these ends, research has concentrated on the nature of the open forest and woodland, plantation timbers, tropical pastures and agricultural crops such as cassava.[11]

Revegetation includes planting nursery stock, direct seeding and planting trees at a density of thirteen hundred per hectare. Fertilizer and trace elements are added to the soil. Management problems include damage or loss of vegetation through fire, wind, flood, drought, termite and rodents.

The ten year plan to 1987 aims to rehabilitate nearly six thousand hectares at Weipa and Andoom as shown in table 38. Table 39 sum-

Table 38. Proposed 1987 rehabilitation targets after bauxite mining at Weipa and Andoom

Fully Vegetated Area Type	(Hectares)	Tailings Dam Area (Hectares)	Residential and Recreation Area (Hectares)
Natural open forest	3,500	600	250
Plantation timber	700		
Grassland pasture	700		
Research	100		
	5,000		

Source: B.A. Middleton, "Land Use Planning, Bauxite Mining Operation, Weipa", in *Management of Lands Affected by Mining,* ed. F.A. Rummery and K.M.W. Howes (Division of Land Resource Management, CSIRO, Australia, 1978).

Table 39. Summary of revegetation completed at Weipa by March 1979

	Hectares	Percentage
Open forest woodland	760	41.7
Plantation timbers	466	25.6
Tropical pastures	175	9.6
Agricultural trials	102	5.6
Research trials	195	10.7
Amenity (gardens, etc)	126	6.9
	1,824	100.0

Source: B.A. Middleton, "Rehabilitation of Bauxite-mined Lands at Weipa", in *Mining Rehabilitation*, ed. Ian Hore-Lacy (Canberra: AMIC, 1979).

marizes the revegetation programme at Weipa to March 1979. Costs in 1978 were reported to be approximately $3,000 per hectare.

The progress of annual revegetation from the commencement of mining is illustrated in figure 14. This plot indicates the relatively low priority placed on revegetation in the ten years after production commenced, but it does not reveal three important aspects of the revegetation programme: since 1957 when the Comalco Agreement Act was passed, public expectation of the level of reconstruction has risen and Comalco has responded with investigation and research in some measure beyond the legal requirements under the Act. Secondly, the approach to revegetation has been almost exclusively that of stabilization by

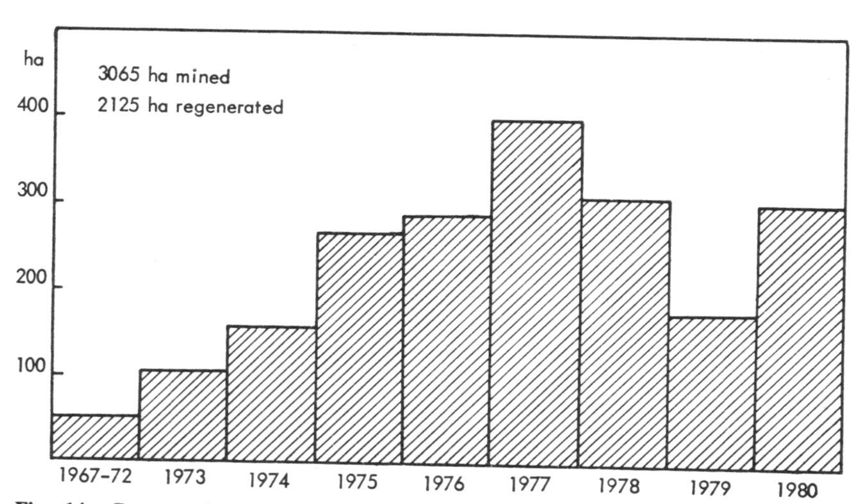

Fig. 14. Regeneration plantings per year at Weipa, 1967–80. Modified after B.W. Mackenzie, *Mineral Economics*, vols. 1 and 2, Notes prepared for Workshop 121/79 (Adelaide: Australian Mineral Foundation, 1979), pp. 42-46.

modified agronomic techniques and reafforestation for timber production with very limited plant ecology. Finally, the lack of direction by the Queensland government for long-term use after mining has obliged the operating company to extend its resources into virtually creating a tropical research station.

In the future, Comalco plans to revegetate about 240 hectares annually to keep pace with the rate of area developed for mining. It would be of great assistance to this work if the needs and aspirations of the local communities could be assessed as an essential ingredient in planning for reconstruction.

Iron Ore in the Pilbara[12]

Western Australia produces 95 per cent of the total Australian output, nearly all of which comes from the Pilbara region. There are four companies involved.

The local topography is of rounded hills and ranges with escarpments and gorges. The ore has a high quartz content (between 3 and 8 per cent at Pannawonica).[13] Soils range from shallow uniform textured gravelly loams in the Hamersley and Chichester Ranges, to cracking clay, red earthy sands and duplex alkaline earths on the Abydos Plain and in the Fortescue River Valley. The climate ranges from semi arid on the north coast to hot arid in the inland areas. Average annual rainfall is variable but usually two hundred to three hundred millimetres is received. The area is prone to tropical cyclones. A variety of vegetation covers the area: low closed mangrove forests grow in coastal areas, further inland are salt marshes and tussock grasslands, and eucalypt woodlands dominate the Hamersley Plateau, though distinctive vegetation associations are found in gorges and valley floors. Previous land use in the area has led to some alteration in the composition of the vegetation. For example, following burning and heavy grazing *Triodia* species are replaced by *Eragrostis* species; *Acacia aneura* often dies as a result of soil changes due to trampling under grazing. On areas denuded by earth moving, it has been suggested that the natural succession to climax may take only five to ten years.[14]

Effects of development and production

The individual mine sites are not very large: the Mount Newman mine area covers about 7,500 hectares. The open cast mining operations have a similar general effect to those of the surface coal and bauxite mines but additionally pose problems of dust control and water quality.

At the Mount Newman mine, the most obvious physical effect is in the alteration of the landscape. The existing landforms — a hill and its surrounding plain — are being converted into a large pit and waste dumps. Problems encountered include dust, instability of dumps and general aesthetics.

Dust is a problem here because of the particular combination of a hot dry climate, the pattern of prevailing winds and the high rate of production and nature of the ore. Dust has been thought to impair functions in native vegetation and animals. Studies are inconclusive but the effects seem to be minimal.[15] The extent of the problem is illustrated in table 40 which shows the amount of particulate matter at various sites. By comparison the USA standard in 1971 was for a mean annual maximum of 75 $\mu g/m^3$ and a maximum of twenty-four hour exposure of 260 $\mu g/m^3$ permitted only once a year.[16] The problem is accentuated by town development close to the mines.

Table 40. Dust concentrations at various sites in Western Australia

Site	Dust Concentration ($\mu g/m^3$)
Port Hedland motor hotel	23
South Hedland townsite west	86
Public Works Dept.	86
Perth suburbs	16-122

Source: R.A. Powell and D.B. Sykes, "Dust surveillance for trend analysis versus dynamic sampling directed towards immediate evaluation", in *Symposium: Control of Dust in the Environment in the Pilbara Iron Ore Industry* (Institute of Engineers, WA Division, 1975).

Not only is dust annoying but it is a health hazard. Normal dust is of such a size (greater than or equal to five microns) that particles are not held or absorbed by lungs, but industrial crushing may reduce the size so that respirable dust is able to permeate lung tissue. There is a relationship between mining and silicosis of the lung reported by McNalty and in the popular press.[17] There are no legal standards in Australia, although the companies generally attempt to suppress dust, and the total dust per tonne of ore shipped has shown a steady decrease over recent years.[18] This does not imply that absolute quantities of dust fallout have decreased. Methods of dust suppression largely involve the use of water spraying. There are thus many water-related effects of mining in the Pilbara. For example, so much use is made of water in dust control that abandoned mine pits will probably become permanent water bodies. Mount Goldsworthy pit, in 1979, was 122 metres below the water table. The effect of increased water on fauna and flora has been reported in

Jones.[19] Increases have been observed in water dependent species, in feral species and in the periodic visits of water-dependent migratory species. Jones also notes, for example, the demise of saline xerophytes.

Indirect physical effects are related to development associated with the mines. Between 1965 and 1975, ten towns were constructed with an associated population increase, and effects included the introduction of exotic species, responses due to changes in water quality and to recreation activities such as picnicking, trail bike riding, and so on. Over twelve hundred kilometres of railway track were built to transport ore to the coast, with construction effects on soils, drainage and vegetation. In 1975 over sixty thousand cubic metres of water and over two hundred kilowatts of power were consumed each day.[20]

Port Hedland handles shipments of ore. The coastal area has been extensively dredged and a thirteen kilometre entrance channel was developed, which has altered the local hydrology and the plant communities. The Mount Newman Mining Co. is currently studying these effects.

Reconstruction

The aim of reconstruction is to develop a maintenance-free area to assist in dust control and waste dump stabilization and to provide an aesthetically acceptable landscape.

The lake areas have tourist and recreation potential. Suggested revegetation is to a moist gorge woodland. In the waste dumps and mined out areas, it is hoped to establish vegetation similar to surrounding areas (*Triodia* tussock grassland).

Small areas of the man-made landforms become available for rehabilitation relatively early in the life of the mining operation (Mount Whaleback mine has a life of forty to fifty years at the present rate of extraction), and these are being used for revegetation experiments. At Mount Whaleback, land is revegetated at the rate of fourteen hectares per year, with smaller amounts at other mines. Areas at the port and mine are being revegetated for dust control and to improve their aesthetic values. The revegetation steps are similar to those for coal and bauxite operations: first the ground is prepared and then vegetated by means of climax seed mixtures, pioneer seed mixtures or by the use of seed already present in the top soil. Irrigation is limited to the establishment phase only. Research has indicated that fertilizer treatment does not improve the establishment of stable vegetation communities. Areas disturbed by railways are being monitored for regrowth and, to accelerate revegetation, are being deep ripped.

Sand Mining in Eastern Australia

Mineral sand deposits contain a variety of minerals, namely ilmenite, zircon and rutile. Western Australia is the major producer, though New South Wales and Queensland account for nearly 30 per cent of total output.

Mineral sand mining projects and potential mines in eastern Australia are located at a few coastal sites from Wyong in New South Wales to near Gladstone in Queensland. The heavy mineral sand deposits are essentially ancient placers of various concentrations produced by near-shore marine, beach and wind processes which operated in the past. These deposits are now found close to the present shoreline or in the sub-surface buried deeply by water- and wind-deposited sand. Thus, they are identifiable parts of the substratas which support many different coastal ecosystems. The ecosystems evolved and function today under soil conditions of low reserves of available plant nutrients and generally poor water-holding capacity. Today, the climate along this coast is humid with mild winters and hot summers.

Effects of development and production

Sand mining on the beach between tide levels leaves little physical evidence, though some short-term impact is made on various littoral fauna. There has been little beach "skimming" in eastern Australia since the early 1960s, and the viability of the industry has been dependent on the large tonnages processed by dredge and bucket wheel using water or (the now discontinued) "dry" mining. These methods have nominal throughputs of up to several thousand tonnes per hour in the mining/concentrating plant.

Whether mining occurs in foredunes, backdunes, high dunes, dry heath or wet heath, it destroys surface vegetation, original topography and soil profiles which may be some tens of metres thick in the high dunes. Efficiency in handling large tonnages and top soil is very important to success in the reconstruction stage which today is closely integrated with the development/production stage.

Surface impacts created by infrastructure such as jetties, tracks, roads, power lines, pipe lines, buildings and stock piles of mining wastes must be added to the direct effects of mining. In most localities, the facilities established by individual mining companies have set the framework for subsequent land uses. On North Stradbroke Island, Queensland, for example, the state and local government officials have chosen to use the facilities where appropriate. In contrast, officials in New South Wales have directed companies to close mining facilities which could have served the public. In neither state is there much evidence of co-ordinated planning to capitalize on the unavoidable effects of mining; though in New South Wales there is a much greater awareness of and some experience in this.

Reconstruction

There is a considerable literature covering many aspects of reconstruction after heavy mineral sand mining over the past twenty-five years or so which reveals a growing maturity from empirical adaptation of agronomic techniques to the early stages of an "ecological" approach. It is also apparent that not all companies within the sand mining industry have adopted and applied these methods, nor has any one company settled on using general ecological principles with appropriate modifications for local environments.

Much of the continuing argument about reconstruction depends on the various interpretations attached to: "revegetation", "rehabilitation", "reclamation" and "restoration". There is an atmosphere of sterility where this dispute is carried out in the popular press. The problem is not yet resolved even at the botanical level, let alone the ecological. Coaldrake stresses that a stabilizing cover of vegetation can be established and maintained but that "it is ecologically impossible on a sensible time-scale to *restore* completely many of the original plant communities".[21]

Some mineral sand mining companies in different areas of eastern Australia, and in different physical environments within those areas, have demonstrated that good quality reconstruction is possible, provided the job is done by a skilled and co-ordinated team. However, most companies have not been able *consistently* to achieve their goals in the field. The difficulties and costs of material handling, variable terrain and climate, unpredictable weather, and problems of co-ordinated management of a capable mining/rehabilitation team, tend to combine and yield a result less than expected and desired. This is not to deny that there are examples of superior reconstruction which have been a success in the short term and should continue to be so in the longer term.

In very general terms, the most recent techniques involved: returning the varied topography to a quasi-natural state; management of topsoil; preserving nutrients and seed stock; control of water and salinity; and stabilization using a range of sand "traps" such as rapidly growing plants, straw mulching, laying of brush and chemical binding of slopes. Natural regrowth is complemented by direct seeding and planting nursery-raised trees in some situations.

The difficulties of reconstruction in coastal areas cannot be overemphasized, yet too much attention has been given to a simplistic "restoration-versus-stabilization" question at the expense of a more fundamental investigation of appropriate land use and social responsibility for mining on the coast. Such an inquiry would involve consideration of the roles, values and attitudes of the public, of company personnel, of state and local government officers, and of politicians.

Uranium at Ranger, Northern Territory

The general background to the Ranger Uranium Environmental Inquiry is discussed in chapter 9.[22] Uranium mining at Ranger commenced in 1981. Government approval has been given for other projects in the Northern Territory and some are awaiting approval. Much of the information in this section, which is concerned with the *potential* effects of uranium mining on the environment, is derived from the Ranger Environmental Inquiry *Second Report*.

Mary Kathleen, which ceased operations in early 1982, was Australia's only producing uranium mine in 1977–78. After consideration of the Ranger Uranium Environmental Inquiry, the government decided to proceed with the mining of uranium. At 30 June 1978, the Australian Atomic Energy Commission (AAEC) estimated a total *recoverable* uranium reserve (including losses in mining and drilling processes) of between 33.7 and 35.2 million tonnes, depending on an acceptable cost of recovery between US $80 and US $130 per kilogram.

The Ranger area comprises the catchments of the East, South and West Alligator Rivers, together with adjacent vacant land in the Wildman River catchment plus Field and Barron Islands. There are five physiographic regions – the sandstone plateaux, lowlands, flood plains, tidal flats and the southern hills and basins. The climate is humid tropical with a summer rainfall maximum, and because of the rainfall pattern there are large variations in stream flow and the main tributaries dry up towards the end of the dry season. A diverse range of plants and animals are adapted to these conditions. During the environmental fact-finding study for the Ranger Inquiry over 950 plant species were recorded and there were at least 4,500 species of insect; about 25 per cent of all recorded Australian freshwater fish are represented in the area and more than one third of the known Australian bird species have been sighted in the area. The vegetation ranges from woodland to forest and scrub communities. The area has been occupied by Aborigines for at least twenty-five thousand years but in the last one hundred years the Aboriginal population has been decimated. Of the four major mines proposed, Koongarra, Ranger I and Nabarlek will be open cut, and Jabiluka will use underground methods. Extraction is by the use of a sulphuric acid/amine solvent process. Mining results in both physical and chemical problems.

Effects of development and production

Mining results in the construction of pits, tailings dams and waste piles and general disturbance at mill sites. For Ranger, Jabiluka and Nabarlek, the estimated areas to be affected by the proposed development are shown in table 41.

The pits range in design size from 70 metres deep by 180 metres wide at Nabarlek to 175 metres deep by 680 metres wide at Ranger I. Direct physical effects specifically include disturbances to drainage systems along communication lines, following the removal of vegetation and the destruction of fauna habitats. This may result in increased erosion and sedimentation of streams. The construction of settlements, processing plants and the planned intensification of pastoral industries will also affect the area. Similar effects have already been considered in relation to coal, bauxite and iron ore extraction. Uranium mining has the *additional* problems of chemical pollution of water and air, and radiation dangers to living organisms — the people and animals — who inhabit or move into the area.

Water pollution is one problem. Estimates of contaminated water releases during mining in a year of average rainfall are 1,690,000 cubic metres in year two to 2,500,000 cubic metres in year 10. It is difficult to estimate the level of contamination. Table 42 illustrates the likely range of contamination during the mining operations, and shows how this compares to estimated "safe levels". These are determined by the concentration that kills 50 per cent of a selected test species in a given period of time (TLm concentration or median tolerance limits). This is then reduced by a safety (or application) factor to estimate a "safe level". The local hardyhead fish (*Craterocephalus majoriae*) was used with a time of four days. (Another measure is the "fish avoidance level" which is generally around one third of the TLm levels.) *Predicted* levels of con-

Table 41. Predicted environmental disturbance after uranium mining in Alligator Rivers region

	hectares
Open pits	260
Waste Rock	468
Tailings Dam	346
Ponds	350
General	499
Total	1,923

Source: D.R. Davy, "Uranium — overview of the Alligator Rivers Area, Northern Territory, Australia", in *Management of Lands Affected by Mining,* ed. F.A. Rummery and K.M.W. Howes (Division of Land Resources Management, CSIRO, Australia, 1978).

Table 42. Comparison of predicted contaminants during mining and estimated "safe levels" in water

Contaminant (μg/l)	Range of concentrations by location (year ten)	"Safe level" (TLm x Application factor)[a]
Copper	0.5 (MC) – 100 (RP)	2
Lead	0.5 (MC) – 300 (T)	2
Zinc	1.0 (MC) – 80 (T)	1
Uranium	0.1 (MC) – 170 (RP)	130

[a] Values for copper, lead and zinc were obtained using hardyhead fish. The uranium figure was obtained using striped grunter and is lower than that for the hardyhead (212 μg/1).

MC – Magela Creek

T – tailings dams

RP – Retention Pond no. 2

Source: Ranger Uranium Environmental Inquiry, *Second Report* (Canberra: AGPS, 1977)

taminants in the study area do not generally (in an average year) exceed the "safe level".

Table 43 shows the estimated content of major contaminants released from the mine during and after mining compared with the natural water quality. Effects on water quality will vary between different parts of the Magela system – on-site drainage, tributaries and flood plains. Drainage on-site may result in chemical and sediment pollution of local billabongs, although dilution of released water should maintain contamination below "fish avoidance level". It is likely though that occasional adverse effects on fish and other organisms in some areas will be experienced. In tributaries such as Gulungal Creek, lead is the most likely toxic component to exceed its "safe level", together with nutrients such as phosphates. Soil erosion may result from tailings dam walls leading to sedimentation and clay deposition. This may act as a trap for metal contaminants, some of which could later become available to organisms. On the Magela flood plain the release of contaminants would be accompanied by considerable dilution. Almost certainly some cumulative effect would be found in the local vegetation communities.

Table 43. Estimated annual contamination from mine site in year of average rainfall

Contaminant	(Unit)	Water from mine site in natural condition	Total released from mine site during year ten	Total annual release after mining ceases
Copper	(kg)	45	130	71
Lead	(kg)	21	80	57
Zinc	(kg)	30	80	57
Uranium	(kg)	6	180	89
Radium	(kg)	0.003	0.050	0.024

Source: Ranger Uranium Environmental Inquiry, *Second Report* (Canberra: AGPS, 1977).

When mining operations cease, the Magela system could be contaminated for some considerable time. Runoff quantity would be similar after mining to that during mining, but removal of ore stockpile and revegetation of waste rock dump and tailings dam would reduce contaminant concentrations in the runoff. The tailings would continue to produce radon at significant levels for around a hundred thousand years or more. The expected rate of radon release could be reduced if the tailings were to be covered by earth or rock.

There are three main types of air pollution likely to cause problems at Ranger. These are sulphur dioxide (SO_2) emitted from the acid plant; radon and radioactive dust from mine workings and mill, and radioactive yellowcake dust from the mill where dried yellowcake is handled.

In considering the potential for air pollution, it is important to consider the amount and method of dispersion of pollutant, which is affected by the meteorological conditions in the area. Wind data collected at Jabiru since 1971 indicate that north and north-west winds predominate during the wet season and east and south-east winds during the dry. Over the year, east and south-east winds are dominant. The proposed regional centre is thus directly downwind from the mine area. The rate of dispersal depends on turbulence. The range of predicted concentrations of sulphur dioxide in the Ranger area is illustrated in table 44 for several sites (pastoral lease, Mudginberri homes, town centre, southern boundary of special mineral lease).

Table 44. Estimated sulphur dioxide concentrations in the Ranger area

Period		Predicted range of sulphur dioxide (SO_2) concentration ($\mu g/m^3$)	US EPA Standard
Annual average		.005–1.28	80
Annual maximum	24 hour	40–100	365
Short term peaks	3 hour	195–680	1,300
	1 hour	540-1,990	–

Source: Ranger Uranium Environmental Inquiry, *Second Report* (Canberra: AGPS, 1977).

The presence of sulphur dioxide is likely to have *some* effects on the ecology of the area but these will probably be minimal outside the mine site. Over the long term, the sulphur dioxide will cease to be produced when mining and milling operations cease. The long term effects of sulphur dioxide on the Aboriginal rock paintings in the areas is uncertain; preliminary tests exposing paint materials to high humidity and sulphur dioxide did not reveal visible damage.

Radium and radon gas are the most important radioactive contaminants from uranium mining and milling, although uranium is also present. Both are present naturally in the environment at low concentrations. The potential hazard is related mainly to the effect on plants and animals, including man, both directly and indirectly via the food chain. Only very approximate estimates of likely doses can be made prior to production. External radiation has been estimated by the Australian Atomic Energy Commission (AAEC) to range from 1.6 to 20 millirems per hour at the mine face. Exposures at tailing surfaces and in the cabins of haul-trucks may be around 1 to 2 millirems per hour. Ranger Uranium Mines Pty Ltd (Ranger) predicted an annual maximum whole body dose for workers of 5 rems. Overseas miners are reported to receive a maximum of 0.5 rems per year.

Ranger obtained predictions that the amount of radon emission could range from 0.54-0.72 curies per day. The AAEC estimates a "realistic" amount of 0.98 curies per day based on production of three thousand tonnes of yellowcake per year. Yellowcake dust is predicted by Ranger to emit 730 microcuries of uranium per day at a production level of three thousand tonnes per annum. However silica particles may be a greater problem than radioactivity in ore dust. Dissolved radium in water supplies is also a potential problem.

Doses of radon received by people and animals off the site are likely to be in the ranges indicated in table 45. The percentage of the dose limit per annum is also indicated.

Table 45. Predicted maximum radon dose at sites near Ranger

	Site of Mining Lease		Town Centre			
Rate of production of yellowcake (tonnes U_3O_8 p.a.)	Lung dose (m rem) (% annual dose limit)	Bone dose (m rem) (% annual dose limit)	Lung dose (m rem) (% annual dose limit)		Bone dose (m rem) (% annual dose limit)	
	Adult	Adult	Adult	Child	Adult	Child
3,000	254	95	58	66	19	91
	(17%)	(3%)	(4%)	(4%)	(0.6%)	(3%)
6,000	291	166	77	90	36	159
	(19%)	(6%)	(5%)	(6%)	(1%)	(5%)

Source: Ranger Uranium Environmental Inquiry, *Second Report* (Canberra: AGPS, 1977)

Reconstruction proposals

Originally, the Jabiluka deposits were proposed to be mined by surface methods and the predicted effects listed above and reconstruction proposals are based on this expectation. However, the company now intends to extract by underground methods and this clearly changes much of the information given here for the Jabiluka mines.

During the mining operations, it is proposed to investigate the propagation of suitable native species, whilst attempting to minimize erosion. Morley considers that cleared areas will rapidly regenerate.[23] After decommissioning, the area will revert to a portion of Kakadu National Park and will have to meet National Park requirements. In the mill area, buildings will be removed and the area revegetated. The waste dumps will be contoured, landscaped and revegetated and open pits will be allowed to fill with water; the leaching of heavy metals and radioactive elements from pit walls is not considered to be a hazard.

The tailings ponds are a serious problem. The tailings are to be held indefinitely in a high security pond. After initial evaporation of water, tailings will be covered by at least one metre of clay or similar low permeability material, plus two metres of overburden and soil. This should reduce radon emanation by 99.7 per cent to a level of about twice the natural radon background concentration. The downslope rock zone will be revegetated and there will be a further buffer zone. The toe drain will be maintained.

LEGISLATION AND CONTROL OF MINING IMPACTS

Control of Mining in General

In Australia constitutional ownership of minerals resides with the various state and territorial governments and these assume the responsibilities of a landlord.[24] The federal government plays a role in the preparation and implementation of a national mineral policy which should ensure that mineral investment and extraction activities make the best possible contribution to regional and national economic and social needs.[25] It is able to influence and regulate development and productivity by its statutory powers in international trade, customs and excise, taxation and loans, and in the environment.

State governments make mineral policy too, though it is rare for them to make such policy explicit except in very general terms of state development projects and employment opportunities. Few state governments question the belief that all and every mining activity must be "good for the State" and actively vie for projects by offering lower taxes and royalties and direct and indirect subsidies. While state governments have not clarified their policies, they have expended major efforts to "win back" territory from the commonwealth and to dispute ownership of territory and jurisdiction, as witnessed by federal and state legislation and High Court challenges concerning submerged lands.

Historically Australian federal and state governments have not been able to fix on comprehensive and integrated mineral policies and have preferred to remain "flexible" to make the most of "opportunities". This uncertainty, particularly at the federal level, creates planning difficulties and has drawn much criticism from the mining industry as well as the public.

Within each state and territory there are mining laws and regulations which cover the prospecting for and working of minerals from areas of leasehold tenure. Implicit in the provisions of mineral tenures are the notions that: (1) minerals are an important resource in modern society, (2) they are of limited occurrence due to geological factors, and consequently, (3) their continued availability should be assured by affording mining some priority as land use.

Stated in such a bland way few people would argue with these notions. However, with today's rapidly shrinking world and expanding population, the argument that mineral resources occur only in specific places and that mining is only a temporary use of the land is also applicable to other land uses. Thus although mining is an important industry it, too, must compete for land and priority like any other form of land use. The aggressiveness of companies and apparent acceptance of the ultimate priority of mining is most tangible in the Mining Warden's court and public enquiries where conflict resolution is attempted.

Mining legislation and regulations were established at various times for the different states, but mainly towards the end of the nineteenth century, and they related to small and labour intensive operations. Despite amendments, most recently in the 1970s, they are proving inadequate for the larger mechanized operations of the second half of the twentieth century. A review to reveal the interdependence and contradictions of mining and non-mining legislation and regulations, against a background of modern concepts of planning and management, is long overdue. An inquiry like that into planning over mineral working in Great Britain is necessary before action.[26] Such an investigation would cover at least three prerequisites for effective management of any resources: adequate technical knowledge, the regulation (and possible resolution) of conflicts in values, and the development of suitable institutions — both to effect conflict regulation and to make technical knowledge effective.[27]

Control of Mining Impacts

In the current atmosphere of positive encouragement for mining, and ambiguity at the federal and state level towards even basic policies of mineral exploitation, it is hardly surprising that legislation and regulation concerning the biophysical impacts of mining have received little attention. Where authorities have attempted to regulate impacts it has usually been through legislation and regulations *outside* the mining Acts.[28] For example, the Queensland Water Quality Council is given powers to control the discharge of acid mine water under the Clean Waters Act 1971–79 and Clean Waters Regulations 1973–79. In 1976 the potential of the federal Customs Act 1901–73 and Customs (Prohibited Exports) Regulations were realized under the terms of the Environmental Protection (Impact of Proposals) Act 1974, and this effectively prevented additional biophysical impacts of mining mineral sands on Fraser Island.

At least five categories of control by state and federal authorities can be recognized: those to resolve land-use conflicts; those which specify environmental states; those which impose planning or pollution control; those which provide the bases for environmental impact statements; and those which impose excise powers. Where mining legislation and regulations specifically refer to potential biophysical impacts, such as the need for sludge abatement, authorities have established in-house techniques to discharge their responsibilities. However, unless these responsibilities are taken seriously the utility of environmental sections in mining legislation is open to question. Alternatively, there are sufficient powers under existing state and federal mining acts and regulations to allow high quality control of mining impacts given the administrative methods and manpower to carry them out.

The New South Wales Department of Mines has made conservation and environmental protection part of its objectives and in 1973 amended some provisions in the Mining Act and set up mechanisms which amount to voluntary environmental planning. This stands in stark contrast to the Queensland Department of Mines which has reduced its consideration of environmental matters under 1979 statewide regulations and also amended the Mining Act to minimize town planning considerations and involvement of the public in the Mining Warden's Court.[29]

Events in Australia have shown it is very naive to anticipate that any government or public service will continue to display benevolence towards protecting the biophysical environment and will formalize this in legislation. There is very little regular monitoring of the ramifications of existing legislation and regulations and it is so easy for a bureaucracy to say "that's good enough". Many of the advances and the heightened awareness achieved by an active public over the last twenty years can very easily be dissipated by piecemeal and retrograde "adjustments".

Ideally, individuals and groups of individuals in institutions, companies and governments should display a concern (if not a respect) for the biophysical environment which would ensure minimal impact from the mining projects we need. That utopian situation has not yet been achieved and the future appears to be one requiring control measures which are both efficient and effective.

Notes

1. J.B. Sless, "Control of sand drift in beach mining", *Journal of the Soil Conservation Service of N.S.W.* 12 (1956): 164-75.
2. A.B. Hicks and J. Hookey, *Fraser Island Environmental Inquiry. Final Report of the Commission of Inquiry* (Melbourne: Advocate Press, 1976); New South Wales State Pollution Control Commission, *Diamond Hill Inquiry*, Report and Findings of the Environmental Inquiry into a Proposed Extractive Industry at Diamond Hill, Kurrajong, 1979; and R.L. Heathcote, *Australia* (London: Longman, 1975), pp. 124-34.
3. Australian Conservation Foundation, *Conservation and Mining*, Special Publication No. 8, Papers of an Australian Conservation Foundation Symposium (North Blackburn, Victoria: Dominion Press, 1973); F.A. Rummery and K.M.W. Howes, eds., *Management of Lands Affected by Mining* (Division of Land Resource Management, CSIRO, Australia, 1978), pp. 129-40.
4. A.H. Parbo, "Conservation and the Mining Industry", Supplement to Australian Institute of Mining and Metallurgy, *Bulletin*, No. 341 (1971).
5. Conzinc Riotinto of Australia Limited, *Submission and Supplementary Submission to the Industries Assistance Commission on Petroleum and Mining Industries Inquiry*, May and November 1975, Appendix E, p. 1.
6. S.S. Clark, "Australian ecosystems and their modification", in *Australia as Human Setting*, ed. A. Rapoport (Sydney: Angus & Robertson, 1972), pp. 39-58.
7. Australian Bureau of Statistics, *Year Book, Australia*, No. 63 (Canberra: AGPS, 1979), p. 369.
8. R.S. Dick, "A map of the climates of Australia: according to Köppen's principles of definition", *Queensland Geographical Journal* 2, 3 (1972): 33-69. Subsequent references to climate draw on this article.
9. J.C. Hannan, "Coal: rehabilitation of coal mines in the Hunter Valley of New South Wales", in *Management of Lands*, ed. Rummery and Howes, pp. 18-33; J.C. Hannan, "Rehabilitation of mined land in the Hunter Valley Coalfield, NSW", in *Mining Rehabilitation* ed. I. Hore-Lacy (Canberra: AMIC, 1979), pp. 49-58.
10. B.A. Middleton, "Land Use Planning, Bauxite Mining Operation, Weipa", in *Management of Lands*, ed. Rummery and Howes, pp. 140-55; B.A. Middleton, "Rehabilitation of Bauxite-mined lands at Weipa", in *Mining Rehabilitation*, ed. Hore-Lacy, pp. 17-22.
11. K.W. Stewart, J. Leggate and D.F. Bevage, *Some Aspects of Mine Rehabilitation in Northern Australia*, collection of papers presented to a regional symposium, Weipa, July 1974; J. Leggate, *Mine Regeneration Policies and Practices at Weipa: A Further Report*, paper presented at AMIC workshop meeting, Adelaide, October 1976; and R.L. Specht, R.B. Salt and S.T. Reynolds, "Vegetation in the Vicinity of Weipa, North Queensland", *Proceedings of the Royal Society of Queensland* 88 (1977), pp. 17-38.
12. H. Jones, "Iron ore mining – effects and rehabilitation" in *Management of Lands*, ed. Rummery and Howes, pp. 155-59; W. Carr, "Rehabilitation of disturbed areas in the Pilbara, W.A.", in *Mining Rehabilitation*, ed. Hore-Lacy, pp. 23-33.
13. M.W. Pryce, "The Mineralogy of the Pilbara Iron ore and associated rocks with particular reference to their quartz content and its measurement in airborne dust", *Symposium: Control of Dust in the Environment in the Pilbara Iron Ore Industry* (Institute of Engineers, Aust., WA Division, 1975), chapter 9, p. 9.0.
14. Carr, "Rehabilitation in the Pilbara", p. 25.
15. Jones, "Iron ore mining".
16. R.A. Powell and D.B. Sykes, "Dust surveillance for trend analysis versus dynamic sampling directed towards immediate evaluation", in *Symposium: Control of Dust*.
17. J.C. McNulty, "Health effects of inhaled mineral dusts within the working environment", in *Symposium: Control of Dust*.

18. Powell and Sykes, "Dust surveillance".
19. Jones, "Iron ore mining".
20. D.R. Kelly, "Planning and the total environment for iron ore mines and ports", in *Symposium: Control of Dust,* p. 1.2.
21. E. Coaldrake, "Beach Sand Mining" in *Conservation and Mining,* Australian Conservation Foundation, Special Publication No. 8, Papers of an Australian Conservation Foundation Symposium, (North Blackburn, Victoria: Dominion Press, 1973), p. 57.
22. Ranger Uranium Environmental Inquiry, *Second Report* (Canberra: AGPS, 1977).
23. A.W. Morely, "Rehabilitation in vein deposit mining — the Jabiluka uranium project", in *Management of Lands,* ed. Rummery and Howes, pp. 87-89.
24. The commonwealth maintains ownership and control of uranium in the Northern Territory.
25. B.W. Mackenzie, *Mineral Economics,* vols. 1 and 2, Notes prepared for Workshop 121/79, Australian Mineral Foundation, Adelaide, 1979, pp. 7-12.
26. Sir R. Stevens, Chairman, Committee of Inquiry, *Planning Control Over Mineral Working* (London: HMSO, 1976).
27. J.A. Sinden, "The problem of resource management", in *The Natural Resources of Australia,* ed. J.A. Sinden (Sydney: Angus & Robertson in association with ANZAAS, 1972), pp. 3-16.
28. D.F. Fisher, *Environmental Law in Australia: An Introduction* (St Lucia: University of Queensland Press, 1980).
29. New South Wales Department of Mines, *Mining and the Environment,* NSW Parliamentary Paper No. 135 of 1972; and Queensland Co-ordinator General's Department, *Impact Assessment of Development Projects in Queensland* (Brisbane: Govt. Printer, 1979).

9 The Uranium Decision

Hugh Saddler and J.B. Kelly

It was observed in chapter 7 that until recent years mining seldom generated much political controversy In the absence of conflict, administration tends to replace politics, and for this reason there is a notable scarcity of scholarly (or any other) analysis of the politics of mining.[1] Apart from notable exceptions like the Eureka Stockade and the New South Wales coal strike of 1949, mining operations tended to create little community conflict, though in the case of large new discoveries or developments, considerable initial enthusiasm may have been aroused. Most mining operations were regarded as normal, laudable and worthy of encouragement. As the Queenstown–Zeehan area of Tasmania still testifies, not even hideous environmental pollution caused any notable protest in early mining days. Mining was, and to a considerable extent still is, a major component of the dominating Australian development-progress-prosperity ethos. That attitude may still prevail but it no longer enjoys general, uncritical community acceptance; it is a long way in more than one sense from "the peaks of Lyell" to the sandhills of Fraser Island.

Of course at various levels politics has always been associated with mining. Most mining operations require some kind of government policy-making with respect to safety measures and other social conditions affecting the mining workforce. Social reformers over a long period prodded reluctant governments to legislate to improve the bad conditions in the coal mines; presumably such issues could be regarded as very early aspects of the politics of mining.

When compared with its earlier relatively apolitical history, mining in recent years has become highly politicized. In part this has been due to the increasing concern of commonwealth governments with mining activities, whereas formerly state governments were the principal regulators of the mining industry. The period of the Whitlam government, 1972–75, coincided with and partly stimulated what might be termed mining politics. Highly political issues concerning mining were not of course invented, or discovered, by the Whitlam government. Issues concerning mining and conservation, pollution, foreign ownership, royalties, and deficiencies and disparities in state control of mining had been growing in intensity for a considerable time, particularly following the minerals boom of the 1960s and the advent of the Gorton government in 1968. As Susan Bambrick has noted:

> Before the 1972 election, the Federal Government had adopted minor restrictions on foreign investment, had made natural gas subject to an export embargo until proved reserves increased, and was subject to pressure to apply export controls to raise the price of coal to Japan. The state governments were being criticised for their low royalty rates and for their lack of environmental concern.[2]

During the 1970s, however, these issues became increasingly important in political terms. Gary Smith has commented as follows:

> Minerals and energy resources have become a major concern of government policy in Australia largely because of their economic importance and their international significance . . . Domestically it appears that the very success of the minerals boom generated an increased concern in Federal political and bureaucratic circles over the distribution of the financial and environmental costs and benefits of this rapid growth, linked to increasing pressure-group activity around quality of life issues such as environmental protection and planning.[3]

There is considerable scope for major analytical studies of the politics of mining during this period: Gary Smith's analysis is an excellent beginning. This chapter focuses on the politics of uranium mining and serves to illustrate at least some of the vital political issues created by mining in the fundamentally changed circumstances of recent years. A case study of uranium aptly illustrates many aspects of the politics involved in mining; at the same time it must be recognized that it is in many ways a special case. Those responsible for decision-making in the area are confronted with an issue which is inherently more vital, more emotive, more complex, more potentially disruptive than any other mining issue which has arisen before or since. It is obviously impossible to separate the issue of mining and exporting uranium from the wider issues of the nuclear power industry and the possibility of a nuclear holocaust resulting from nuclear war. The industry raises vital issues such as nuclear arms proliferation, risks of contamination of the environment through reactor breakdown, and the immense problems created by the difficulties of safe storage of nuclear fuel residues. The first Ranger Report, though dealing specifically with the question of mining and exporting uranium, was largely concerned with these wider issues.

The history of uranium mining in Australia dates from the second world war. It therefore spans a period during which governments of both major political party groups were in power: the Curtin–Chifley Labor governments, Liberal–Country Party governments 1949–72, the Whitlam Labor government and since 1975 the Fraser Liberal–National Country Party government. All of these governments were concerned in one way or another with policy-making regarding the mining of uranium. This study traces only briefly the developments prior to the advent of the Whitlam government in 1972; because the decision-making process regarding the mining, processing and exporting of uranium on a large scale has coincided with the incumbency of the Whitlam and Fraser governments, the bulk of the analysis deals with this period.

Although the Whitlam government came to power with a series of specific, basic objectives regarding development of minerals and energy, it retained the general Australian objective of promoting and encouraging mining and processing with the aim of increased exports and economic development. In pursuing its specific objectives, this government placed much more stress on a series of six factors than their predecessors: reducing and discouraging foreign ownership of Australian mineral resources; protecting the total environment from damage caused by mining; development of local processing of minerals; securing a fair price for Australian minerals on world markets; ensuring that Australia's own future energy and industrial requirements were met; and protecting the Australian Aborigines against harm caused by inadequate control of mining operations.[4]

These objectives, viewed as a whole, constitute many of the main elements of what is now the politics of mining. Most of them are involved in the decision-making process concerning uranium mining, which renders an account of this industry most suitable for illustrative purposes. The policy process inaugurated by the Whitlam government placed particular stress on environmental protection and Aboriginal land rights. Both factors figured prominently in the uranium decision-making process and receive considerable attention in this study.

This chapter is essentially a case study written within the comparatively new sub-discipline of public policy studies. The aim of illustrating the politics of mining by study of the uranium mining decision required an account of the history of uranium mining, of the development of public attitudes to the problem and the reaction of governments to changing circumstances. Developing this material into an analysis relevant to the study of public policy meant adopting an orientation to the decision-making process influenced by the relevant special approaches and procedures developed in the literature on the study of public policy. The classic analytical framework developed for policy study — policy formation, formulation, implementation and evaluation — was used to guide the research. The chapter is written under these main headings.

The "group model" is a well-known analytical tool in policy studies and proved particularly well suited to analysis of the uranium debate, in which group participation was a notable feature. The trade unions played, and are still playing, a vital role in determining whether uranium will be successfully mined or not, and this role is briefly analysed here. Apart from the unions, the debate has produced other important contributions from proponents and opponents of uranium mining and the use of nuclear power, and their attempt to contribute to the decision-making process is also analyzed.

There is a great need for policy case studies so that common factors in policy making can be located and a body of theory about policy making built up, particularly in Australia where there has been a marked lack of research on policy written from the specialized standpoint of public policy theory. A notable recent exception is the valuable collection of case studies of policy at the state level, edited by Roger Scott. The studies are all theory-oriented and use the group model or theory to interpret the data assembled.[5] From the standpoint of the present study the most immediately relevant of these essays deals with bauxite mining and its impact on jarrah forests in Western Australia.[6] The essay illustrates very effectively an important contemporary aspect of the politics of mining: the clash between a development-minded government intent on encouraging mining apparently irrespective of cost and ecologically-oriented groups and citizens striving to establish the need to protect the environment — specifically the jarrah forests — from the impact of bauxite mining, or at least of expansion of the industry. The study shows a government involved in the politics of mining treating opposition to mining with contempt and making unilateral decisions without any impartial advice. Advice from the bureaucracy involved did not appear to be impartial and that coming from groups supporting mining scarcely could be.

The contrast to the policy-making procedure adopted at the national level regarding uranium mining is marked. Policy formulation in this case was based on two reports from the very able and obviously impartial Ranger Uranium Environmental Inquiry. But both the bauxite and the uranium mining studies indicate that some mining issues are of such a controversial nature that they are highly likely to provoke community division and confrontation caused by conflict between the values of economic development and the values of environmental protection. Important philosophical, social and political attitudes and values are shown to be involved in both issues.

On the basis of this analysis it is suggested that in Westminster-type political systems the most suitable decision-making procedure regarding such issues is investigation and recommendations by a high-calibre royal-commission type inquiry, followed by a non-party vote in parliament. Alternatives such as decision by the government with or without an advisory referendum are considered unsuitable.

Thus, to summarize, the study has two major aims:

1. by analysis of the uranium-mining issue and its background to provide an illustration of the links between politics and mining; and
2. to analyze government decision making in a controversial and politically sensitive area.

POLICY FORMATION: THE YEARS TO 1972

Although the metallic element uranium was isolated and named as long ago as 1789, it remained little more than a laboratory curiosity until world war two. In 1938 it was discovered that, under certain conditions, if a nucleus of the uranium isotope, uranium-235, were bombarded with neutrons it would split (fission) into two roughly equal halves, at the same time releasing two or more further neutrons and a large quantity of energy. The possibility of releasing energy by means of a nuclear chain reaction became apparent.[7] While scientists foresaw the possibility of controlling this energy release for peaceful purposes, the circumstances of war made an uncontrolled release, in the form of the atomic bomb, the first priority. As is well known, the effort to make such a weapon succeeded in 1945.

Almost immediately the nuclear arms race between the USA and the USSR was launched, and the British government decided to build its own nuclear weapons independently of the USA. The USA and the UK wanted large amounts of uranium to make nuclear weapons and Australia was one source of supply. Commonwealth control of uranium mining was initially facilitated by the Atomic Energy (Control of Materials) Act, 1946, which was subsequently replaced by the Defence (Special Undertakings) Act, 1952 and the Atomic Energy Act, 1953. Inducements offered included large cash rewards for discoveries of uranium ore and the high price paid by the combined US/UK purchasing agency. During the 1950s uranium was mined at Radium Hill in South Australia, Rum Jungle and the South Alligator Rivers region in the Northern Territory and Mary Kathleen in Queensland.

However, by the early 1960s stockpiles of nuclear warheads were deemed to be sufficient, and the military demand for uranium collapsed. The Mary Kathleen project was mothballed in 1963 and when Rum Jungle was permanently shut down in 1971 there were no operating uranium projects in Australia. Exports from Rum Jungle had ceased in 1965 and the remaining production was stockpiled by the Australian Atomic Energy Commission (AAEC). Total Australian production up to that date was about 9,100 tonnes of yellowcake (U_3O_8).[8]

This commission was established under the Atomic Energy Act, 1953. At that time many countries were beginning to show an interest in civil nuclear power development, which eventually led in 1966–67 to a surge of commercial orders (about sixty) for nuclear power stations in the USA.[9] In Australia the commonwealth government announced a new export policy for uranium, and a number of companies, doubtless expecting a strong demand for uranium within a few years, began to explore for uranium ore. In 1970 the discoveries were announced of the Ranger, Nabarlek and Koongarra deposits in the Northern Territory. By 1973 a further large deposit at Jabiluka had been announced, as well as deposits at Yeelirrie in Western Australia and Beverley in South Australia. These discoveries were so large that by 1975 Australia had over 20 per cent of the world's reasonably assured low cost resources of uranium outside the centrally planned economies.

During 1972 Peko-Wallsend and EZ Industries (the Ranger partners), Queensland Mines (Nabarlek) and Mary Kathleen Uranium negotiated sales contracts totalling about 9,000 tonnes at the relatively low price which prevailed at the time.[11] During October and November, immediately prior to its electoral defeat, the McMahon government granted export approval for these contracts.

POLICY FORMULATION

The Labor Government, 1972–75

Until that time, policy on uranium and nuclear energy had enjoyed invariable bipartisan support. For example the Labor opposition supported the Atomic Energy Act in 1953 and continued to support the AAEC which had been established by the Act. At its National Conference in 1971, the ALP adopted a policy to "stimulate the growth of nuclear technology, establish uranium enrichment plants and nuclear power stations".[12] On taking office at the end of 1972, however, the new Labor government made two changes to Liberal–Country Party policy on nuclear matters. Firstly, the responsible Minister R.F.X. Connor refused to approved further export contracts for uranium. Secondly, the new government ratified the Treaty on Non-Proliferation of Nuclear Weapons (NPT), an action which the previous government had refused to take. Nuclear weapons proliferation will be discussed further below.

It was almost two years before Connor fully explained the reasons for his policy, in a ministerial statement on 31 October 1974:

> Whilst I have questioned the propriety of the approvals of uranium exports contracts immediately prior to December 1972, I have stated on numerous occasions that this Government will ensure that the commitments under those contracts are met. But the Government has not been prepared to approve further export contracts because of the unsatisfactory nature of the market.[13]

The Minister went on to explain his plans for government participation, in joint venture with the private companies which had made the discoveries, in the exploitation of uranium reserves in the Northern Territory.

By "the unsatisfactory nature of the market", Connor meant the low price for uranium which prevailed during 1972 and 1973. He felt confident that the price would rise and was vindicated by subsequent events.

The price passed through a historic minimum (in real terms) in 1973 and by later 1974 a rapid rise was well under way. The price eventually peaked in early 1976 at a level more than six times higher (in real terms) than the 1973 price.[14]

The desire to increase the participation of Australian public instrumentalities in the exploitation of the country's mineral wealth was an important part of the Labor government's overall policy. Though radical by Australian standards, public participation was by the 1970s well on the way to becoming the norm in most other mineral producing and exporting countries. In the oil industry, not only OPEC countries, but also France, Norway, Canada and the UK had established significant public equity. Public corporations had a major role in the uranium industries of four of the six largest producers, Canada, France, Niger and Gabon.

In Australia, attempts to implement public participation were bitterly opposed by the conservative parties and the mining industry. Legislation to give effect to the policy was blocked by the Senate and only passed after the double dissolution election of May 1974. The failure of attempts to circumvent opposition in obtaining loans to finance the new policy was a major cause of the eventual downfall of the Labor government in November 1975. Given the vehemence of the opposition and the complexity of the task, it is perhaps not surprising that it took Connor some time to formulate his plans for uranium. For the same reasons his secretive and abrasive manner is understandable, though, equally, there can be little doubt that these personal characteristics increased the difficulties experienced by him and the government in bringing the policies to fruition.

The powers given it by the Atomic Energy Act made the government's administrative and legislative difficulties in securing public participation less formidable for uranium than for other minerals. The proposal announced in October 1974 was for the Australian Atomic Energy Commission to be the vehicle for public participation in the Ranger project, which was the most advanced of all the new uranium prospects and was also, at the time, thought to be the largest (subsequent exploration indicated that the Jabiluka deposit is probably larger). The commission was to be the sole agent for future export sales, after commitments under the contracts negotiated in 1972 had been met, and was also to be the exclusive holder of rights to explore for uranium in the Northern Territory. The policy announcement also emphasized that the government's policy was "to treat and fabricate Australia's minerals in Australia to the greatest practicable extent". Accordingly, it would be giving "full consideration of the technology to be used in a uranium enrichment plant to be built in Australia".[15]

Simultaneously with these developments, other policies, relating to Aboriginal land rights and environmental protection, were being implemented. Within a few days of taking office, the Labor government froze all applications for mining and mineral exploration on Aboriginal reserves. Queensland Mining was directly affected by this decision because its Nabarlek deposit is within the Arnhem Land reserve. Shortly afterwards a judicial commission under Mr Justice Woodward was appointed to advise how (not if) land rights should be granted in the Northern Territory. The Aboriginal Land Rights Commission brought down two reports in 1973 and 1974. Legislation giving effect to the commission's recommendations had been prepared but had not been passed when the Whitlam government fell in 1975. When, eventually, the Aboriginal Land Rights (Northern Territory) Act was passed in December 1976, its consequences included making the sites of the Ranger and Koongarra deposits subject to a land claim. An additional and most important element was thus formally introduced into the policy-making process.

The effect on that policy-making process of Labor's environmental concerns occurred more quickly. The Environment Protection (Impact of Proposals) Act 1974 laid down a formal procedure for environmental impact assessment and decision making for proposals over which the commonwealth government has constitutional jurisdiction. The uranium prospects in the Alligator Rivers region were subject to the Act both by virtue of their export-orientation and because of their location in the Northern Territory. The proponents of a development subject to the Act were (and still are, since at the time of writing the Act has been neither amended nor repealed) required to submit a draft environmental impact statement to the Minister of Environment and make copies available to the public. The Minister may, at his or her discretion, order that a public inquiry into the proposals be held.

During 1974 and 1975 opposition to the nuclear power industry and to the mining and export of uranium to supply that industry began to develop within the community and among rank and file members of the ALP. It would be quite mistaken to view this opposition as an attempt by outside groups to infiltrate and influence the ALP. From the time of its upsurge, the environmental movement had looked to the ALP, being the party of reform, as the political party most likely to respond favourably on environmental issues. The record in office of the Labor government had, on the whole, vindicated such expectations, and it was therefore only natural that opponents of uranium mining should seek to promulgate their views within the ALP.

It would be equally mistaken to regard concern about nuclear power as an essentially alien import, brought in from the USA and Europe. In the early 1970s, the great majority of Australians knew almost nothing about nuclear power and much of what was known emanated directly from the AAEC (acting, it should be said, under the section of the Atomic Energy Act which requires the commission "to collect and distribute information relating to uranium and atomic energy" S.17(1) (k)). As information from other sources gradually became available it began to appear that nuclear energy was not necessarily the safe, clean energy its proponents liked to portray. At a somewhat later date, in November 1976, Bill Hayden, at the time an opposition frontbencher, spoke for many people both inside and outside the ALP when he said: "I once felt and expressed the view that we should mine and export uranium. . . . My position has changed in the face of what I regard as more profound and worrying information which is submitted by this report" (the *First Report* of the Ranger Uranium Environmental Inquiry — see below).[16]

The debate over French atmospheric testing of nuclear weapons at Mururoa Atoll in the early 1970s and the vigorous opposition of the Labor government in 1973 and 1974 to the testing were also important in raising public awareness of and concern about nuclear matters.

The results of these processes first came to general public attention at the biennial national conference of the ALP in February 1975. A number of delegates expressed opposition to uranium mining, though Connor secured a comfortable majority in support of his position. In June of that year, however, three state branches of the ALP passed motions which opposed the Connor policy. The Victorian branch called for an outright ban on uranium exports. New South Wales and South Australia called for a delay in mining and export until independent studies were produced to show that it would be safe to do so.

By this time the process of establishing a public inquiry into the Ranger project under the Environment Protection (Impact of Proposals) Act had been launched. The public announcement of the appointment of the presiding commissioner, Mr Justice Fox, and two other commissioners was made in July 1975. When the matter was first discussed a few months previously it had been envisaged that the inquiry would be a much smaller and more low key affair, conducted by hearings commissioners, who were public servants within the Department of the Environment. This was the format used for the inquiry into sand mining on Fraser Island, which was established at that time. It seems probable that tensions within the Labor Party were an important influence on the decisions to hold such a major inquiry (Mr Justice Fox was at the time of his appointment to the inquiry the senior judge of the ACT Supreme Court). Since a decision by the government on exports of uranium from new mines was automatically postponed until after the inquiry had reported, further development of these projects was again delayed.

However, one uranium mine was opened, or rather reopened, during 1975. This was Mary Kathleen, which had contracted in 1970 and 1971 for the supply of uranium to companies in West Germany, the USA and Japan, starting in 1975. The need for capital to reopen the mine necessitated a financial restructuring of the company holding the lease, Mary Kathleen Uranium, in 1974, and the federal government took the opportunity to underwrite the new issues, through the agency of the

Atomic Energy Commission. As expected, public interest in the issue was negligible, with the result that the original parent company, Conzinc Riotinto of Australia, held 51 per cent of the new Mary Kathleen Uranium and the government 41.6 per cent.[17] Public judgement was vindicated by the losses which the company made for several years after the reopening.

A Vital Step in Policy Formulation: the Ranger Uranium Environmental Inquiry

As already noted, the inquiry was officially established in July 1975. It started taking evidence in September and produced two reports, the first of which was released in October 1976 and the second in May 1977. The first report dealt with what have been termed the generic issues of nuclear power, that is to say those issues which are raised by Australia's involvement in the nuclear fuel cycle but do not necessarily affect Australia directly, including likely demand for uranium and the economic benefits from its export, reactor safety, radioactive waste disposal and nuclear weapons proliferation. The second report dealt with consequences of the Ranger project for the natural and social environments of the Alligator Rivers region.

Many of the more than three hundred witnesses who gave evidence were members of the general public appearing in their own time and at their own expense. For the most part these witnesses were opposed to uranium mining and their evidence concerned the generic issues.[18] Despite their general lack of formal technical expertise, much of the evidence was of a very high quality, and many of the witnesses were able to participate in the give and take of cross examination on fully equal terms with the formally qualified experts. On some topics, evidence given by supposedly expert witnesses was manifestly inferior to that from members of the public. That people without technical expertise are able to understand technical issues in sufficient depth to present informative, technically accurate evidence should not be surprising. After all, just such an assumption underlies the appointment as commissioners of three people with expertise in only some of the many fields covered in evidence.

Among proponents of mining, most evidence on the generic issues was given by the AAEC. Being a partner in the Ranger venture, the AAEC was a proponent of the development, but it was also (and still is) the main source of technical expertise on nuclear science and technology in Australia and of advice to government on these matters. The situation was not a novel one for the AAEC; it had performed these at least potentially conflicting roles from the time of its establishment. Moreover the strong personality and political skills of its chairman from 1956 to 1972,

Sir Philip Baxter, had effectively enabled the AAEC to make nuclear policy.[19] Moreover, the future corporate morale of the AAEC depended on its growing involvement in uranium mining. The termination of the Jervis Bay nuclear reactor project in 1972 had left the commission floundering without a major focus of activity;[20] Mr Connor's policies offered the prospect of a new and expanding corporate *raison d'être*. The new government at the end of 1975 was opposed to public participation in the uranium industry and in February 1976 instructed the commission to dispose of its equity in Mary Kathleen Uranium.[21] In fact, however, the consistent unprofitability of the Mary Kathleen operation meant that no buyer could be found.

The influence of the AAEC on the Ranger Inquiry was by no means as dominant as might have been expected. On no topic did the commission offer the only evidence of substance; during the course of the inquiry it became obvious that in the Australian community there were a large number of people able to make informed and constructive comments on nuclear policy. By its powers under the Impact of Proposals Act, the inquiry was also able to base its findings on documents submitted as exhibits, and could thus draw on a vast body of international literature. On some topics the AAEC itself possessed no independent expertise and was entirely dependent on the same literature. For example, the important question of the relative economics of nuclear as against other methods of electricity generation was examined only by members of the public. The AAEC stated, towards the end of the inquiry, that they had no economists on their staff at that time.

If properly conducted, a public inquiry can contribute greatly to the process of developing public policy by taking a fresh and independent look at contentious issues. The Ranger Inquiry was able to pursue its own investigations uninhibited by constraints arising from current or previous official policies. The commissioners, their advisers and staff all came from outside the bureaucratic/business decision-making network that had been involved in the issue up to that time. The broad terms of reference, together with the liberal though firm interpretation by the commissioners, gave witnesses scope to raise important matters which might, on a narrow interpretation, have seemed irrelevant to the main concerns of the inquiry. It was significant that on some important questions the inquiry reached conclusions which were diametrically opposed to the conventional wisdom in official circles in Australia at the time, but were subsequently corroborated by official sources outside Australia.[22]

Although the Environment Protection (Impact of Proposals) Act gives a public inquiry under the Act independence from the government of the day in the conduct of its hearings and research, an inquiry requires funds from the government to meet the expenses of staff salaries, accommodation, and so on. Whether and to what extent a commission appointed under this Act or under the Royal Commissions Act could continue to carry out its functions if deprived of funds by the government is an interesting and subtle legal problem, but one of some practical significance. Early in 1976 the new government, by cutting off funds, prematurely terminated a number of important commissions of inquiry, such as the Royal Commission on Petroleum. Press reports in January of that year claimed that the prime minister wished to end the Ranger Inquiry in a similar manner. However, the presiding commissioner, Mr Justice Fox, issued a formal statement which drew attention to the statutory independence of the inquiry and subsequently the inquiry was allowed to run its full course without interruption to its funding.

Because of its independence and its wide terms of reference the Ranger Inquiry was able to examine all aspects of uranium mining in the Alligator Rivers in a holistic manner, recommending an integrated decision scheme which made conscious trade-offs in important areas of conflicts over land use in the Alligator Rivers area, which had been simmering since the mid-1960s, well before the uranium discoveries were made. While the recommendations of the inquiry may not have fully satisfied any of the conflicting parties, the Aboriginal, national park, mining, pastoral and tourism interests, they were far more satisfactory than the *de facto* situation in 1975.[23]

One important and distinctive feature of the Ranger Inquiry was the influence of its general approach, as distinct from its recommendations, on the subsequent evolution of uranium policy. It seems to be widely believed that the task of a public inquiry is to make definite recommendations for or against a proposal. The fatuous comparison between the Ranger Inquiry and the Windscale Inquiry into nuclear fuel reprocessing in Britain, made by three engineers closely associated with the Australian uranium mining industry, that "It is refreshing to note that the report of the Windscale Inquiry concludes with definite recommendation [*sic*]" is typical.[24] (Would they have been equally refreshed if the inquiry had recommended against proceeding with the nuclear fuel reprocessing plant?)

In contrast, the Ranger commissioners did not see the making of recommendations as their sole objective. They did not reach unequivocal conclusions on every issue raised by a comprehensive examination of the nuclear fuel cycle. Instead, they indicated the directions of the various arguments put forward and left the reader to reach his or her own conclusions. There can be no doubt that the Australian public is better informed about uranium as a result of the publication of the two Ranger Inquiry reports. The public education function, which does not require that recommendations be made, should be seen as an important task of public inquiries on contentious topics.

With respect to the link between nuclear energy, uranium export and the proliferation of nuclear weapons, the Ranger Inquiry had an additional reason for not making unequivocal recommendations. Arguments were advanced as to why immediate export might discourage proliferation and also why a few year's moratorium on export might achieve that objective. It was further argued that matters "beyond the ambit of the Inquiry" were involved and that the choice between these two policies should be made on the basis that "Australia should take the course which is deemed to be the most effective and most practical in order to bring a favourable response from other states in relation to the proliferation problem".[25]

If the holding of a public inquiry is to be considered worthwhile and if its deliberations are to influence the policy-making process then the inquiry must point towards decisions that are both fair and seen to be fair. (It may of course happen that governments do not move in the direction to which an inquiry points.) The Ranger Inquiry took scrupulous care to be fair and to be seen to be fair to and by all shades of interested opinion. As a result, the inquiry's reports seem to have been widely accepted in Australia and also abroad as fair and reasonable documents, a fact of great significance when the government came to develop policy.

The debate over uranium and nuclear power has been characterized not only by strongly held views but by a conviction, particularly among opponents of nuclear development, that important ethical issues are involved. A number of witnesses argued that the question of whether or not to export uranium was, in essence, a question of the values which should prevail in Australian society. They suggested that, if Australia and the world were to have a long-term harmonious future, society must turn away from its present dominant values, which they saw as selfish, acquisitive and exploitive. The role of technology in modern industrial society was seen as the central element in its overall character, and the changes which they were advocating would require a fundamental change in the nature and function of technology. Nuclear power, they argued, was a paradigm of modern technology, and its renunciation would be an essential element of a change toward environmentally harmonious life styles.

The commissioners recognized that values were involved in making judgements about the evidence. On several occasions during hearings the presiding commissioner indicated his views in the course of a dialogue with witnesses. The position adopted was summarized in the *First Report*: "While the conclusions we arrived at must of necessity be our own, we have, where appropriate, endeavoured to apply the standards and values generally accepted in our society, as we understand them. To do otherwise would be to express purely personal opinions, and this would be both wrong and unhelpful."[26]

It is clear that the commissioners recognized the political function of the inquiry and the value-laden nature of some of the most important findings and recommendations. However, the assumption that there exists a set of "values and standards generally accepted in our society" raises other questions. Some critics would argue that no such set exists; others would hold that there are such "values and standards" which, however, belong not to society as a whole but to a dominant group or groups within society, possessing the power to impose their views on the rest. Such fundamentalist critics would also query the assumption that a group of individuals, appointed as commissioners, could completely put aside their own values and substitute for them the prevailing values of society or of "the ordinary man". Unfortunately, such questions go well beyond the limited scope of this chapter, and must therefore be left undiscussed here.

The enactment by the commonwealth government of Aboriginal land rights legislation for the Northern Territory has already been mentioned. Section (11) of the Act gave the Ranger Inquiry power to make a determination concerning Aboriginal ownership of the land in the Alligator Rivers region. An additional series of hearings was therefore held, after the *First Report* were included in the *Second Report*, where they were linked to other recommendations about land use referred to above.

Policy Formulation by the Fraser Government

On a number of occasions after taking office at the end of 1975 the Liberal–National Country Party government of Malcolm Fraser stated that a decision on whether to approve the mining and export of uranium would not be made until after the Ranger Uranium Inquiry had released its final report. Nevertheless, both supporters and opponents of the government had little doubt that the government favoured export. In particular Doug Anthony, who at the time was Minister for National Resources and Trade, publicly indicated that he strongly favoured uranium export and nuclear power development.

Few were surprised, therefore, when the government announced in August 1977, three months after the publication of the second and final report of the Ranger Inquiry, that it would approve mining and export of uranium. What was of more interest about the announcement was the effort which had obviously been devoted to justifying the decision and to articulating formal administrative arrangements for uranium development.

On the first matter, the government's position had been greatly strengthened in April 1977 by the announcement from the USA's new Carter administration of a policy to prevent the proliferation of nuclear weapons. Among other measures, the policy sought to discourage spent fuel reprocessing (which, because it involves separating out plutonium, provides a route to nuclear weapons capability), by providing "adequate and timely supplies of nuclear fuels", that is, uranium, to countries without their own supplies. It was clearly indicated that Australia could assist the USA in this objective.[27] In May 1977 the prime minister made a statement on the non-proliferation policy which his government would pursue and this was amplified in the August statement of approval for uranium mining. In addition, the government took the unusual step of appointing Mr Justice Fox, once the Ranger Inquiry had been completed, as Ambassador-at-Large for nuclear non-proliferation matters.

On paper, the government policy articulated strong opposition to nuclear weapons proliferation and a clear interest to try to prevent it. In practice, since 1977, however, the policy has been progressively weakened. As Indyk has argued, this has occurred because the non-proliferation objectives have turned out to conflict with the commercial objective of selling uranium.[28] Moreover the Carter policy has been effectively abandoned by the USA in face of the combined opposition of most other countries which supply and use nuclear energy technology.

The United States policy had a major influence on the government's decision to emphasize the non-proliferation aspects of its uranium mining policy, but there can be little doubt that the importance which the Ranger Inquiry attached to non-proliferation was also influential.

This influence of the Ranger Inquiry's report can also be deduced from the way in which the government dealt with many other issues and from the emphasis on how closely the recommendations of the inquiry were being followed. Among other matters, the government announced that it would grant title over the land in the Alligator Rivers area, including the Ranger mine site, to the Aboriginal people, establish a national park over much of the area and set up the complex machinery for environmental protection. However, a recommendation that development of the various prospective mines in the regions be developed in sequence controlled by the government was rejected.

Considerable care (and presumably expense) was devoted to the presentation of the government's policy. This took the form of six ministerial statements (by the Prime Minister and the Ministers for National Resources, Foreign Affairs, Environment, Aboriginal Affairs and Health), which were delivered in parliament and simultaneously released in printed form, accompanied by a packet of supporting documents.[29] The care lavished on presenting the policy indicates a recognition that the Australian community was deeply divided on the issue of uranium mining. So too does the attention which the Prime Minister in particular devoted to addressing specific matters raised by the opponents of uranium mining, such as the problem of radioactive waste disposal and the relevance of nuclear energy to meeting world energy needs.[30]

Perhaps most surprising of all was the government's decision to retain its participation in the Ranger project, as negotiated by the previous Labor administration. As Mr Fraser pointed out in his statement ". . .

the Government is most conscious that the Memorandum of Understanding between the Commonwealth and the Ranger partners . . . would not have been the Government's preferred approach to mineral development". Making an obvious reference to ALP policy (see below), he then went on: "We believe, as a matter of principle, that the repudiation by our Government of contracts entered into by a previous Government would be quite wrong".[31]

Another consideration which may have influenced this aspect of the government's policy was the wish to avoid the lengthy delay to the start of the Ranger project which would probably have occurred had the government decided to sell its equity. Subsequent events support this supposition. In 1979, when construction at Ranger was about to start, the government announced that it wished to dispose of its equity in the Ranger project.[32] The successful tenderer was a new public company, backed by Peko-Wallsend and EZ, called Energy Resources of Australia (ERA). Late in 1980, during the stock market boom, ERA was floated with striking success.

POLICY IMPLEMENTATION

Implementation of the government's uranium policy was a lengthy procedure. The machinery for environmental protection required the enactment by the commonwealth parliament of six separate pieces of legislation, which took until June 1978.[33] One of these Acts established the Office of the Supervising Scientist, a Co-ordinating Committee and the Alligator Rivers Region Research Institute to study the effects of uranium mining on the environment of the region, to develop standards and procedures for environmental protection and to supervise the implementation of environmental requirements applying to uranium mining activities. Another of the Acts established a national park in the region under the management of the Australian National Parks and Wildlife Service.

A third Act granted title over land in the Alligator Rivers region to the traditional Aboriginal owners, as recommended by the Ranger Inquiry. This meant that the mining companies were required, under the Aboriginal Land Rights Act, to negotiate an agreement concerning the conditions under which mining should proceed with the Northern Land Council, which represents the traditional owners. The agreement for the Ranger project was signed in November 1978 and subsequently one was signed for the Nabarlek project. Considerable pressure was placed on the Northern Land Council and the traditional owners to give their assent to the Ranger project. The role of the commonwealth government was particularly noteworthy. Far from acting as an independent arbiter, neutral as between Aboriginal and mining interests, the government, which at the time held a 50 per cent equity in the project, actually conducted the negotiations on behalf of the mining companies. Furthermore, Aboriginal communities normally employ commercial decision-making procedures following independent and lengthy discussions. The haste with which the Ranger agreement was concluded made it impossible to follow such a procedure. The character of the agreement that resulted is well described by Dr H.C. Coombs: "The Government and the mining company have therefore signatures as required by the Land Rights Act to validate an agreement. But whether it was agreement freely entered into by the Aboriginal parties, an agreement to which they feel honourably committed, is quite another matter."[34]

Once an agreement with the traditional owners had been secured, the government was able to issue mining leases for Ranger and Nabarlek, which it did early in 1979. Contrary to the strong recommendation of the Ranger Inquiry, the Ranger lease was issued under the Atomic Energy Act, rather than the Northern Territory Mining Act.[35] The environmental requirements applying to the two projects take the form of conditions attached to the leases. Work at the two mine sites started shortly thereafter, at the beginning of the 1979 dry season.

Concurrently with these events, which specifically concerned the uranium projects in the Alligator Rivers region, a number of other pieces of the overall policy were being put in place. The Uranium Export Office was established in 1978, to set conditions for and monitor uranium export contracts. A large part of the Department of Trade and Resources also has responsibility for regulating uranium development and export.

Under the government's policy, approval to export uranium depended on the conclusion of a safeguards agreement with the importing country. The Department of Foreign Affairs was heavily involved in these negotiations and agreements were concluded with a number of potential customer countries during 1979 and 1980.

The other general commonwealth control over uranium development is via the foreign equity limitations applying to individual projects, which are administered by the Treasury. As originally enunciated, the foreign investment guidelines required 75 per cent Australian equity in uranium projects (compared with 50 per cent for other classes of investment). However a significant watering down of these guidelines (termed flexibility) occurred in 1979 when it was decided to approve the Yeelirrie project in Western Australia, in which Esso and West German enterprise Urangesellschaft hold a combined interest greater than 25 per cent.[36] Esso later withdrew from this project in May 1982.

During 1980 preliminary development work started on several smaller uranium projects in South Australia and Queensland. From 1977 Pancontinental Mining was involved in negotiation and planning for development of its huge Jabiluka project in the Alligator Rivers region, and reached an agreement with the Northern Lands Council in 1982. One factor inhibiting the early development of any of these projects is the current low level of international demand for uranium. While reports indicate that the Ranger and Nabarlek projects have succeeded in obtaining contracts, mainly from Japan, covering most of their output, it does not follow that other potential producers will be equally successful.

Another government action late in 1978 was the appointment of twelve wise men and women to constitute the Uranium Advisory Council. The appointment of such a body was recommended by the Ranger Inquiry, "to advise the government . . . with regard to the export and use of Australian uranium, having in mind in particular the hazards, dangers and problems of and associated with the production of nuclear energy".[37]

Not surprisingly, the cost of all this government activity is quite considerable. Estimates for the 1980–81 financial year included $6.9 million for environmental protection in the Alligator Rivers regions (including the operation of the Office of the Supervising Scientist), $10.0 million for construction of infrastructure at Jabiru (the town to service the Alligator Rivers uranium projects) and $0.9 million for the Uranium Export Office, the Uranium Advisory Council and other regulatory activities. The funds for environmental protection represent some 6.6 per cent of total commonwealth expenditure on environmental protection budgeted for 1980–81.[38]

These estimates do not include the costs of administrative activities relating to uranium within the various departments involved, which include Trade and Resources, Foreign Affairs, National Development and Energy, Treasury, Prime Minister and Cabinet, Home Affairs and Environment, and Health. However they do include reimbursement of environmental and administrative expenses incurred by the Northern Territory government on behalf of the commonwealth. With the passage of the Northern Territory (Self-Government) Act 1978 a number of legislative and administrative responsibilities relating to the Northern Territory were passed from the commonwealth to the territory. Although the commonwealth retains responsibility for uranium developments in the Northern Territory, the move towards statehood for the territory has tended to complicate further the administrative arrangements for the government's uranium policy.

THE CONTINUING DEBATE

The government's decision to proceed with uranium mining was not supported by the Australian community as a whole. The announcement of the decision in August 1977 did not mean the end of opposition to uranium mining.

The continuing opposition comes from a variety of sources. One of the most important is the Australian Labor Party, which at its biennial National Conference held in July 1977 formally adopted a policy of opposition to uranium mining and export. The new policy committed the party when in government to impose a moratorium on uranium mining in Australia and to repudiate commitments regarding mining or export made by a non-Labor government. The wording of the policy, particularly the adoption of a moratorium, relied heavily on the findings and recommendations of the Ranger Inquiry's *First Report*.

The ALP made its opposition to uranium mining a major issue in its general election campaign in December 1977. The then Premier of South Australia, Don Dunstan, was featured in television advertisements in which he made it plain that, having previously favoured uranium mining, he had changed his mind as a result of material he had read, which included the reports of the Ranger Inquiry. Dunstan had been the mover of the resolution for the new uranium policy at the National Conference. Interestingly, the government parties said very little about uranium. The heavy defeat which Labor suffered in the election led mining interests and some members of the Liberal and National Country Parties to claim that the electorate had decisively endorsed the government's uranium policy. However, it was quickly pointed out that the total primary votes for the ALP and the Australian Democrats, which also opposed uranium mining, virtually equalled the primary votes cast for the government parties and additional small groups supporting uranium mining.

No move to change the ALP's policy on uranium was made at the 1979 National Conference. Although some of the leadership may have wished to make a change it was fairly obvious that the 1977 policy was strongly supported by rank and file party members. In fact, the policy was made more comprehensive by the insertion of an additional clause, which called for a prohibition of the establishment of other stages of the nuclear fuel cycle in Australia, including nuclear power stations.

Many state ALP politicians in states with operating or prospective uranium mines have tended to regard the party's policy on uranium as an electoral liability. This feeling was reinforced by the outcome of the South Australian state election in September 1979. The conservative parties claimed that the policies of the state Labor government had stifled economic development and that the giant Roxby Downs copper/uranium prospect would, if it were allowed to be exploited, bring economic salvation. It is of course difficult to say how much voters were swayed by these particular arguments, but the result of the election was a heavy and unexpected defeat for Labor.

At the federal general election in October 1980 none of the major parties placed much emphasis on their uranium policies. Among leading ALP figures, only Tom Uren, who had for long led the opposition to uranium mining within the party, spoke about uranium.

Another important source of opposition to the government's uranium policy has been the trade union movement. Of course there is a special relationship between the trade unions and the ALP and some of the most articulate opponents of uranium mining within the ALP were from the union wing of the party. However, the trade union movement itself has distinctly different functions and concerns from the ALP and has its own procedures to determine its own policies, which often differ from those of the ALP.

Trade union involvement in the uranium issue surfaced dramatically in May 1976. The Australian Railways Union (ARU) placed a ban on the movement of freight to and from Mary Kathleen. A railwayman at Townsville followed this policy and was promptly suspended, whereupon the ARU held a twenty-four hour national stoppage. The issue was temporarily resolved through the intervention of the ACTU, on the basis that all freight except uranium exports should be handled until the reports of the Ranger Inquiry were released. After that time the union movement as a whole would determine a policy on uranium.

This happened at the biennial congress of the ACTU held in September 1977, shortly after the government had announced its policy. After a lengthy debate the conference resolved that the government be given two months to hold a referendum on the mining and export of uranium and that, if it refused, the union movement should place a total ban on uranium mining and export, subject to endorsement by the rank and file union membership.

The role of the then president of the ACTU in formulating this motion was important. Bob Hawke had made it clear that he personally favoured uranium mining and that his main concern was not the uranium issue itself but the importance of avoiding a disastrous split in the union movement. The policy achieved the latter objective, but had few other virtues. The Prime Minister had already stated bluntly that the government would not hold a referendum. Moreover there was no clear procedure for obtaining rank and file endorsement and some big unions refused to participate at all. Finally, in February 1978, the ACTU convened a special national conference to consider the uranium policy, at which it was decided that uranium export contracts existing at that time would be fulfilled but that further mining should be banned until the union movement was satisfied that a solution had been found for all the hazards associated with nuclear energy. This policy was re-endorsed at the 1979 congress.

In 1980 Hawke was succeeded as ACTU president by Cliff Dolan, who had been one of the leading opponents of uranium mining within the trade union movement. He has admitted that the ACTU policy has so far had little effect on construction of the Ranger and Nabarlek projects.[39] ACTU policy is not binding on affiliated unions and most of the unions whose members are involved in the construction activities have been unwilling or unable to prevent the work being done.

It is the important place of the ALP and the trade union movement within Australian society that makes their opposition to uranium mining significant. However, these organizations have followed rather than led the opposition within the wider community. The organizational and intellectual cutting edge of the opposition has been provided by an array of environmental pressure groups, including the Movement Against Uranium Mining, the Campaign Against Nuclear Power and Friends of the Earth. These organizations are for the most part rather loosely structured and have relied mainly on voluntary labour with a few very poorly paid employees. They have nevertheless been able to mobilize wide community support, as indicated by the number participating in public rallies, particularly during 1977 and 1978. Opposition to uranium mining has spread to some rather unexpected places; for example, the correspondence columns of *Engineers Australia,* the magazine of the Institution of Engineers, Australia, have revealed a lively debate among members of the generally conservative engineering profession.

The pro-uranium arguments have for the most part been made by the mining companies, the AAEC and a few academics in nuclear science and engineering. From 1974 to 1978 most of the mining companies with uranium interests formed an organization called the Australian Uranium Producers' Forum. In 1977 the forum launched an extensive (and expensive) advertising campaign in the mass media. This was apparently judged to have been a failure and the forum was replaced by the Uranium Information Centre, a much more low-key organization which has produced and distributed information for schools, media kits and similar material. Since 1977 the pro-uranium argument has of course enjoyed the support of the massive resources of the commonwealth government.

Although opposition to uranium mining appeared to diminish during 1980, it would be a mistake to assume that community debate is about to end. During that year the possibility of establishing a uranium enrichment plant and nuclear power station in Australia began to be widely canvassed. International experience indicates that plans for power stations in particular can be expected to generate far wider and more intense opposition than uranium mining. As Saddler has previously argued, it is likely that any definite moves to build nuclear power stations will greatly increase the opposition to all related activities, including uranium mining.[40]

Despite the lengthy decision-making process which has already occurred, it would be a mistake to assume that the uranium question is settled. Much of the opposition to uranium mining springs from deeply held convictions about the problems of modern industrial societies and the way in which Australian society should evolve in the years ahead. These opinions and values are not likely to be changed by assurances from technical experts as to the safety of and financial benefits to be derived from the mining and use of uranium.

EVALUATION OF THE URANIUM DECISION-MAKING PROCESS

Evaluation of the quality of the uranium decision — wise, foolish, justified, misguided, dangerous — is not feasible in the context of this study. The whole problem of nuclear power is enormously difficult to evaluate in any precise, quantitative, "scientific" fashion. The Ranger Inquiry, after about a year and a half of intensive and wide-ranging investigation by highly-trained analysts, recorded no findings which could be interpreted as a definite recommendation for or against either uranium mining or nuclear power in general. It is, however, possible to evaluate the decision in terms of the appropriateness of the decision-making process.

How decisions are made is obviously a vital aspect of the process of evaluation — one which seems to be strangely neglected in the public policy literature. For example a referendum on uranium mining could have reversed the decision that was made by the Australian government or decided that there should be an indefinite moratorium on mining. (The Swedish electorate recorded a majority referendum vote in 1980 in favour of phasing out of the nuclear power industry after twenty-five years.)

In Australia's partial "Westminster" system of government, at least five decision-making options, set out below, were available to the government regarding the uranium-mining problem:

- decision by parliament on party lines
- decision by parliament on party lines but guided by a select parliamentary committee
- decision by parliament guided by the outcome of an advisory referendum
- decision by parliament on party lines guided by a royal commission type of public inquiry (the method actually used)
- decision by parliament, guided by a royal commission type of public inquiry, but with a non-party vote on the issue.

A full-scale evaluation of the procedure adopted would involve fairly extensive analysis of each of the options listed, which again is outside the scope of this chapter.[41] The merits of the decision-making procedure adopted will be assessed by brief comparison with the other alternatives considered feasible. The major conclusion of this study is that the decision-making procedure adopted regarding Australia's uranium issue — which, as the foregoing analysis has established, was an intensely controversial one — was generally very well suited to the nature of the issue, but that the policy-deciding process could have been improved at the parliamentary level.

Australia's hybrid Westminster-federal system of government, operating in a context of rigid party politics, ensures that, whilst the government's majority holds firm in the House of Representatives and in the Senate, the executive and not parliament controls the content of government policy. Debates therefore are mostly ineffective in changing policy and, partly as a result, are frequently poor in quality. The debates between 1974 and 1977 in the federal parliament concerning the uranium issue illustrate the point most effectively. Their generally poor quality indicates that a sensitive, complex, social issue is not likely to receive adequate treatment in parliamentary debates where government and opposition hold opposed policies on the issue. During the 1977 uranium debate Don Chipp pointed out that at one stage the speaker, the prime minister, the leader of the opposition, and all bar two ministers were absent from the House and that attendance varied between eleven and nineteen.[42] In a very useful analysis of the 1974–77 House of Representatives debates, Wood has concluded:

> Discussion of issues of uranium mining *per se* often was less prevalent than political interplay: e.g. claims and counterclaims of deliberate distortion and misrepresentation of the Fox Report, of the academic literature and of each other's speeches . . . personal abuse and smearing by association (the NCP with the multi-national mining corporations and the ALP with environmental fringe extremists and militant unionists), attempts to portray differences of opinion between senior party spokesmen and others in the same party. . . . Attendance in the House, outside Question Times appears often to have been low. . . .[43]

It is thus rather alarming to contemplate what the content of the uranium decision might have been if it had been made in the usual manner after a ritual debate and on the basis of the government's normal advisory resources, that is to say if the Ranger Environmental Inquiry had not been held. As has been shown, the work of the commission was, in general, admirably thorough, impartial, balanced and cautious which was exactly what was needed to provide the government with adequate advice on such a complex issue. The commissioners had the ability, the time and the resources needed to cope with the daunting challenge of providing advice on such an issue. It is highly unlikely that the government could have obtained suitable assistance with the relevant policy making from any source other than a highly competent public inquiry of the kind appointed.

An *ad hoc,* or select, parliamentary committee is listed above in the options available. In spite of considerable improvements in recent years, the record of Australian parliamentary committees is generally not impressive, especially in the House of Representatives. Committee work can be interrupted by the end of parliamentary sessions or, more disrupting still, by dissolution of parliament, or of the House of Representatives. The pressures of electoral duties and often of work on various committees place considerable demands on the resources of parliamentarians. In the Australian parliament the value of committees also tends to be reduced by the general inability of Australian politicians to transcend the ubiquitous effects of party confrontation, conflict and pointscoring. This is an additional obstacle in the way of effective committee work. Australian parliamentary committees are chaired by government members. Sensitive social issues like uranium mining and nuclear power make special demands for impartiality and balance in policy making. Intrusion of party factors militates against realization of these needs. In the later stages of the uranium decision-making process the major party groups had taken up opposed policy stances regarding the issue. Had a select committee been examining the issue, its work would have been automatically terminated by this development.

Opposed party policy regarding uranium mining is also of vital significance concerning the third option listed above – decision by parliament (effectively the government) following an advisory referendum. It seems reasonably certain that any referendum held after July 1977, when the ALP adopted a policy of opposition to uranium mining and export, would have been held in a context of party campaigning on confrontational lines. Such a development in a referendum campaign means that large numbers of electors are likely to vote on party lines. Whether they give the issue the independent analysis on its merits postulated by the theory of direct democracy is coincidental and clearly unlikely. It is impossible in this context to explore the case for and against use of the referendum device.[44] Its use may well be fully justified in some circumstances, such as resolution of issues of national sovereignty. In the case of policy decisions on sensitive social issues, the proposition is advanced, necessarily without adequate analysis, that referendums are better avoided if at all possible. Issues like prohibition, conscription, capital punishment, abortion, environmental protection, nuclear power, race relations and so on, are likely to relate to deeply felt values, moral attitudes and convictions held by many citizens. That is a basic reason why projected political changes concerning such issues are so controversial. Some, like abortion, conscription and capital punishment, relate to beliefs about the sanctity of life and the concept of violence. The uranium-nuclear issues involves judgements about radiation, thermonuclear war, genetic mutations and, in a broad sense, for some people, the viability, justice, and worth of the whole western social system.

It is suggested, therefore, that the most satisfactory option available to democratic governments faced with such an issue is to have it thoroughly analyzed by the best available investigators in a public inquiry, and have parliament decide the issue adopting the normal procedure, which is to follow the recommendations submitted. A referendum has obvious and admirable democratic features, but it would be difficult to argue persuasively that it is likely to produce the impartiality, balance and judgement necessary for rational decision-making on sensitive, complex, controversial issues. A referendum can open the way to massive propaganda campaigns by the interests involved. If one group has disproportionately large resources, this can be the deciding factor. In the case of uranium this factor would have operated; the anti-uranium promotional groups were, and are, notoriously penurious, whereas the mining companies have extensive resources. In a campaign distorted by party and interest intervention, the opportunities for distortion of the issue by irrational and irrelevant campaigning are numerous. The likelihood of chance, apathy, or ignorance determining the issue is high.

On the basis of the above brief outline of the decision-making alternatives available regarding the uranium-mining issue, it is thus concluded that reliance by the government on the advice of a high calibre, royal commission type of public inquiry was in general a fortunate development. But it is further postulated that the best available alternative for governments faced with complex political-social-ethical issues such as uranium mining and/or nuclear power development, is the final one listed above — decision by parliament with a non-party ("free") vote after submission of and debate on recommendations by a public inquiry of the calibre of the Ranger Commission. Non-party votes are rare in Australian parliaments but they can be and have been used. Such a decision-making procedure would have freed parliament to consider the recommendations of the Ranger Inquiry on their merits without the point-scoring techniques of confrontational party politics. Assuming that on such issues the parliamentary majority would assent to the main thrust of the inquiry's recommendations (as seems likely), a valuable consensus would be established in a particularly difficult policy-making area.

CONCLUSION

The importance and difficulty of the general problem of effective decision making on issues of intense political controversy warrants much more attention than it has received. There is surely an onus on governments, political analysts, and others particularly concerned, to search for the most appropriate decision-making procedure available. Evaluation of the uranium-mining issue has involved consideration of quite broad issues of political decision making. If public policy studies are to progress beyond isolated analyses of particular issues with no wider theoretical significance, such an approach is necessary. The uranium-mining issue lent itself admirably to this purpose. It is as good an example as could be found of the problems of decision making in areas of intense political controversy. The study illustrates and emphasizes the political problems which mining can produce, and has significance for decision making regarding development of mineral resources other than uranium. Bauxite, oil exploration, sand mining and development of aluminium smelters would be examples. Proposed establishment of uranium enrichment plants and of a nuclear power industry would be even more relevant.

Notes

1. Australian libraries do not even have a category for "politics of mining".
2. Susan Bambrick, "The Australian Mining Industry", *Current Affairs Bulletin* (1 May 1977): 4.
3. Gary Smith, "Minerals and Energy", in *From Whitlam to Fraser,* ed. Alan Patience and Brian Head (Melbourne: Oxford University Press, 1979), pp. 233-35. The nature of some of the more important economic and environmental issues has been analyzed in chapters 7 and 8.
4. Ibid., p. 236.
5. Roger Scott, *Interest Groups and Public Policy: Case Studies from the Australian States* (Melbourne: Macmillan, 1980).
6. Owen Hughes, "Bauxite Mining and Jarrah Forests in Western Australia", in *Interest Groups and Public Policy,* ed. Roger Scott.
7. For an account of the fundamentals of nuclear science and technology, see Ranger Uranium Environmental Inquiry, *First Report* (Canberra: AGPS, 1976).
8. R.K. Warner, "The Australian uranium industry", *Atomic Energy in Australia* 19, 2 (1976), pp. 19-31.
9. I.C. Bupp and J.C. Derian, *Light Water: how the nuclear dream dissolved* (New York: Basic Books, 1978), p. 49.
10. Organisation for Economic Co-operation and Development Nuclear Energy Agency, *Uranium: Resources, production and demand* (Paris, 1975).
11. Ranger Uranium Environmental Inquiry, *First Report,* p. 64.
12. Australian Labor Party, *Uranium: a fair trial* (Canberra, 1977), p. 2.
13. Australia, House of Representatives, *Debates* 1974, no. 91, p. 3167.
14. M. Radetzki, *International uranium: economic and political instability in a strategic commodity market* (Institute for International Economic Studies, University of Stockholm, 1980).
15. Australia, House of Representatives, *Debates* 1974, no. 91, p. 3167.
16. Australia, House of Representatives, *Debates* 1976, no. 102, p. 2996.
17. Australian Atomic Energy Commission, *Annual Report 1974*–75 (Sydney).
18. For an account of the principal issues raised see H. Saddler, "Australian uranium — the environmental issue", in *Australia and Japan: nuclear energy issues in the Pacific,* ed. S. Harris and K. Oshima (Canberra and Tokyo: Australia-Japan Economic Relations Research Project, 1980).
19. On the AAEC see A. Moyal, "The Australian Atomic Energy Commission: a case study in Australian science and government", *Search* 6 (1975): 365-84. For an account of the career and views of Sir Philip Baxter see B. Martin, *Nuclear knights,* (Canberra: Rupert Public Interest Movement, 1980).
20. Moyal, "Australian Atomic Energy Commission".
21. Australian Atomic Energy Commission, *Annual Report 1975*–76 (Sydney).
22. H. Saddler, "Public participation in technology assessment with particular reference

to public inquiries", in *Proceedings of Workshop on technology assessments* Department of Science (Canberra: AGPS, 1978).
23. H. Saddler, "Implications of the battle for the Alligator Rivers: land use planning and environmental protection" in *Northern Australia: options and implications,* ed. R. Jones (Canberra: Research School of Pacific Studies, Australian National University, 1980).
24. D.T. Woods, L.T. Nicholls and L.G. Kemeny, "The Ranger Uranium Environmental Inquiry and the Environment Protection (Impact of Proposals) Act 1974", *Environmental Engineering Conference, Sydney* (Institution of Engineers Australian National Conference Publication No. 78/5, 1978), pp. 25-30.
25. Ranger Uranium Environmental Inquiry, *First Report,* pp. 181, 185.
26. Ibid., p. 175.
27. United States, *Presidential Documents,* Jimmy Carter, 13, 15 (1977), p. 507. See also Saddler, "Australian Uranium – the environmental issue" and M. Indyk, "Safeguarding nuclear energy in the Pacific: the role of Australia", in *Australia and Japan,* ed. Harris and Oshima.
28. Indyk, "Safeguarding nuclear energy".
29. All the material was published as a package, under the title *Uranium – Australia's decision* (Canberra: AGPS, 1977).
30. Saddler, "Australian Uranium".
31. Statement by the Prime Minister the Rt Hon. Malcolm Fraser, in *Uranium – Australia's decision,* p. 10.
32. Australian Financial Review, 7 August 1979, p. 1.
33. Supervising Scientist for the Alligator Rivers Region, *First Annual Report 1978–79* (Canberra: AGPS 1979); Saddler, "Implications of the battle for the Alligator Rivers".
34. H.C. Coombs, "Impact of uranium mining on the social environment of the Aborigines in the Alligator Rivers Region", in *Social and environmental choice: the impact of uranium mining in the Northern Territory,* ed. S. Harris (Canberra: Centre for Resource and Environmental Studies, Australian National University, 1980), p. 127.
35. Saddler, "Implications of the battle for Alligator Rivers".
36. Treasurer, *Press release,* 10 June 1979.
37. Ranger Uranium Environmental Inquiry, *First Report,* p. 186.
38. See Statement No. 3 attached to Budget Speech, Australia, House of Representatives, *Debates* 1980, Weekly Hansard no. 12, pp. 128-278.
39. "Winds of change at ACTU", interview with Cliff Dolan by Michael Gordon, *Chain Reaction,* No. 22 (Summer 1980–81).
40. Saddler, "Australian Uranium".
41. A larger study of decision making on controversial social issues is a research project of one of the present authors, J.B. Kelly.
42. M. Wood, unpublished Master of Administrative Studies paper (Canberra: Australian National University, 1978), p. 10.
43. Ibid., p. 11.
44. A fairly recent and valuable analysis is included in David Butler and Austin Ranney, eds., *Referendums: A Comparative Study of Practice and Theory* (Washington, DC: American Enterprise Institute for Public Policy Research, 1978).

10 The "Mining Theme" in Australian Literature*

Reba Gostand

> "That's nothing, once you could find,"
> (So the old-timers say)
> "Sapphires big as your nail;
> We used to throw them away — . . ."
>
> James McAuley[1]

The lure of vast mineral wealth; the hardships and dangers of life for the solitary fossicker, puddler, prospector, or gouger, and later for the company mine worker and his family; the moral, social, economic, industrial and political problems of that life; the effects of mining upon personal relationships, the health of miners and their families, the environment, and the earlier utilizers of the land such as the Aborigines and the pastoralists: these are themes that, from the earliest days of mineral discovery, have interested many Australian writers.

The physical and later the technological activities that are involved in extracting minerals or gemstones from the earth — the processes of mining — provide material for some of the early verse-writers and for novelists like Rolf Boldrewood, Katharine Susannah Prichard, and Harold C. Wells, but on the whole it is not the actual processes that have been of primary interest to creative writers: of far greater significance in individual works and as a shaping influence on the nature of Australian literature in general has been the "image" of mining.

To take this term in the narrower sense, the mining "image" has provided an important source of metaphor in Australian literature: certain mining activities (such as fossicking, tunnelling, refining ore) and the products that are mined (gold, opal, diamonds, coal, tin, copper, iron, and so on) have frequently been used as descriptive terms, ranging from the simple epithets of Dorothea Mackellar's "opal-hearted country"[2] or Marc Radzyner's:

> A fool's-gold moon
> touches the river currents
> with mica gleams and threads of metal lace . . .[3]

---

* This survey is confined to creative literature — poetry, fiction and drama. The extensive 'factual literature' of mining, such as history, biography and autobiography, would require an essay in its own right.

to the more oblique suggestion of opal in Christopher Brennan's "fire made solid in the flinty stone".[4] They have been used as similes, as in Robert D. FitzGerald's "He . . . can wear an air like mining booms, like profits pile on pile"[5]; or developed into an extended metaphor as in John Manifold's description of poetry:

> Think of an ore new-fossicked, sparse, and crude,
> Stamped out and minted it will buy your food,
> Cajole a mistress, soften the police,
> Raise a revolt or win ignoble peace,
> Corrupt or strengthen, sunder or rejoin;
> Words are the quartz, but poetry's the coin.[6]

In R.F. Brissenden's entire poem "Firedamp", he compares the death of the young poet Michael Dransfield "Under that falling rock/And in that stifling dark"[7] of drug addiction, to the sensations of men caught underground by firedamp. Mining images have also been used as structural devices or as central symbols in a work of literature, one of the most effective examples being the motif of the miner buried alive, stated in the Proem to Book 1 ("Australia Felix") of Henry Handel Richardson's trilogy *The Fortunes of Richard Mahony,* and developed throughout the three novels as a linking image of Mahony's emotional response to the marital and financial pressures of his life.

It is the "image" of mining in the broader sense, however, that has been most influential in the shaping of Australian literature and in the literary reflection of aspects of our national persona or self-concept. Certain ethical and materialistic values, emotional attitudes and responses to experience, personal and social behavioural patterns, associated with the *idea* and the practice of mining have provided basic thematic material for much of our fiction, verse, and drama: concepts like labour, wealth, power, manipulation; adventure, excitement, courage, danger, violence, death; independence and anti-authoritarianism, idealism and self-interest; mateship, personal relationships, divided loyalties, treachery; hardship, fear, sorrow, loneliness, disappointment, despair (happy humans leading a blameless or uneventful life do not make for sustained literary interest). These provide themes not of course restricted to the literature of mining, but mining and the mining life have given some of our major writers and many minor writers a powerful context for their expression.

Of particular interest to reader and literary historian alike is the way in which this thematic material has changed in direction and emphasis since the early nineteenth century. Man's relation to the land, for instance, has always been a signficant theme in Australian literature, as has the concept of mateship: in the literature of mining both of these have their place, and both have undergone significant changes. The early diggers, working with their hands in a close physical relation with the soil, rock, and streams in their search for alluvial gold, retain something of the feeling of personal rapport with the land that is such a feature of the "bush" literature of Australia. With the introduction of more sophisticated machinery when the mining companies move in, this personal connection between man and mine/land is gradually lost in the growing depersonalization of labour. Increasing technological complexity introduces many different kinds of mine worker plus a class system of foremen, engineers, analysts, surveyors, managers, owners, and — as manipulators on the "outside" — speculators and shareholders. The old feeling of solidarity, the mateship of the diggers based on an identity of aim and a similarity of life and work styles, is replaced by the selective mateship of unionism and class interest.

With changes of technology, mining itself becomes less prominent as a theme, and attention moves towards the social, industrial, and political issues related to mining: industrial issues following the growth of consortiums are dominant in the literature of the nineteen forties and fifties; in the sixties and seventies there is a swing towards violence and sensationalism — literary responses characteristic of many gold mining stories of the nineteenth century — but now with the emphasis on personal, financial, or political power games. Most striking of all is the movement away from entertainment to polemic: while there are still authors who use mining themes and settings at a purely narrative level, an increasing number — especially novelists and poets — are using this material to argue a socio-political or environmental case.

In the following pages I will trace these changing patterns of interest in the work of some of our better-known authors. Note that only works from which an extract has been quoted are footnoted; other recent publications mentioned are listed in the Select Bibliography.

Although gold is not the first mineral to be mined here, it is "the magic power of gold"[8] that first interests colonial singers and writers, and throughout the nineteenth century few authors give much attention to the literary possibilities of any other type of mining; but with the turn of the century a spate of novelists and playwrights explore opal, silver-lead, iron, coal, copper and tin mining, and since the 1930s asbestos, nickel, oil, and uranium have been added to compete with gold as foci of interest. Contemporary poets seem to be chiefly concerned with environmental implications, and with man's use of the products of mining to destroy himself and his world — a theme that has also produced a number of novels and at least one play. Authors of children's books are turning to mining themes and settings too, sometimes simply as the basis for adventure yarns, sometimes as the context for a moral or environmental message. Of particular interest to the literary historian is the very recent development of a body of Aboriginal writing, touching on themes of the displacement and destruction of the Aborigines and on questions of land ownership and utilization; an outlet for this new "nationalism" is provided in the journal *Aboriginal and Islander Identity*, as well as in published fiction and collections of verse.

In the old bush songs and ballads of the early nineteenth century can be found almost every theme later taken up by some of Australia's major writers, as well as by popular authors both home-grown and from outside who, like Hammond Innes in *Golden Soak* (a novel that was filmed in Australia for television in 1979), have set stories of romantic adventure, violence, and graft against the background of the Australian mining fields. In the work of the early versifiers is all the tragedy, pathos, comedy, and farce of life in a solitary prospector's camp or on the crowded diggings. Here is the ubiquitous *Anon:*

I thought a good home could be found
At the diggins-oh.
But soon I found I got aground
At the diggins-oh.
The natives came one day,
Burnt my cottage down like hay,
With my wife they ran away
To the diggins-oh.

I built a hut with mud
At the diggins-oh.
That got washed away by flood
At the diggins-oh.
I used to dig, and cry
It wouldn't do to die,
Undertakers charge too high
At the diggins-oh.[9]

These old songs and ballads record the promise of instant riches that brought men to Australia from almost every country of the world; Charles R. Thatcher sums up the hopes of one such specimen:

I came out here like many more
To pick up lots of gold;
If I greased my boots 'twould stick to them,
At home I had been told; . . .'[10]

Sometimes the lure of riches is allied to an urgent need for anonymity or escape, as a later balladist E.G. Murphy ("Dryblower") suggests in his account of the prevalence of Smiths at Coolgardie; even later the Pilbara provides a similar haven for some of the characters in John Powers's play *The Last of the Knucklemen.*

The various national "types" thronging to the diggings are characterized — in fact often exaggerated to the point of caricature — through speech or behavioural patterns. Particular social or moral traits associated in the popular imagination with particular races are constantly stressed — Irish thirst and love of brawling, Scottish parsimony, English greenness or gullibility, American business acumen verging on graft,

Chinese inscrutability and cunning — as British, American, and Oriental immigrants (legal and illegal) rub shoulders with stalwart Canadians, massive Negroes, earnest Scandinavians, phlegmatic Germans, and volatile Greeks and Italians. A man's nationality all too often typecasts him beyond redemption in the black-and-white world of the early versifiers, and affects the characterization of many later short story writers, novelists, and dramatists. Some writers use this kind of national shorthand for parody, sending up speech habits or idiosyncrasies of diction, a favourite comic device of both Joseph Furphy and Rolf Boldrewood:

> "Waal, Mrs Mullockson," says he, "so you've pulled up stakes from Bendigo City and concluded to locate here. How do you approbate Turon?"[11]

or making the racist attitudes implicit in such typecasting the target for satire or irony.

Indian hawkers and Afghan camel drivers bring stores and other necessities to the fields — Prichard, in her novels of the West Australian gold fields, records the high feelings that result when man and beast foul the precious waterholes along the route to Coolgardie — and in time these itinerants are followed by the precursors of Big Business like the quartermaster in M. Barnard Eldershaw's *A House is Built:*

> ". . . Just think what those fields will be like, men from everywhere just rushing in unprepared — no food, no necessities — water'll be dearer than gold soon, I'll be bound — and no thought of how to get them. That's where we'll come in . . ."[12]

With such a diversity of peoples at the diggings, it is hardly surprising that racial feelings at times run high. The Chinese are always considered "fair game", especially by the comic writers, although a group of Chinese fossickers very nearly get the better of Michael Doyle over some valuable bricks, in Dyson's "A Golden Shanty". But the outcome of racial tension is not always comic. Mary Gilmore recalls the Lambing Flat hanging of Chinese diggers in her poem "Fourteen Men", and a sympathetic treatment of the twentieth century race riots on the gold fields is given by Prichard in *Golden Miles* and by Cyril E. Goode in two of his stories, "The Man with the Nordic Complex" and "Maker of Stilettos":

> Before anyone could raise a hand he grabbed one of the toys and plunged it into a second-generation Italian. The remaining five stilettos were not nearly enough for the fight that was soon under way.[13]

The literature offers a fascinating reflection of the changing conditions of life in the mining environment. At first it is largely a male society, emphasizing the values of mateship, honesty, courage and endurance. Boldrewood defines "dividing mates":

> The Major and Cyrus had by chance become mates in the colony of Victoria, where we first met them, and by the merest hazard joined forces with us. Since then we have journeyed together. Quitted moderate goldfields where nothing more than an easy liberal livelihood was to be had for the stern hazards of a new rush, at a moment's notice. Here, "dividing mates", as the mining phrase is, one half of the

> party, when times were bad, working at bush or other labour, in order to provide food and raiment, tools and lodging for the whole, while the other pair tried the mining ventures of the locality, on the chance of striking, at any moment, a fortune, small or great, to be loyally and equally divided into four parts.[14]

"Humpy" Bannon, in Dyson's "A Zealot in Labour", is endurance personified. But also, for those men who want them, there are the compensations of a good booze-up at the grog-shop or shanty pub, or the blandishments of the whores who soon move into the fields.

In the early stages the digger's ambition is to make his pile and go "home", usually back to the country of his birth, but sometimes – as it is for old Peter McKenzie in Lawson's "Payable Gold" – his home is a mortgaged cottage in the city far away at the coast, where a lonely wife struggles to care for the children.

The population of any mining area tends to be both shifting and shiftless, although evidence from the literature is conflicting on the latter point – some writers stressing the lawless elements which provide good material for stories ranging from simple claim-jumping and gold-coach robbery to more sensational exposures of robbery with violence or brutal murder; others, like Prichard or Boldrewood, insisting that the diggers are always an honest, decent, law-abiding (if occasionally rumbustious) lot, and that it is not until the big companies move into the fields that moral and ethical standards begin to wilt, and ruthless exploitation, share manipulation, and personal, economic, and political power struggles take over. Randolph Bedford regrets the changes that have come over the Silver City (Broken Hill) since "The Days of '84", and Alex Buzo dramatizes some fast operators from Stallion Oil in his play *Tom:*

> STEPHEN: It's just that this whole thing's very important for Stallion, and very important for me. Oil men judge you according to how good you are in a crisis, how well you measure up when your fellows are in danger. It's a kind of camaraderie, a brotherhood of adventurers. Mind you, they'll stab you in the back without a qualm in the least.
> SUSAN: I've seen that happen.[15]

During the early rushes tent settlements spring up quickly – Boldrewood, Prichard, and many other novelists describe these makeshift "towns" in some detail – and in time wives and families join their men in the bush camp or the shanty town nearby. From a tendency to concentrate on the fossicker or prospector the literary interest gradually widens to include the lonely wife battling the terrors of the bush or of Aboriginal attack, trying to care for sickly or dying children many miles from medical or other help, or suffering the heartbreak of widowhood through the death of her man in a drunken brawl or a disaster at the mine. Some of the short stories explore the opposite side of this picture: the long-suffering, patient husband coming off shift exhausted, to cope with a sick, drunken, flighty, or lazy and selfish wife, as in Lawson's "The Story of the Oracle" and Goode's "Alexander the Great and a Dog";

or they take a comic look at the "hard cases" like Mother Middleton in Lawson's "The Golden Graveyard".

As gold mining becomes more widely established and quantities of gold are carried by stage-coach from the fields to the cities, bushrangers make ever more frequent appearances in the literature, and plays like George Darrell's *The Sunny South,* or novels written or set in the nineteenth century like Boldrewood's *The Miner's Right* and *Robbery Under Arms,* Will Lawson's *When Cobb and Co. Was King* and Frank O'Grady's *Wild Honey,* add the excitement of robbing the gold escort to their tally of adventure. Opals are another commodity particularly susceptible to hijacking (Prichard's *Black Opal,* Nicholas Miles's *Opal Fever,* Kenneth Cook's *The Man Underground*). Dire need, or sometimes simply too much temptation, leads even good men into theft (there are always illicit gold or opal buyers waiting to do business, as in Henrietta Drake-Brockman's play "Hot Gold", Prichard's mining novels, Casey's *Downhill is Easier,* or Stuart Gore's *Down the Golden Mile*) and sometimes through mischance also involves them in murder. Such ironies are a constant bonus to literature.

Although the lure of a wealthy strike so often calls men to set out with a cheerful bravado to "make a push for that new rush/A thousand miles away. . . . In spite of blacks and unknown tracks"[16], throughout many of the verses, plays, and fiction a terrible fear stalks — the fear of sudden death in a cave-in or a flooded mine; the fear of lingering death from an accident down the shaft:

> BUTCHER: They're not answering anymore. Cass? [*then*] WHY?
> *BLACKOUT.*
> *Throughout the intervening darkness: we hear Butcher tapping urgently, refusing to give up.*
> BUTCHER: Can you hear anything. Cass? [*No reply; he taps again.*] Help me out, Cass. [*No reply; he taps again.*] There's nothing, is there, Cass? For fuck's sake . . . CASS? [*He taps frantically again. Then breathless pause, then:*] WHY?[17]

the fear of the "Talking Ground":

> But from out of the great black void of the stope the voice of the earth continued to protest. Whispers grew into muffled groans, and occasionally the hollow rattle of a little falling stone, diving down a rock wall, hitting the broken rock, and spinning and leaping to the bottom of a rill, made an exaggerated clatter in its higher key . . .[18]

and above all the fear of the "dust":

> I cursed the mines, in my mind, and the silent stone that surrounded Tom and Don and myself and all the rest of us. A treacherous mass that blunted steel and resisted all your efforts to overcome its solidness. Until one day it gave a grunt and a mouth opened in it and grinned at you, before the lot of it came down and snuffed you out. Or it played a waiting game. Let you tear it about with fracteur and cart it and haul it and crush it. But it left its poison, the poison you couldn't see, and could only taste in your mouth in the mornings, in the still air.[19]

The mine itself is almost a physical presence in some of Dyson's and Casey's stories.

Ghosts haunt the diggings and the surrounding bush:

> "Someone had skinned a dead calf during the day and left it on the track, and it gave me a jump, I promise you. It looked like two corpses laid out naked. I finished the whisky and started up over the gap. All of a sudden a great old-man kangaroo went across the track with a thud-thud, and up the siding, and that startled me. Then the naked, white, glistening trunk of a stringybark tree, where someone had stripped off a sheet of bark, started out from a bend in the track in a shaft of moonlight. . . . There was a Chinaman's grave close by the track on the top of the gap. An old Chow had lived in a hut there for many years, and fossicked on the old diggings, and one day he was found dead in the hut, and the Government gave someone a pound to bury him. When I was a nipper we reckoned that his ghost haunted the gap, and cursed in Chinese because the bones hadn't been sent home to China. . . ."[20]

One of the occupational hazards is madness, as a number of short story writers (for instance, Dyson in "The Trucker's Dream" and "After the Accident") testify. But always, against the fear, shoring up the weakening walls of the mind, there is great courage and determination:

> ". . . There's more men down there. I must go back."
>
> "Stop that, Mac. You've got all the bloody gas you can stand. It would be suicide." Strong hands held him firmly.
>
> "I don't care. I know the pit, an' I know where they'll be. Get out of my bloody way."[21]

There is dumb, undemanding endurance of hardship, loss, or loneliness that is epitomized in the pathos of Lawson's "His Father's Mate" and the patient, weary toiling of J.M. Marshe's "Old Fossicker Jack".

A number of stories – and a large proportion of Prichard's three gold field novels – consist of old miner's yarns: memories of old mates (Lawson's "An Old Mate of Your Father's"); old strikes, in both senses; old journeys and rushes:

> ". . . The Palmer, in 1873, that was the start of it for me, and it's been a lust in my blood ever since. Palmer, and the Coen, Nebo, Croyden, and dozens more in Queensland, I was on 'em all, and then it was the Kimberley Rush, in '86. Hall's Creek and Mt. Dockerel and a score of nameless gullies, and south then to Nullagine, and all across the Pilbarra, and down to the Ashburton Top Camp, and on to Nannine. The Mainland, Cue, Day Dawn, Mt. Magnet, Youanme, Lake Way, Darlot, Leonora, ah, it's a long list o' camps, an' the fever still burns in a man. Bayley's Find, Hannan's Find, Kurnalpi and Kanowna, and always short of water, and living like a dog, and if a man got gold, he put it into searchin' for more elsewhere. Always the country further on, further out, luring us forward. Hard times and high hopes, and . . . well, now a man's caught in the web of it, a slave to it, and he hugs his chains."[22]

Old jokes too. The tall story, characteristic of the sardonic outback tradition of Australian humour, flourishes in the get-rich-quick atmosphere and amongst the muscular outback Supermen of the mining communities: examples include Max Brown's story of the radioactive prospector in "Wheelbarrow and the Whirlwind"; the abortive theft of gold bullion in Wally Wynne's "Fairy Gold", or the prowess of his "Old George" the carpenter at the Great Croesus; and the trials of strength in Gavin Casey's "The Hairy Men from Hannigan's Halt". The whole heroic breed is effectively summarized in Ian Mudie's send-up of the legendary Australian in "They'll Tell You About Me":

> I fought at Eureka and Gallipoli and Lae;
> and I was a day too early (or was it too late?) to discover Coolgardie,
> lost my original Broken Hill share in a hand at cribbage,
> had the old man kangaroo pinch my coat and wallet,
> threw fifty heads in a row in the big game at Kal. . . .[23]

Other yarns concern old treacheries (Dyson's "Dead Man's Lode") or tales of one-up-manship or of revenge, sometimes comic and sometimes quite deadly serious in their outcome.

Nostalgia for a long-past way of life is a significant mood in much Australian literature, and many examples can be found in the stories and verse dealing with the early rushes, although at times the nostalgia is mingled with irony: memories are sometimes shown to be deceptive. Farce is another popular mode that comes into its own with the later short story writers: macabre farce as in Lawson's "The Golden Graveyard" and Ted Mayman's "Gold Comes to Sunrise" for instance, or hilariously indigestible as in Casey's "Rich Stew".

Many ambivalent attitudes that we have come to regard as characteristically "Australian" are already established in the earlier literature: a sympathy for the underdog, which expands on occasion into an admiration for the daring and courage of the bushranger, seen for instance in *Robbery Under Arms*, but undercut ironically in this comment from Frank O'Grady's romantic adventure *Wild Honey*:

> He didn't expect to find any money in Thunderbolt's plant, but there'd be gold for sure and the rightful owners of that could never be traced. They'd be only Chows, anyhow. Thunderbolt always robbed the Chows when he stuck up a goldfield. Chows had no rights. That's how Thunderbolt got so popular with the diggers.[24]

This sympathy for the lawbreaker is one of the manifestations of the anti-authoritarianism we see again and again in attitudes to law and order.

Other ambivalent attitudes include the puritanical objection to theft or cheating (one's mate) — seen in *Black Opal, The Sunny South,* and *Opal Fever,* to mention only three examples — that seems to be in no way incompatible with a hearty enjoyment of the successful con trick worked on a stranger or foreigner (mateship clearly operates very selectively); while by the nineteen thirties, as David Hutchison notes, gold-stealing is a "popular pastime, one of several ways of pitting oneself against the distant bosses"[25] — a point illustrated by the operation of the *ad hoc* syndicate in Ted Mayman's "The Gold Steal":

> One day, to everyone's astonishment, the underground manager finding time hanging heavily on his hands on the surface, went underground. He found the three

> ringleaders, with a couple of offsiders, having the time of their lives in the level's "Jeweller's Shop". After a hastily improvised display of anger and wrath, in which he threatened the entire shift with the sack and gaol, they adjourned to the pub to drink it over.[26]

Then there is the distrust of and contempt for – mingled with a certain obsequiousness towards – the aristocracy as represented by the Governor or any of the other English "toffs" who come to the fields:

> "Wa-al!" said one of our Yankee friends, "wat 'yur twistin' your necks at like a flock of geese in a corn patch? How d'ye fix it that a lord's better'n any other man?"
>
> "He's a bit different, somehow," I says. "We're not goin' to kneel down or knuckle under to him, but he don't look like any one else in this room, does he?"[27]

So much for Australian egalitarianism – even on the diggings Australia has never been a classless society, as Richard Mahony would feelingly verify, and the wives of managers and other executives in the mining townships of Prichard's novels, Vance Palmer's *Golconda,* Wells's *The Earth Cries Out,* Betty Collins's *The Copper Crucible,* or Kay Brown's *Knock Ten* would confirm.

Another ambivalence is seen in the pride in personal strength and masculine virility that expresses itself on the one hand in the dominance of the puritan work ethic – stressed in Lawson, Prichard, Casey, Wells and Palmer – and on the other hand in the exalting of masculine prowess at "holding a skinful", subduing a woman, or being ever ready to "put up the mitts" and take on one and all in fight:

> Then the swing doors of the bar bulged out and someone inside shot a bloke into the gutter. . . . He was roaring defiantly when he shot through the door, and he kept on roaring as he hung on to the post.
>
> It was Don Bell.
>
> "Who says a ticket makes any difference t' me?" he yelled. "Good as ever I was. Lotta bloody rot. Who says I ain't? Come out 'ere and I'll bloody well kill 'im."[28]

Attitudes to law and order, and to their upholders the troops and police ("traps", "joes"), that characterize much of the literature and are still a significant characteristic of Australian social attitudes, probably reflect both the original convict background of settlement and the adventurous and hardy nature of the old diggers. On the early diggings the licensing laws and the manner in which they are policed stir both the spirit of fair play and the spirit of revolt to boiling point; Victor Daley's poem "A Ballad of Eureka" (1901) commemorates an event that has become a national symbol. The affair of the Eureka Stockade has been dramatized both as a straight play for stage or radio (by for instance Louis Esson, Leslie Haylen, and Richard Lane) and as a musical (Kenneth Cook); used as a central theme or a major incident in numerous short stories and novels (by such writers as James Middleton Macdonald, Edward Dyson, H.H. Richardson, Leonard Mann, Henrietta Drake-Brockman, and Eric Lambert); and has been sung and sung again in verse, until the word Eureka has become a synonym for any fight against a tyrannous authority. A recent example of this appropriation, as a cultural symbol, of a significant incident from the social history of mining can be seen in the title of *The Black Eureka,* Max Brown's factual account of striking Aborigines in Western Australia. Similarly, the name of Peter Lalor is often invoked as a symbol, while — to move away from Eureka for a further example of appropriation — Lasseter's lost reef occurs frequently in verse as a symbol of the unattainable, of treasure-mirage, and has been used as a theme for fiction and even for a musical.

Hard though the early digger's life is, the camps and fields are not without their entertainment. Along with the chauvinist pleasures of fighting, drinking, and womanizing, there is plenty of other fun to be had in Chinaman-baiting, or in singing and dancing — Henry Lawson, in "The Songs They Used to Sing", recalls many favourite songs and tells how "the most glorious voice of all belonged to a bad girl"[29] or in gambling of every kind from two-up (a favourite of Prichard's miners) to fan-tan (popular amongst the Chinese, as Boldrewood notes), cock fighting, and of course horse racing. Bets are accepted on anything: a man's speed and dexterity with a shovel (Casey's "The Hairy Men from Hannigan's Halt"), his ability to move a pressure-gauge by lung-power (Wally Wynne's "High Pressure"), his capacity to drink a hundred stubbies in four hours (Ivan in Kenneth Cook's *The Man Underground*). *The Sunny South* opens its third act in a mining camp where the diggers are celebrating Queen Victoria's birthday with sack races, a wrestling match, and the "three card monte man and his buttoner . . . Now you see him — Then you don't!"[30] More recently, Powers's play *The Last of the Knucklemen* reaches its dramatic climax in a tense poker game, when Methuselah loses the thousand dollars he has saved up so that he can at least die comfortably in the city. On a lighter note, Goode records the tense excitement of "The Last Cockroach Race":

> They were neck-and-neck for awhile and nearly everyone forgot about The Spotted Dog's number Three out near the centre of the table. You could have ruled a line across their noses. Then number Three propped dead in its tracks and looked around wild eyed; its owner trotted to the top of the table and started shouting advice. It seemed to understand and slapped on a bit of speed, outpacing the other two on the rails . . .[31]

Entertainers, singly or in troups, visit the fields. Songsters like Charles R. Thatcher (who is credited with many of the old bush songs that survive today) and glamorous dancers like Lola Montez (whose spider-dance inspires diggers to throw nuggets of gold to her on the stage) are in great demand, but performers of Shakespearian plays, or of popular successes like J.C. Williamson's legendary *Struck Oil,* also attract big audiences. In later days the delights of a night at the cinema, or a walk down town to look at the shops, are added — but these are usually regarded by the miners as a concession to the little woman.

Work hard, live hard, drink hard, play hard: these themes remain central to the literature of mining experience even today, but changing modes of viewing that experience and its effects on the people concerned are clearly evident in twentieth century Australian writing. The best of our contemporary writers have moved a long way from the earlier black-and-white clear-cut relationships and experiences of the versifiers and first short story writers or novelists. Much of the earlier literature focuses, for instance, on the virtues of loyalty and affection between a man and his mate or a man and his wife and family, themes that are certainly not neglected by writers who follow; here, for instance, is Lawson's treatment in "They Wait on the Wharf in Black", a yarn of typical Lawsonian pathos that Mitchell's mate tells, about an elderly Coolgardie digger returning steerage to Sydney to his motherless children:

> "You'd best take what money we have in the camp, Tom; you'll want it all ag'in the time you get back from Sydney, and we can fix it up arterwards. . . . There's a couple o' clean shirts o' mine — you'd best take 'em — you'll want 'em on the voyage. . . . You might as well take them there new pants o' mine, they'll only dry-rot out here — and the coat, too, if you like — it's too small for me, anyway. You won't have any time in Perth, and you'll want some decent togs to land with in Sydney."
> "I wouldn't 'a' cared so much if I'd 'a' seen the last of her," he said, in a quiet, patient voice, to us one night by the rail. "I would 'a' liked to have seen the last of her."[32]

The earlier literature also records the darker nature of man, and especially the obsessive passion for gold or opal that can lead him to treachery and murder. H.H. Richardson evokes this sense of lust:

> A passion for the gold itself awoke in them an almost sensual craving to touch and possess; and the glitter of a few specks at the bottom of pan or cradle came, in time, to mean more to them than "home", or wife, or child.[33]

This image of lust and greed is extended through her powerful metaphor of the land itself "lying stretched like some primeval monster in the sun, her breasts freely bared" watching "with a malignant eye, the efforts made by these puny mortals to tear their lips away". The sensationalism inherent in this theme of obsessive passion is exploited in any number of verses (like Bernard McElhill's "A Bush Secret" or J.E. Liddle's "Harry Dale and Olaf Cubb"), short stories (Dyson's "Dead Man's Lode"), plays (Lionel Shaw's one-act "Red and Gold") and novels (*The Miner's Right,* Prichard's mining novels, and more recently *Downhill is Easier, Down the Golden Mile,* and *Opal Fever*). Perhaps because the search for wealth is his real love, romantic entanglement for the digger or gouger, especially in the nineteenth century literature, tends to be given comic treatment (as in Charles R. Thatcher's song, "Moggy's Wedding", or Darrell's cheerful melodrama *The Sunny South*), or later, and particularly in popular fiction, to be sentimentalized (Kay Brown's *Knock Ten* and Helen Wilson's *Bring Back the Hour*).

As the mining townships grow in size, more complex personal, economic, and social problems become apparent. In most of the nineteenth century literature the woman's life in the camp or the mining township is told from the man's-eye-view. The unending battle to keep the red dust out of house and washing, and the struggle to find enough food during the hard times when the seam peters out or, later, when the men are out on strike or unable to find work, are seen as woman's chief preoccupations, and it would appear that, however hard her life may be, if a woman has her man and her kids this is considered compensation enough of days of weary drudgery and nights of lonely waiting for her man to stagger home from a session with his mates. That the averge man might continue to see this as "the woman's lot" is evidenced in Casey's *Downhill is Easier:*

> "We're all right," Baldy agreed. "I've always had work, an' the missus has always had me an' the kids. That's what a woman needs up here — plenty o' kids. If they haven't got a family they've got nothing to do but feel sorry for themselves because of the dust an' heat, an' the way they reckon their husbands neglect 'em."[34]

Since the middle of the twentieth century, however, Casey himself and writers like Prichard, Wells, Palmer, Wilson, Collins, Brown and Dean are beginning to probe a little more deeply into the psychological effects of the life upon wives and children, but there is still plenty of room for a more sympathetic and systematic exploration of this field.

Perhaps instrumental in pointing the way towards the deepening of psychological interest are some of Henry Lawson's sad and gentle stories (especially the stories about Joe Wilson and his wife, although these concern bush life, not mining): these might well have influenced Gavin Casey's sensitive and compassionate probing of the greys and browns of half-formulated feelings, dissatisfactions, hopes and disappointments, regrets and fears, the silences and the sourness that can creep into circumscribed lives whatever the environment. In Casey's best work (the collection of stories about gold miners and their lives originally published under the title of *It's Harder for Girls,* though now reprinted as *Short-Shift Saturday and Other Stories;* and his novel *Downhill is Easier,* which utilizes many of the themes from his short stories) he explores the relationships of man and man, man and woman, parent and child, man and the mining environment, writing with understanding and occasional touches of humour and irony of youngsters growing up in a mining area; of the anxious search for work during the Depression years; of miners on holiday down at the coast where they feel lost in an unfamiliar environment; of man/mate/wife jealousies; of watching a mate dying from "dust". His best-known story "Short-Shift Saturday" is an acknowledged classic.

The influence of the environment on the miners' children is touched on by Wells in *The Earth Cries Out:* Miss Wakely, who teaches eight-year-olds at Cessnock, explains to a new teacher from the city the effect of the colliery whistles on even quite young children:

> ". . . When you've been in the coal-fields as long as I have you'll come to understand why they tighten up and become tense at whistle-time. Especially when the pits are working badly or at holiday-time, as it is now . . . the three whistles through the day

> are worst of all. . . . That means a man has been hurt, you know. When three whistles blow at Aberdare during the day I always know which girls' fathers work at that colliery. They are no good for the rest of the day. I usually let the ones concerned go early. It's no good keeping them. They can't settle to work. . . ."[35]

Geoffrey Dean's "The Town That Died", set in a mining town where the mines have closed, perceptively traces the effect of this closure on the lives of the people, and especially on the youngsters just growing up to a local world that offers them absolutely nothing, so that they fill the void of their days with fantasies and slip easily into crime for kicks. In *Kangaroo Court* John Jost shows the effects of inertia, of the insecurity and anxieties of a copper mining community where the mine has been temporarily closed by a landslide, and the divisions that arise over the company's limited work offer. Helen Wilson's "The Scholarship" is the story of a woman's determined fight for her son's future.

Most of the literature to date has been concerned with life in the older mining areas. There is a whole new field waiting to be explored in the social problems posed by life (especially for the women) in the outback suburbias of the newly established mining towns – towns like Moranbah, Dysart, or Blackwater in Queensland. It is surprising that writers have as yet shown so little interest in the potential of these settings.

Twentieth century writing evidences a deepening interest in the theme of racial relations; one new twist of this theme is that of some of the lonely wives finding, in the more sophisticated attentions of Greek or Italian workers, a welcome change from their ocker husbands (Dean's "Strangers' Country"). Of still more literary and socio-cultural interest is the treatment of the relationship between the Aboriginal people and the whites who have moved into tribal territory, shattering the indigenous social structure by introducing the Aborigines to liquor and prostituting their women. Prichard weaves this theme through her gold field trilogy:

> The blacks had steered clear of the white man's camp while the season was good, hunting kangaroos, bungarras and emus on the plains and in the far-away hills. They came into the Gnarlbine Rock and their well in Coolgardie when the dry season began, to find them commandeered by the white men, almost empty with the drain of so many men and beasts. There were still soaks and rock holes they knew of in the outlying country; but the secret of these places was well guarded.
>
> When the kangaroos and even crows were lying dead round the rock holes, and the natives learned that the white men had food, they congregated round the camp like flies, living on any garbage the storekeepers threw out: butcher's offal and the dregs in tins and bottles. Wine and tobacco! You could buy a native for either after a while. They would come from the bush, thin and hungry, with no more than a hair string belt round their middles, and the women, quite comely when they were young, stark naked. In no time, the same crowd would be skulking about in odds and ends of cast-off clothing, begging food and tobacco, bartering their women. . . .[36]

Opal Fever, Kangaroo Court, and Donald Stuart's *Yaralie* also touch on the damaging moral and cultural effects that the influx of miners and the growth of mining towns have had on the Aboriginal population, and a number of poets have also written of the displacement or destruction of the Aborigines, although often their statement of concern is a generalized one, not specifically linked with mining activity. Judith Wright in one of her best-known poems "Bora Ring" laments that:

> The song is gone; the dance
> is secret with the dancers in the earth,
> the ritual useless, and the tribal story
> lost in an alien tale.[37]

but in "Two Dreamtimes", a much later and much more obviously didactic poem dedicated to Aboriginal author Kath Walker, she writes:

> Raped by rum and an alien law,
> progress and economics,
>
> are you and I and a once-loved land
> peopled by tribes and trees;
> doomed by traders and stock exchanges,
> bought by faceless strangers.[38]

Similar themes are also beginning to receive expression from a growing number of Aboriginal writers. Monica Clare makes two brief references to the loss of her people's land to mining companies in her autobiographical novel *Karobran,* the first novel written by an Aboriginal woman to achieve publication. Jack Davis sings an ironic "Mining Company's Hymn":

> The Government is my shepherd,
> I shall not want.
> They let me search in the Aboriginal reserves
> which leads me to many riches
> for taxation sake. . . .[39]

but Kath Walker takes an even more militant attitude:

> The miner rapes
> The heart of earth
> With his violent spade.

and she calls on the "gentle black man" to

> Make the violent miner feel
> Your violent
> Love of land.[40]

Banumbir Wongar's collection of short stories, *The Track to Bralgu,* deals far more fully and satirically with the effects of uranium mining on the land-oriented Aboriginal culture.

Closely related to the theme of the destruction of the original inhabitants and their traditional way of life is that of the destruction of the land itself. Dyson's "A Zealot in Labour" offers a fairly early statement of what one man can accomplish:

> The ruin effected looked like the work of many men. The muddy, yellow stream had been diverted from its course several times within half a mile, and all along the banks were torn down, great cuttings made, piles of gravel heaped up, dams built,

races dug. But the ravisher was there – a lone man, gouging his way into a bank at the head of the flat where it met the hill, looking a mere midge amongst the destruction he had wrought with his two good hands.[41]

Bernard O'Dowd in his long poem "The Bush" (1912) lists amongst his "roadside crimes unnamed" the "sloughing pock-marks of the gold-disease" and the "sludgy creek".[42] Boldrewood, Richardson, Prichard and Palmer are amongst the novelists who create equally vivid pictures – indeed, Richardson's evocative description of the raped landscape and of the all-pervasive sound of mining activity in the Proem to "Australia Felix" is close to prose-poetry. A.D. Hope, writing of the miracle of evolving nature in his poem "Conversation with Calliope", uses an image suggested by an old Tasmanian copper mining area to describe a bare, treeless, rock-exposed landscape:

the dead moon-landscape made
By an abandoned copper mill. . . .[43]

Pollution and the destruction of the landscape, and of the fauna dependent on it, are major concerns of Judith Wright's poetry. In "Remembering Michael" she recreates powerfully the visual and chemical pollution caused by the refineries and smelters for the treatment of ore:

Smelters, refineries, crouched among smoking phalluses
smoking six colours,
cylinders, blocks, cubes
swung at our left as we turned to the airfield.
Behind them a hell of a sunset
enraged their outlines,
a stumping circling dance of black geometrical dwarves
among black geometrical towers;
below them the chemical waters
attacked those leaking shores.

That wild copper mare of a sunset
reared right out of this world.
Flanks stabbed by chimneys,
acidic smoke in its veins,
kicking the sun's last arc
to death.[44]

The shift of literary interest in this century has not only been in the direction of a more subtle analysis of the human lives and a growing concern over the effects of mining on the environment; there has also been a refocusing of attention by writers on to the industrial aspects of life in a mining community. Although there are still fossickers picking over the old gold or opal areas (Goode's "Perhaps it was the Water", Miles's *Opal Fever*), in most mining areas the puddlers, fossickers, prospectors, and gougers have been superseded by the big consortiums, buying up the claims and leases (Prichard's trilogy, Palmer's *Golconda* and *Seedtime*), introducing more sophisticated machinery and paid workers, and giving plenty of scope for analysis of conflicting interests between labour and capital:

". . . The bedrock question is whether this field is going to be properly developed. Those gougers say that doesn't matter a damn; everything's jake with them; they've got hold of the bulk of the leases and they can make some sort of a living scratching at the hill's top. So any real progress is to be held up for the sake of fifty or sixty men who're in the position of small property-owners. But how does that square with your interests? You're not paid to represent them but the men working for wages."

"And you're paid," grinned Donovan, "to represent the bosses, the big moneyed jokers who'd like to get a mortgage on this field but can't."[45]

The appearance of union organizers on the fields is recorded by novelists before the end of the nineteenth century (George Garnet's *A Barrier Bride: A Story of the Broken Hill Strike,* 1898, is one example), and unionism increases in importance as a theme or sub-theme. Wells writes about the early struggles to organize the coal mining unions and to give them a more effective voice, as Geordie recalls for young John's benefit how both owners and workers pursued self-interest, the men "playin' one owner off against the other when trade was good"[46] to the detriment of the industry and the delay of a nationwide improvement in conditions. Unfortunately, Wells's novel — like Prichard's gold mining trilogy — often sacrifices literary to factual interest, and in its attempt to give a fully detailed picture of mining life and of union efforts to attain solidarity and political muscle becomes too obviously didactic. A recent novel to incorporate the sub-theme of divided economic and industrial interests is Jost's *Kangaroo Court:*

Rodgers frowned: "Look I hope we don't blue over this. I've got strict instructions from the union. It says accept immediately."

"Accept?" Murphy was on his feet immediately. "What's with the fucking union? Has it joined the bosses against us? What about severance? Holiday pay? You ain't mentioned that. And anyway, we all know there's fuck-all chance of the twenty odd blocks who miss out on jobs now of getting other ones quickly. Why doesn't the union call us all out and force that fucking mining company to expand the mine? We turn up for work rain hail or shine and they reckon they can toss us to the wolves any fucking time they feel like it. . . . They're conning us. I wouldn't be surprised if they wanted to automate the mine and are using the landslide as an excuse."[47]

The political as well as economic and industrial possibilities of mining are also receiving more and more attention. An unusual early fictional treatment is C.H. Chomley's *Mark Meredith: a Tale of Socialism* (1905), in which the mines — like the canefields and some factories — are used as places of banishment. Political idealism and political expedience are recurring motifs in Prichard's gold field trilogy, and here again her deeply felt convictions lead her into a great deal of didactic writing that is interesting to social historians but often tedious to the general reader. Prichard writes at length in *The Roaring Nineties* about the struggle of the alluvial diggers to retain their rights; the political skulduggery of government legislation to protect property and class interest, resulting in an alignment of government and the police with capital against the workers; the bitterness when Alf deserts his old mates to become a mine manager and so joins the enemy. In *Golden Miles* and *Winged Seeds* she adds the politics of war and war profiteering as her documentary novels cover the first and second world wars and the Spanish civil war, tracing the effects of war on mining production in Australia and on the changing population and shifting loyalties of the fields, as conflicts arise over the issues of foreign workers, returned soldiers, the growth of the Labor movement, the rise of the International Workers of the World (IWW) in Australia and abroad, and communism versus fascism; and always preaching solidarity through a strong trade unionism, the need for the workers to own the mines, and the retention of mining profits in Australia.

Ron Tullipan's social realist novel *March into Morning* is another work concerned to make its points loud and clear:

> ". . . Sure, we got the kind of government that would nationalise up to a point, and when they got a going concern they'd step down and the next party elected would sell out cheap to foreign investors. Until the people gets sick of that and decides to go to other extremes. They get frightened then and start nationalising everything that ain't payin' dividends again. . . . If you understand it and speak up again it, you're a traitor to your country, not to the international investors, but to yourself. Talk about a proper people's government set up on the principle of the Miners' Federation and I'll say nationalise the lot."
>
> A farmer looked down upon the small weed of a man and charged, "If your miners put on too many demonstrations of solidarity with the people involved, the government will find ways and means of taking away their influence, even if it means closing the mines down and buying coal from abroad. The government would consider that less costly than having people stirred up against the present order of things."[48]

Wells is also interested in tracing the relation between the politics of mining and the politics of war, and Wells, Prichard, Palmer, Casey, Collins, Brown and Jost devote varying proportions of their novels to accounts of strikes and strike-breaking; the way government and mine owners and police combine to introduce and protect scab labour; the effects of strike action on the workers and their families; and in most cases also incorporate accounts of the long and bitterly-fought battles for improved health and safety measures that are added to the original union battles for a decent living wage.

Although conflicts and struggles like these are legitimate material for literary treatment, it is no easy matter for a writer to avoid the pitfalls of didacticism and emotive writing — of pleading a socio-political case as a substitute for creating a work of literature. A similar didactic spirit affects the literary quality of some of the recent verse concerned with mining, whether it is the unashamed special pleading of "Blood on the Coal":

> Come down and breathe the dank air, the foul air, the rank air;
> Fill up your lungs with coal dust, disease dust, for proof;[49]

or the passionate statement of some of Judith Wright's later poetry, where the strength of her personal concern for the world is vigorously attested, but its expression falls far short of the fine poetry in some of her earlier work like "The Two Fires", in which she handles her theme with greater subtlety and poetic richness.

The politics of power is another area of interest to many writers. The wealth to be won from mining, wealth that was once merely a passport back "home" to the old country and to a life of ease and refinement, is in much of the recent literature a passport to power – and power is all too often corrupt. Cartels and deliberately engineered strikes, speculation and dealing in leases and shares, the manipulation of government officials: the excitement and the rewards of these new modes of gambling are intoxicating to the power-hungry, and these variations on the mining theme have been taken up by novelists like Prichard, Palmer, Casey and Cook, and dramatists like Alex Buzo, Patrick White and Cliff Green.

Speculation in shares, and the rise to wealth and political influence firstly by petty theft and later by shady deals and war profiteering, are (of course) well documented by Prichard. Cook's *The Man Underground* uncovers graft and corruption in the opal mining township of Ginger Whisker:

> "There's a friend of mine coming along here tonight I want you to meet."
> I made an indeterminate noise hopefully expressing polite interest.
> "He's the man you and I are going to make Federal Member for Ginger Whisker." . . .
> "I see," I say: "So you have Ginger's only newspaper and only radio station and you have . . . what is it, six months to the next Federal elections? Yes. I see your point."[50]

Buzo's *Tom* is a troubleshooter from an oil rig, but the play itself is concerned with sexual and economic power games, not with oil drilling *per se*; White's *Big Toys* makes peripheral use of the uranium issue as a part of his three characters' sexual, economic, and political power games; and Cliff Green in *Burn the Butterflies* brings the moral dilemma of nuclear politics to television drama.

The lure of mineral wealth is stronger than ever, but today the ground rules are seen to be changing. Exploitation is more than ever the name of the game, and it also extends into the realm of sexual exploitation – the mining environment is still fundamentally a male chauvinist society, as *The Man Underground* and *Kangaroo Court* so amply demonstrate, and this chauvinism reaches even to its city outposts, as illustrated in *Big Toys* and *Tom*. Tom, having caught his friend Stephen at two A.M. climbing in through a window to retrieve an incriminating file, puts his cards on the table:

> TOM: I didn't go to Whale Beach tonight. I carried out some investigations. Baldock was out – I was luckier than you. His sister was asleep, like Susan.
> *He glances at the file.*
> Aluminium . . . asbestos . . . Alkapolite Holdings Proprietary Limited . . . housing . . . roofing . . . guaranteed percentage. A photograph of Miss Anthea Baldock underneath Mr John Morelli of Stallion. This file is the personal property of G. Baldock Esquire. Very generous of you to accept an inflated tender from your own private company.[51]

Our writers have not been slow to join the rush to these new fields. Indeed a few bleak visionaries are already looking beyond the sensational present to the even more sensational future, where they see man using the products of mining — and particularly uranium — to destroy the world. Ric Throssell's play *The Day Before Tomorrow* is concerned with the break-down of civilization as it is reflected in the moral disintegration of an ordinary family, in a city devastated by nuclear explosion. Popular authors like Nevil Shute (*On the Beach*), Jack Danvers (*The End of It All*), and Colin Mason (*Hostage*), look at the aftermath of nuclear holocaust or global nuclear war, Danvers combining it with germ warfare just to make certain of total destruction. But interestingly enough it is the poets who seem to be most vocal on this theme. Judith Wright, whose passionate concern for the environment and for humanity sometimes outweighs her poetic judgement, can in her best work on these themes create superb poetry out of polemic, as for instance when she expresses the pressure to use the destructive power man has forged:

> The will to power destroys the power to will.
> The weapon made, we cannot help but use it;
> it drags us with its own momentum still.[52]

The poems written in Australia about Hiroshima, Nagasaki, the atom bomb, and the effects of radiation would require a separate chapter to themselves, while the number of poems which incorporate nuclear fission or the atom bomb into their structure as metaphors is even greater. Treatment varies from the matter-of-fact, ironic statement of expatriate poet Peter Porter:

> The Polar DEW has just warned that
> A nuclear rocket strike of
> At least one thousand megatons
> Has been launched by the enemy
> Directly at our major cities . . .[53]

or Judith Wright's "Eve to Her Daughters":

> It was warmer than this in the cave;
> there was none of this fall-out . . .[54]

to the compassion of Val Vallis's "Flotsam":

> . . . Dead earth of Nagasaki, bleeding in
> The surf, what dreams are taking flight from Death?[55]

or Colin Thiele's "Radiation Victim":

> This ancient incandescence fanned and freed
> To leap the air invisibly until
> Your mortal breast, ignited and ablaze,
> Shrinks to its blackened ashes silently.[56]

From her first book *The Moving Image* (1946), in which her final poem "Dust" speaks of how "The remnant earth turns evil"[57], to her recent *Fourth Quarter* (1976), Judith Wright returns in poem after poem to the thought that people, too, are an endangered species:

> Bombs ripen on the leafless tree
> under which the children play . . .[58]

and that man who "runs down destruction like a hound" is "the instrument of this planet's death"[59]. Perhaps her "shadow-sister"[60] Kath Walker should be allowed the last word in this short survey of directions – directions that range from pure entertainment to urgent polemic and prophecy – taken by the "mining theme" in Australian literature:

> Lay down the woomera,
> Lay down the waddy,
> Now we got atom-bomb,
> End *every*body.[61]

Notes

1. James McAuley, "Tailings", in *Australian Poetry 1970,* ed. Rodney Hall (Sydney: Angus and Robertson, 1970), p. 59.
2. Dorothea Mackellar, "My Country", in *Poets of Australia,* ed. George Mackaness (Sydney: Angus and Robertson, 1946), p. 288.
3. Marc Radzyner, "Factory and River", in *Poetry Australia* no. 48 (1973), ed. Grace Perry (Sydney: South Head Press, 1973): 53.
4. Christopher Brennan, "Fire in the heavens, and fire along the hills", *The Verse of Christopher Brennan,* ed. A.R. Chisholm and J.J. Quinn (Sydney: Angus and Robertson, 1960), p. 115.
5. Robert D. FitzGerald, "Currencies", *Product: Later Verses by Robert D. FitzGerald* (Sydney: Angus and Robertson, 1977), p. 59.
6. John Manifold, "A Hat in the Ring", *Collected Verse,* (St. Lucia: University of Queensland Press, 1978), p. 40.
7. Robert F. Brissenden, "Three Poems for Michael Dransfield: III. Firedamp", in *Australian Poetry 1973,* ed. J.M. Couper (Sydney: Angus and Robertson, 1973), p. 66.
8. Charles R. Thatcher, "Look Out Below!", in *Old Bush Songs and Rhymes of Colonial Times,* ed. Douglas Stewart and Nancy Keesing, enlarged and revised from the collection of A.B. Paterson (Sydney: Angus and Robertson, 1957), p. 97.
9. *Anon.,* "The Diggins-Oh", *Old Bush Songs,* p. 95.
10. Thatcher, "The Green New-Chum", *Old Bush Songs,* p. 101.
11. Rolf Boldrewood, *Robbery Under Arms* (1881; reprinted London: Oxford University Press, 1949), p. 325. (Note the landlady's mining name.)
12. M. Barnard Eldershaw, *A House is Built* (1929; reprinted Sydney: Australasian Publishing Company, 1965), p. 190.
13. Cyril E. Goode, "Maker of Stilettos", *Stories of Strange Places* (Melbourne: The Hawthorn Press, 1973), p. 19.
14. Rolf Boldrewood, *The Miner's Right* (1890; reprinted London: Macmillan and Co., 1927), p. 25.
15. Alex Buzo, *Tom* (Sydney: Angus and Robertson, 1975), p. 25.

16. *Anon.*, "The Old Palmer Song", *Old Bush Songs,* pp. 94-95.
17. Bill Reed, *Cass Butcher Bunting* (Melbourne: Edward Arnold (Aust.) Pty Ltd, 1977), p. 69.
18. Gavin S. Casey, "Talking Ground", *It's Harder for Girls, and Other Stories* (Sydney: Angus and Robertson, 1944), p. 51.
19. Casey, "Short-Shift Saturday", *It's Harder for Girls,* pp. 164-65.
20. Henry Lawson, "The Chinaman's Ghost", *The Stories of Henry Lawson,* ed. Cecil Mann, Second Series (Sydney: Angus and Robertson, 1964), p. 102.
21. Harold C. Wells, *The Earth Cries Out* (Sydney: Angus and Robertson, 1950), p. 140.
22. Donald Stuart, "To Each a Dream", *Morning Star Evening Star: Tales of Outback Australia* (Melbourne: Georgian House, 1973), pp. 103-4.
23. Ian Mudie, "They'll Tell You About Me", *The Blue Crane* (Sydney: Angus and Robertson, 1959), pp. 4-5.
24. Frank O'Grady, *Wild Honey* (Sydney: Angus and Robertson, 1961), p. 129.
25. David Hutchison, "Kalgoorlie of the Thirties", review of *View from Kalgoorlie,* ed. Ted Mayman, in *Westerley* 4, (1969): 94.
26. Ted Mayman, "The Gold Steal", in *View from Kalgoorlie,* (ed. Ted Mayman (Perth: Landfall Press, 1969), p. 52.
27. Boldrewood, *Robbery Under Arms,* p. 350.
28. Casey, "Short-Shift Saturday", *It's Harder for Girls,* p. 182.
29. Lawson, "The Songs They Used to Sing", *The Stories of Henry Lawson,* First Series, p. 265.
30. George Darrell, *The Sunny South,* ed. Margaret Williams (Sydney Currency-Methuen, 1975), p. 32.
31. Goode, "The Last Cockroach Race", *Stories of Strange Places,* p. 26.
32. Lawson, "They Wait on the Wharf in Black", *The Stories of Henry Lawson,* First Series, pp. 413-14.
33. Henry Handel Richardson, *The Fortunes of Richard Mahony:* "Australia Felix" (1917; reprinted Melbourne: William Heinemann Ltd, 1951), p. 16.
34. Gavin S. Casey, *Downhill is Easier* (Sydney: Angus and Robertson, 1945), p. 66.
35. Wells, *The Earth Cries Out,* p. 113.
36. Katharine Susannah Prichard, *The Roaring Nineties* (London: Jonathan Cape, 1946), p. 35.
37. Judith Wright, "Bora Ring", *The Moving Image* (Melbourne: The Meanjin Press, 1946), p. 12.
38. Judith Wright, "Two Dreamtimes", *Alive: Poems 1971-72* (Sydney: Angus and Robertson, 1973), p. 24.
39. Jack Davis, "Mining Company's Hymn", *Jagardoo* (Sydney: Methuen, 1978), p. 32.
40. Kath Walker, "Time is Running Out", *My People* (Brisbane: Jacaranda Press, 1970), p. 95.
41. Edward Dyson, "A Zealot in Labour", *The Golden Shanty: Short Stories by Edward Dyson,* sel. Norman Lindsay (Sydney: Angus and Robertson, 1963), p. 49.
42. Bernard O'Dowd, "The Bush", *The Poems of Bernard O'Dowd: Collected Edition,* introd. Walter Murdoch (Melbourne: Lothian Publishing Co. Pty Ltd, 1941), p. 193.
43. A.D. Hope, "Conversation with Calliope", *Collected Poems 1930-1970* (1966; reprinted with additional poems, Sydney: Angus and Robertson, 1972), p. 192.
44. Judith Wright, "Remembering Michael", *Fourth Quarter and Other Poems* (Sydney: Angus and Robertson, 1976), pp. 24-25.
45. Vance Palmer, *Golconda* (Sydney: Angus and Robertson, 1948), p. 107.
46. Wells, *The Earth Cries Out,* p. 51.
47. John Jost, *Kangaroo Court* (Sydney: Angus and Robertson, 1979), p. 133.
48. Ron Tullipan, *March into Morning* (Sydney: Australasian Book Society, 1962), p. 244.
49. John Graham, "Blood on the Coal", *Blood on the Coal, and Other Poems for the People* (Sydney: Current Book Distributors, 1948).

50. Kenneth Cook, *The Man Underground* (Melbourne: The Macmillan Company of Australia Pty. Ltd., 1977), p. 110.
51. Buzo, *Tom,* p. 53.
52. Judith Wright, "Weapon", *Shadow,* in *Collected Poems 1942-1970*(Sydney: Angus and Robertson, 1971), p. 285.
53. Peter Porter, "Your Attention Please", in *Australian Poetry 1963,* ed. Gerald A. Wilkes (Sydney: Angus and Robertson, 1963), p. 68.
54. Judith Wright, "Eve to Her Daughters", *The Other Half* (Sydney: Angus and Robertson, 1966), p. 26.
55. Val Vallis, "Flotsam", *Songs of the East Coast* (Sydney: Angus and Robertson, 1947), p. 29.
56. Colin Thiele, "Radiation Victim", *Meanjin* XIX 4 (1960): 428.
57. Judith Wright, "Dust", *The Moving Image*, p. 33; "Two Songs for the World's End: I. Bombs ripen on the leafless tree", *The Gateway* (Sydney: Angus and Robertson, 1953), p. 52.
58. Judith Wright, "Two Songs for the World's End: I. Bombs ripen on the leafless tree, *The Gateway* (Sydney: Angus and Robertson, 1953), p. 52.
59. Wright, "The Slope", *Alive,* p. 46.
60. Wright, "Two Dreamtimes", *Alive*, p. 24.
61. Walker, "No More Boomerang", *My People*, p. 33.

SELECT BIBLIOGRAPHY TO CHAPTER 10

This bibliography lists a sample only of Australian creative literature on the "mining theme". Publishing dates indicate the editions I have used — later paperback editions of some titles are available.

Verse

Although almost every Australian poet has used mining imagery in at least one poem, and many poets have written one or more poems on a mining theme, the only poets to use mining themes *extensively* are some of the earlier balladists, *Bulletin* contributors (Dyson has been included as an example of the *Bulletin* writers), and in the last few decades the major poet Judith Wright. I have added Daley, Davis, Graham and Walker as examples of smaller output in special fields of interest.

Daley, Victor. "A Ballad of Eureka". *Creeve Roe: Poetry by Victor Daley*. Ed. Muir Holburn and Marjorie Pizer. Sydney: The Pinchgut Press, 1947, pp. 139-46.

Davis, Jack. *Jagardoo*. Sydney: Methuen, 1978.

Dyson, Edward. *Rhymes from the Mines and Other Lines*. Sydney: Angus and Robertson, 1898.

Graham, John. *Blood on the Coal, and Other Poems for the People*. Sydney: Current Book Distributors, 1948.

Stewart, Douglas and Keesing, Nancy, eds. *Australian Bush Ballads*. Sydney: Angus and Robertson, 1955, reprinted 1957.

______ *Old Bush Songs and Rhymes of Colonial Times*. Enlarged and revised from the Collection of A.B. Paterson. Sydney: Angus and Robertson, 1957.

Walker, Kath. *My People*. Brisbane: The Jacaranda Press, 1970.

Wright, Judith. *Collected Poems 1942–1970*. Sydney: Angus and Robertson, 1971.

______ *Alive: Poems 1971–72*. Sydney: Angus and Robertson, 1973.

______ *Fourth Quarter and Other Poems*. Sydney: Angus and Robertson, 1976.

Fiction

Prospectors, miners, or mining life/activities make brief appearances in novels by such important Australian writers as James Tucker (*Ralph Rashleigh*), Catherine Helen Spence (*Clara Morison*), Henry Kingsley (*The Hillyars and the Burtons*), Katharine Susannah Prichard

(*Coonardoo*), M. Barnard Eldershaw (*A House is Built*), Peter Mathers (*Trap* and *The Wort Papers*), Xavier Herbert (*Poor Fellow My Country*), and Thomas Keneally (*The Cut-Rate Kingdom*). The list below, however, has been limited to the Boldrewood and Dyson classics (as representatives of the nineteenth century) and a selection of twentieth century novels and short stories, in all of which mining provides the setting or plays a major role in the narrative, and all of which are currently in print or else readily available in libraries.

Bedford, Randolph. *Aladdin and the Boss Cockie*. Illus. Percy Lindsay. Sydney: Bookstall, 1919.
______ *Billy Pagan, Mining Engineer*. Sydney: Bookstall, 1911.
______ *The Silver Star*. Sydney: Bookstall, 1917.
______ *The Snare of Strength*. London: Heinemann, 1905.
Boldrewood, Rolf. *Robbery Under Arms*. Serialized in the *Sydney Mail* in 1881; first published in book form by Remington, London, 1888. Reprinted London: Oxford University Press, 1949.
______ *The Miner's Right*. 1890. Reprinted London: Macmillan and Co., 1927.
Brown, Kay. *Knock Ten: a Novel of Mining Life*. Sydney: Wentworth Books, 1976.
Brown, Max. "Wheelbarrow and the Whirlwind". *West Coast Stories*. Ed. H. Drake-Brockman. Sydney: Angus and Robertson, 1959, pp. 134-40.
Casey, Gavin S. *Downhill is Easier*. Sydney: Angus and Robertson, 1945.
______ *It's Harder for Girls, and Other Stories*. Sydney: Angus and Robertson, 1944. Reprinted in paperback as *Short-Shift Saturday and Other Stories*. Sydney: Angus and Robertson, 1976.
Collins, Betty. *The Copper Crucible*. Brisbane: Jacaranda Press, 1966.
Cook, Kenneth. *The Man Underground*. Melbourne: The Macmillan Company of Australia Pty Ltd, 1977.
Danvers, Jack. *The End of It All*. London: Heinemann, 1962.
Dean, Geoffrey. *Strangers' Country and Other Stories*. Tasmania: Cat and Fiddle Press, 1977.
Dyson, Edward. *The Golden Shanty: Short Stories by Edward Dyson*. Selected by Norman Lindsay. Sydney: Angus and Robertson, 1963. Dyson's mining stories, written in the 1870s and 1880s, were collected in *Below and on Top,* Melbourne: G. Robertson, 1898.
Foster, David. *Moonlite*. Melbourne: The Macmillan Co. of Australia Pty Ltd, 1981.
Gore, Stuart. *Down the Golden Mile*. London: Heinemann, 1962.
Goode, Cyril, E. *Stories of Strange Places*. Melbourne: The Hawthorn Press, 1973.
Jost, John. *Kangaroo Court*. Sydney: Angus and Robertson, 1979.

Lambert, Eric. *Ballarat*. London: Frederick Muller Ltd, 1962.
______ *The Five Bright Stars*. Melbourne: Australasian Book Society, 1954. Fictionalizes Eureka.
Lawson, Henry. *The Stories of Henry Lawson*. Ed. Cecil Mann. 3 vols. Sydney: Angus and Robertson, 1964.
Lawson, Will. *When Cobb and Co. was King*. Sydney: Angus and Robertson, 1947.
Mann, Leonard. *Human Drift*. Sydney: Angus and Robertson, 1935.
Mayman, Ted, ed. *View from Kalgoorlie*. Perth: Landfall Press, 1969. A collection of short stories by Ted Mayman, Gavin Casey and Wally Wynne.
Miles, Nicholas. *Opal Fever*. London: Peter Davies, 1972.
Palmer. Vance. *Golconda*. Sydney: Angus and Robertson, 1958.
______ *Seedtime*. Sydney: Angus and Robertson, 1957.
Prichard, Katharine Susannah. *Black Opal*. Sydney: Caslon House Publishers, 1946.
______ *Golden Miles*. Vol. 2 of trilogy. London: Jonathan Cape, 1948.
______ *Potch and Colour*. Short stories. Sydney: Angus and Robertson, 1944.
______ *The Roaring Nineties: a Story of the Goldfields of Western Australia*. Vol. 1 of trilogy. London: Jonathan Cape, 1946.
______ *Winged Seeds*. Vol. 3 of trilogy. Sydney: Australasian Publishing Company, 1950.
Rienits, Rex. *Eureka Stockade*. London: Convoy Publications, 1949.
Richardson, Henry Handel. *The Fortunes of Richard Mahony*. 3 vols. First published separately in 1917, 1925 and 1929 by William Heinemann Ltd. Reprinted with introduction by Leonie Kramer. Ringwood: Penguin Books, 1971.
Stow, Randolph. *Tourmaline*. London: Macdonald and Co., 1963.
Stuart, Donald. *Morning Star Evening Star: Tales of Outback Australia*. Melbourne: Georgian House, 1973.
______ *Yandy*. Melbourne: Georgian House, 1959.
______ *Yaralie*. Melbourne: Georgian House, 1962.
Tullipan, Ron. *March into Morning*. Sydney: Australasian Book Society, 1962.
Wells, Harold C. *The Earth Cries Out*. Sydney: Angus and Robertson, 1950.
Wilson, Helen. *Bring Back the Hour: a Romance of the Australian Iron Fields*. Sydney: Angus and Robertson, 1977.
______ *The Golden Age*. London: Robert Hale, 1959.
______ *The Skedule and Other Australian Short Stories*. Sydney: Angus and Robertson, 1979.
Wolfe, Dusty. *The Brass Kangaroo*. Melbourne: Allara Publishing Co. Pty Ltd, 197 .
Wongar, Banumbir. *The Track to Bralgu*. London: Jonathan Cape, 1978.

Drama

There are very few extant printed Australian plays dealing with mining themes: early scripts rarely survived beyond performance; scripts for radio plays or television plays or series (*Rush, Golden Soak*), or for musicals (*Lola Montez, Lasseter, Crushed by Desire*) were seldom published.

Buzo, Alex. *Tom*. Sydney: Angus and Robertson, 1975.
Cook, Kenneth. *Stockade: a Musical Play of the Eureka Stockade.* Ringwood: Penguin Books, 1975.
Darrell, George. *The Sunny South.* Ed. Margaret Williams. Sydney: Currency-Methuen, 1975. Nineteenth century melodrama.
Drake-Brockman, Henrietta. "Hot Gold: a Goldfields Comedy". *Men Without Wives and Other Plays.* Sydney: Angus and Robertson, 1955, pp. 74-172.
Esson, Louis. "The Southern Cross". *The Southern Cross and Other Plays.* Melbourne: Robertson and Mullens, 1946, pp. 1-70. Based on the Eureka Stockade.
Green, Cliff. *Burn the Butterflies.* Sydney: Currency Press, 1979. Television drama.
Haylen, Leslie. *Blood on the Wattle: a Play of the Eureka Stockade.* Sydney: Angus and Robertson, 1948.
Lane, Richard. *Stockade: a Play for Radio.* Unpublished manuscript. Hanger Collection, Fryer Library, University of Queensland, n.d.
Powers, John. *The Last of the Knucklemen.* Ringwood: Penguin Books, 1974.
Reed, Bill. *Cass Butcher Bunting.* Melbourne: Edward Arnold (Aust.) Pty Ltd, 1977.
Shave, Lionel. "Red and Gold". *Five Proven One-Act Plays.* Sydney: The Australasian Publishing Co. Pty Ltd, 1948, pp. 23-47.
Throssell, Ric. *The Day Before Tomorrow.* Sydney: Angus and Robertson, 1969.
White, Patrick. *Big Toys.* Sydney: Currency Press, 1978.

General Bibliography

This general bibliography includes books, articles and other material on the aspects of mining in Australia covered in this book. Not all the sources cited were used in preparing the relevant chapters; in particular it should be noted that some recent material which is included was not available when the chapters were written. On the other hand some sources that were used (particularly very specialized material and works not primarily about mining) are not listed, references to these being given in the footnotes to each chapter.

The essential aim has been to provide a list of material which readers can use to pursue the general topics covered in the book. Clearly it has been necessary to be selective, on occasions somewhat arbitrarily so. The rapid development of the industry quickly renders material out of date; accordingly note should be taken of the date of publication of items. Material more than a few years old has been included only where it has been considered to be of lasting significance or, alternatively, presents a useful analysis of the situation at the time. Articles from daily or weekly newspapers have not been included, nor have articles or papers from specialist publications such as the *Miner* newspaper, the *Mining Review,* journals produced by the Bureau of Mineral Resources and state government mines departments and the Australian *Mining and Petroleum Law Journal.* However, these journals contain much that is relevant. The journal *Resources Policy* also contains several articles in addition to those cited below which are generally relevant to the subject of some of the chapters.

BOOKS

Alexander J. and Hattersley R. *Australian Mining, Minerals and Oil.* 2nd ed. Sydney: David Ell Press, 1981.

Auhl, I. and Marfleet D. *Australia's Earliest Mining Era: South Australia 1841-1851.* Adelaide: Rigby, 1975.

Austin. J.B. *The Mines of South Australia, Including also an Account of the Smelting Works in that Colony.* Adelaide: C. Platts *et al.,* 1863.

Bambrick, S. *Australian Minerals and Energy Policy.* Canberra: Australian National University Press, 1979.

______ *The Changing Relationship: The Australian Government and the Mining Industry.* Melbourne: Committee for the Economic Development of Australia, 1975.

Barnett, D.W. *Minerals and Energy in Australia.* Stanmore, NSW: Cassell, 1979.

Bate, W. *Lucky City: The First Generation at Ballarat, 1851–1901.* Carlton, Vic.: Melbourne University Press, 1978.

Blainey, G. *The Peaks of Lyell.* 3rd ed. Carlton, Vic.: Melbourne University Press, 1967.

______ *The Rise of Broken Hill.* Melbourne: Macmillan, 1968.

______ *Mines in the Spinifex: The Story of Mount Isa Mines.* Sydney: Angus and Robertson, 1970.

______ *The Steel Master: A Life of Essington Lewis.* Melbourne: Macmillan, 1971.

______ *The Rush That Never Ended: A History of Australian Mining.* 3rd ed. Carlton, Vic.: Melbourne University Press, 1978.

Bosson, R. and Varon, B. *The Mining Industry and Developing Countries.* New York: Oxford University Press, 1977.

Brealey, T.B. and Newton, P.W. *Living in Remote Communities in Tropical Australia: The Hedland Study.* Canberra: CSIRO Division of Building Research, 1978.

Bridges, R. *From Silver to Steel: The Romance of the Broken Hill Proprietary.* Melbourne: George Robertson, 1920.

Burke, K. *Gold and Silver: Photographs of Australian Goldfields from the Holtermann Collection.* Ringwood, Vic.: Penguin, 1973.

Butler, S., Raymond, R. and Watson-Munro, C. *Uranium on Trial.* Sydney: Horwitz Group Books, 1977.

Carroll, B. *Australia's Mines and Miners: An Illustrated History of Australian Mining.* South Melbourne: Macmillan, 1977.

Casey, G. and Mayman, T. *The Mile That Midas Touched.* Adelaide: Rigby, 1964.

Charlton, R. *The History of Kapunda.* Melbourne: Hawthorne Press, 1971.

Clacy, Mrs. C. *A Lady's Visit to the Gold Diggings of Australia in 1852–53.* Ed. Patricia Thompson. Melbourne: Lansdowne Press, 1963.

Coghill, I. *Australia's Mineral Wealth.* Melbourne: Sorrett Publishing, 1971.

Crough, G.J. *Foreign Ownership and Control of the Australian Mineral Industry.* Sydney: Transnational Corporations Research Project, Faculty of Economics, University of Sydney, 1978.

Duffield, R. *Rogue Bull: The Story of Lang Hancock, King of the Pilbara.* Sydney: Collins, 1979.

Elliott, M., ed. *Ground for Concern: Australia's Uranium and Human Survival.* Melbourne: Penguin, 1977.

Ellis, M.H. *A Saga of Coal: The Newcastle Wallsend Coal Company's Centenary Volume.* Sydney: Angus and Robertson, 1969.

Fisher, D.F. *Environmental Law in Australia: An Introduction.* St Lucia: University of Queensland Press, 1980.

Fitzgibbons, A. *Ideology and the Economics of the Mining Industry.* Published by the Author, 1978.

Flett, J. *The History of Gold Discovery in Victoria.* Melbourne: Hawthorne Press, 1970.

______ *Dunolly: Story of an Old Gold Diggings.* 2nd ed. Melbourne: Hawthorne Press, 1974.

______ *A Pictorial History of the Victorian Goldfields.* Melbourne: Rigby, 1976.

Gibbs, R.M. *The History of Mining in South Australia.* Kingswood, SA: Pedlow Books, 1979.

Gollan, R. *The Coalminers of New South Wales: A History of the Union, 1860–1960.* Parkville, Vic.: Melbourne University Press, 1966.

Harris, S. ed. *Social and Environmental Choice: The Impact of Uranium Mining in the Northern Territory.* Canberra: Centre for Resource and Environmental Studies, Australian National University, 1980.

Hastings, P. and Farran, A., eds. *Australia's Resources Future.* Melbourne: Nelson, 1978.

Hicks, A.B. and Hookey, J. *Fraser Island Environmental Inquiry.* Final Report of the Commission of Inquiry. Melbourne: Advocate Press, 1976.

Holthouse, H. *Gympie Gold.* Sydney: Angus and Robertson, 1973.

______ *River of Gold: The Story of the Palmer River Goldrush.* Sydney: Angus and Robertson, 1967.

Hore-Lacy, I. and Hubery, R. *Nuclear Electricity; An Australian Perspective.* Melbourne: Australian Mining Industry Council, 1977.

Hore-Lacy, I., ed. *Mining Rehabilitation.* Canberra: Australian Mining Industry Council, 1979.

______ *Broken Hill to Mount Isa: The Mining Odyssey of W.H. Corbould.* Melbourne: Hyland House, 1981.

Hughes, H. *The Australian Iron and Steel Industry, 1848–1962.* Parkville, Vic.: Melbourne University Press, 1964.

Jay, C. *Dollars for Minerals and Energy.* Melbourne: Committee for the Economic Development of Australia, 1979.

Keesing, N., ed. *Gold Fever: The Australian Gold Fields, 1851 to the 1890s.* Sydney: Angus and Robertson, 1967. (Also published as *History of the Australian Goldrushes by Those Who Were There.* Hawthorne, Vic.: Lloyd O'Neill, 1971).

Kennedy, B. *Silver, Sin and Sixpenny Ale: A Social History of Broken Hill, 1883–1921.* Carlton, Vic.: Melbourne University Press, 1978.

Kennedy, K.H. *The Mungana Affair: State Mining and Political Corruption in the 1920s.* St Lucia: University of Queensland Press, 1978.

______ *Readings in North Queensland Mining History.* Vol. 1. Townsville: History Department, James Cook University of North Queensland, 1980.

Knight, C.L., ed. *Economic Geology of Australia and Papua New Guinea.* 3 vols. Parkville, Vic.: Australasian Institute of Mining and Metallurgy, 1975.

Korzelinski, S. *Memoirs of Gold-digging in Australia.* Trans. Stanley Robe. St Lucia: University of Queensland Press, 1979.

Lang, A.G. and Crommelin, M. *Australian Mining and Petroleum Laws: An Introduction.* Chatswood, NSW: Butterworths, 1979.

MacKenzie, B.W. *Mineral Economics.* Vols. 1 and 2. Notes Prepared for Workshop 121/79, Australian Mineral Foundation. Adelaide, 1979.

Madigan, Sir Russel. *Of Minerals and Man.* Parkville, Vic.: Australasian Institute of Mining and Metallurgy, 1981.

Mawson, P. *A Vision of Steel: The Life of G.D. Delprat, C.B.E., General Manager of B.H.P., 1898–1921.* Melbourne: Cheshire, 1958.

McAndrew, J. ed. *Geology of Australian Ore Deposits.* Melbourne: The Australasian Institute of Mining and Metallurgy, 1965.

McKern, R.B. *Multinational Enterprise and Natural Resources.* Sydney: McGraw Hill, 1976.

Monaghan, J. *Australians and the Gold Rush: Californians and Down Under, 1849–54.* Berkeley, Calif.: University of California Press 1966.

Morley, I.W. *Black Sands: A History of the Mineral Sand Mining Industry in Eastern Australia.* St Lucia: University of Queensland Press, 1981.

Paull, R. *Old Walhalla: Portrait of a Gold Town.* Carlton, Vic.: Melbourne University Press, 1963.

Phillipson, N. *Man of Iron.* Melbourne: Wren, 1974.

Prider, R.T., ed. *Mining in Western Australia.* Nedlands, WA: University of Western Australia Press, 1979.

Pryor, O. *Australia's Little Cornwall.* Adelaide: Rigby, 1962.

Quaife, G.R., ed. *Gold and Colonial Society, 1851–1870.* Stanmore, NSW: Cassell, 1975.

Raggatt, H.G. *Mountains of Ore.* Melbourne: Lansdowne, 1968.

Reynolds, J. *Men and Mines: A History of Australian Mining, 1788–1971.* Melbourne: Sun Books, 1974.

Robinson, W.S. *If I Remember Rightly: The Memoirs of W.S. Robinson 1876–1963.* Ed. G. Blainey. Melbourne: Cheshire, 1969.

Rummery, R.A. and Howes, K.M.W., eds. *Management of Lands Affected by Mining.* Melbourne: CSIRO, Division of Land Resources Management, 1978.

Saddler, H. *Energy in Australia: Politics and Economics.* Sydney: George Allen and Unwin, 1981.

Scott, A. *Central Government Claims to Mineral Revenues.* Centre for Research on Federal Financial Relations, Australian National University, Canberra, 1978.

Serle, G. *The Golden Age: A History of the Colony of Victoria, 1851–61*. Carlton, Vic.: Melbourne University Press, 1968.
Shaw, A.G.L. and Bruns, G.R. *The Australian Coal Industry*. Carlton, Vic.: Melbourne University Press, 1947.
Sherer, J., ed. *The Gold Finder of Australia: How He Went, How He Fared and How He Made His Fortune*. Ringwood, Vic.: Penguin, 1973. (First Published 1853.)
Stevenson, G. *Mineral Resources and Australian Federalism*. Centre for Research on Federal Financial Relations, Australian National University, Canberra, 1976.
Stone, D.I. and Mackinnon, S. *Life on the Australian Goldfields*. Sydney: Methuen, 1976.
Sykes, T. *The Money Miners: Australia's Mining Boom 1969–1970*. Sydney: Wildcat Press, 1978.
Titterton, E.W. and Robotham, F.P. *Uranium: Energy Source of the Future?* Melbourne: Nelson, in association with the Australian Institute of International Affairs, 1979.
Trengove, A. *'What's Good for Australia . . .': The Story of B.H.P.* Stanmore, NSW: Cassell, 1975.
UNESCO. *Man and the Environment: New Towns in Isolated Settings*. Proceedings of a Seminar Held at Kambalda. Canberra: Australian Government Publishing Service, 1976.
Whitmore, R.L. *Coal in Queensland, the First Fifty Years: A History of Early Coal Mining in Queensland*. St Lucia: University of Queensland Press, 1981.
Williams, C. *Open Cut: Work and Marriage in an Australian Mining Town*. Sydney: George Allen and Unwin, 1981.
Wills, N.R. *Economic Development of the Australian Iron and Steel In-Industry: An Examination of the Establishment of the Industry, Its Development, Present Distribution, Resources and Importance in the Australian Economy*. Sydney: n.p., 1948.

ARTICLES AND PAPERS

Appleyard, G.R. "Foreign and Local Investment in Mining in Australia: Some Quantitative Factors". *Australian Economic Review* 3 (1978): 46-55.
Aschmann, H.H. "Amenities in the New Mining Towns of Northern Australia". *North Australia Research Bulletin* 5 (1979): 1-41.
Australian Conservation Foundation. *Conservation and Mining*. Special Publication No. 8, Papers of an Australian Conservation Foundation Symposium. North Blackburn, Vic.: Dominion Press, 1973.
Ball, D.J. "Australia and Nuclear Non-proliferation". *Current Affairs Bulletin* 55, no. 11 (1979): 16-30.

Bambrick, S. "Mining: The Problems for Australia". *Australian Quarterly* 45, no. 1 (1973): 64-77.

______. "Mineral Processing in Australia". *Australian Quarterly* 47 no. 1 (1975): 65-76.

______. "The Australian Mining Industry: Can it Lead Economic Recovery?" *Current Affairs Bulletin* 53, no. 12 (1977): 4-12.

______. "Australia's Mineral Commodity Marketing". *Resources Policy* 6, no. 2 (1980): 166-78.

Baxter, P. "The Case for Nuclear Power". *Current Affairs Bulletin* 53, no. 3 (1976): 18-23.

Bell, H.F. "The Consequences of Mining Development for the Australian Economy". *Economic Papers* 12 (1958): 37-62.

Bersten, M. "Mining Since 1926: Change and Adaptations". *National Bank Monthly Summary* (December 1976): 18-22.

Blainey, G. "Herbert Hoover's Forgotten Years". *Business Archives and History* 3, no. 1 (1958): 53-70.

______. "Gold and Governors". *Historical Studies* 9, no. 36 (1961): 337-50.

______. "The Cargo Cult in Mineral Policy". *Economic Record* 44, no. 108 (1968): 470-79.

Brain, P.J. and Gray, B.S. "The 'Gregory Thesis', Protection Policy and Factors Affecting the Size of the Australian Manufacturing Sector". In Kasper, W. and Parry, T.G. eds. *Growth, Trade and Structural Change in an Open Australian Economy*. Kensington, N.S.W.: Centre for Applied Economic Research, University of New South Wales, 1978.

Brealey, T.B. and Newton, P.W. "Migration and New Mining Towns". In Burnley, I.H., Pryor, R.J. and Rowland, D.T., eds. *Mobility and Community Change in Australia*. St Lucia: University of Queensland Press, 1980.

______. "Mining and New Towns". In Holmes, J. and Lonsdale, R., eds. *Rural Australia: Problems and Prospects*. New York: Scripta/Wiley, 1982.

Camilleri, J. "Uranium Exports: Commercial Incentives Versus Nuclear Dangers". *Australian Outlook* 30, no. 1 (1976): 120-35.

Carnegie, R.H. "The Australian Mining Industry: 1975". *Economic Papers* 51 (1976): 1-12.

Catley, B. and McFarlane, B. "Minerals and Multinationals". *Arena* 50 (1978): 39-46.

Caves, R.E. "Policies Towards Australia's Resource-based Industries". In Crawford, Sir John and Okita, S. *Australia, Japan and Western Pacific Economic Relations*. A Report Presented to the Governments of Australia and Japan. Canberra: Australian Government Publishing Service, 1976.

Corrighan, T. "The Political Economy of Minerals". *Journal of Australian Political Economy* 7 (1980): 28-40.

Davies, M. "Bullocks and Rail: The South Australian Mining Association". *Australian Economic History Review* 17, no. 2 (1977): 150-65.

Derrick, S., McDonald, D. and Rosendale, P. "The Development of Energy Resources in Australia: 1981 to 1990". *Australian Economic Review* 3 (1981): 13-55.

Dowell, R. "Resources Rent Taxation". *Australian Journal of Management* 3, no. 2 (1978): 127-46.

Dunn, J.A. "Minerals in the Development of Australia". *Economic Papers* 12 (1958): 5-18.

Fitzgibbons, A. "Mining and the Future Structure of the Australian Economy". *Australian Quarterly* 45, no. 2 (1973): 86-94.

Garnaut, R. "Resource Trade and the Development Process in Developing Countries". In Krause, L.B. and Patrick, H., eds. *Papers and Proceedings of the North Pacific Trade and Development Conference.* San Francisco: Federal Reserve Bank of San Francisco, 1978.

Gregory, R.G. "Some Implications of the Growth of the Mineral Sector". *Australian Journal of Agricultural Economics* 20, no. 2 (1976): 71-91.

Hamley, B.L. "Financing Australia's Mineral Developments". *National Bank Monthly Summary* (May 1977): 8-11.

Hampson, D.C. "Australia's Uranium". *Resources Policy* 6, no. 2 (1980): 3-52.

Harris, S. "Economics of Uranium Mining in Australia". *Current Affairs Bulletin* 54, no. 11 (1978): 17-30.

______. "Resources Policies in Australia". *Resources Policy* 6, no. 2 (1980): 179-91.

Heatley, A. "Aboriginal Land Rights in the Northern Territory". In Scott, R., ed. *Interest Groups and Public Policy: Case Studies from the Australian States.* Melbourne: Macmillan, 1980.

Hughes, O. "Bauxite Mining and Jarrah Forests in Western Australia". In Scott, R., ed. *Interest Groups and Public Policy: Case Studies from the Australian States.* Melbourne: Macmillan, 1980.

Indyck, M. "Australian Uranium and the Non-proliferation Regime". *Australian Quarterly* 49, no. 4 (1977): 4-35.

Leggate, J. "Mine Regeneration, Policies and Practices at Weipa: A Further Report". Paper Presented to Australian Mining Industry Council Workshop, Adelaide, 1976.

Mason, P. "Nuclear Decisions". *Australian Quarterly* 50, no. 2 (1978): 7-21.

Mauldon, F.R.E. "The Decline of Mining". *Annals of the American Academy of Political and Social Science* 158 (1931): 66-76.

McColl, J.D. "The Mining Industry and the Natural Environment". *Resources Policy* 6, no. 2 (1980): 153-65.

McLeod, J. "The Gregory Chant". *Australian Quarterly* 49, no. 2 (1977): 92-105.

McQueen, H. "Gone Tomorrow?: The Aluminium Cycle". *Arena* 55 (1980): 56-80.

Moyal, A. "The Australian Atomic Energy Commission: A Case Study in Australian Science and Government". *Search* 6 (1975): 365-84.

Newton, P.W. "Present and Future Settlement Systems in Sparsely Populated Regions". Paper presented to United States/Australia Seminar on Sparsely Populated Regions, Flinders University, Adelaide, 1978.

Newton, P.W. and Brealey, T.B. "Remote Communities in Tropical and Arid Australia". In Golany, G., ed. *Desert Planning: The International Experience*. New York: Architectural Press, 1981.

O'Mara, P., Carland, D. and Campbell, R. "Exchange Rates and the Farm Sector". *Quarterly Review of the Rural Economy*. 2, no. 4 (1980): 357-67.

Reece, B. "The Mining Boom, Capital Market Reform and Prospects for Housing in the 1980s". *Occasional Paper,* Planning Research Centre, University of Sydney, October 1980.

Saddler, H. "Australian Uranium – The Environmental Issue". In Harris, S. and Oshima, K., eds. *Australia and Japan: Nuclear Energy Issues in the Pacific,* Australia-Japan Economic Relations Research Project, Canberra and Tokyo, 1980.

______. "Implications of the Battle for Alligator Rivers: Land Use Planning and Environmental Protection". In Jones, R., ed. *Northern Australia: Options and Implications,* Research School of Pacific Studies, Australian National University, Canberra, 1980.

Sless, J.B. "Control of Sand Drift in Beach Mining". *Journal of the Soil Conservation Service of New South Wales* 12 (1956): 164-75.

Smith, B. "Long Term Contracts in the Resource Goods Trade". In Crawford, Sir John and Okita, S. *Australia, Japan and Western Pacific Economic Relations.* A Report Presented to the Governments of Australia and Japan. Canberra: Australian Government Publishing Service, 1976.

______. "Bilateral Monopoly and Export Price Bargaining in the Resource Goods Trade". *Economic Record* 53, no. 141 (1977): 30-50.

______. "Australian Mineral Development, Future Prospects for the Mining Industry, and Effects on the Australian Economy". In Kasper, W. and Parry, T.G., eds. *Growth Trade and Structural Change in an Open Australian Economy.* Kensington, NSW: Centre for Applied Economic Research, University of New South Wales, 1978.

______. "Australia's Mineral Production and Trade: A Case Study of a Resource Rich Developed Country". In Krause, L.B. and Patrick, H., eds. *Papers and Proceedings of the Ninth Pacific Trade and*

Development Conference. San Francisco: Federal Reserve Bank of San Francisco, 1978.

Smith, G. "Minerals and Energy". In Patience, A. and Head, B., eds. *From Whitlam to Fraser*. Melbourne: Oxford University Press, 1979.

______. "Forming a Uranium Policy: Why the Controversy?" *Australian Quarterly* 51, no. 4 (1979): 32-50.

Snape, R.H. "Effects of Mineral Development on the Economy". *Australian Journal of Agricultural Economics* 21, no. 3 (1977): 147-56.

Stewart, K.W., Leggate, J. and Bevege, D.F. *Some Aspects of Mine Rehabilitation in Northern Australia*. Collection of Papers Presented to a Regional Symposium, Weipa, 1974.

Stoeckel, A.B. "Some General Equilibrium Effects of Mining Growth on the Economy". *Australian Journal of Agricultural Economics* 23, no. 1 (1979): 1-23.

Temple, R. "The Dangers in Large-scale Use of Nuclear Power". *Current Affairs Bulletin* 52, no. 2 (1976): 4-17.

Trebeck, D. "Rural Industries and Inter-Sector Equity and Efficiency". *Australian Quarterly* 49, no. 4 (1977): 50-65.

Vale, J.V. "Financing Australian Mineral Developments, The Past Decade and the Next Decade". *Economic Papers* 61 (1979): 35-45.

Walker, J. "The Trade Unions and the Uranium Issue". *Current Affairs Bulletin* 55, no. 1 (1978): 18-30.

Warner, R.K. "The Australian Uranium Industry". *Atomic Energy in Australia* 19, no 2 (1976): 19-31.

Woods, D.T., Nicholls, L.T. and Kemeny, L.G. "The Ranger Uranium Environmental Inquiry and the Environment Protection (Impact of Proposals) Act of 1974". Paper Presented to Environmental Engineering Conference. Sydney: Institution of Engineers, Australian National Conference Publication No. 78/5, 1978.

Yarwood, A.T., ed. *Historical Studies: Eureka Supplement*. 2nd ed. Carlton, Vic.: Melbourne University Press, 1965.

OFFICIAL PUBLICATIONS, REPORTS AND OTHER

Australian Mining Industry Council. *Minerals Industry Survey* (Annual).

Australian Mining Industry Council. *Mining Taxation and Australian Welfare*. Canberra: Australian Mining Industry Council, 1974.

Co-ordinator General's Department, Queensland. *Impact Assessment of Development Projects in Queensland*. Brisbane: Government Printer, 1979.

Feros, V.G. Social Survey of Blackwater, 1976. Unpublished Master of Urban and Regional Planning Thesis, University of Queensland, 1977.

Fitzgerald, T.M. *The Contribution of the Mineral Industry to Australian Welfare*. Report to the Minister for Minerals and Energy. Canberra: Australian Government Publishing Service, 1974.

Fraser Island Environmental Inquiry. *First Report,* 1975, *Second Report,* 1976. Canberra: Australian Government Publishing Service.

Industries Assistance Commission. *Report on the Petroleum and Mining Industries*. Canberra: Australian Government Publishing Service, 1976.

Industry and Commerce, Department of. *Major Manufacturing and Mining Investment Projects*. Canberra: Australian Government Publishing Service (bi-annual).

Institute of Engineers, Australia (WA Division) and Australian Institute of Mining and Metallurgy (Perth Branch). *Control of Dust in the Pilbara Iron Ore Industry*. Report of a Symposium, 1975.

Joint Coal Board. *Black Coal in Australia*. Sydney, 1980.

Louthean, R., ed. *Register of Australian Mining*. Nedlands, WA: Ross Louthean Publishing, annual.

Mandeville, T.D. *The Impact of The Weipa Mine on the Queensland Economy*. Melbourne: Comalco Limited, 1980.

Mercer, J.C.B. The Australian Mining Town. Unpublished Master of Urban and Regional Planning Thesis, University of Queensland, 1979.

Mineral Resources, Bureau of. *Australian Mineral Industry Annual Review*. Canberra: Australian Government Publishing Service.

______ *Australian Mineral Industry: Production and Trade 1842–1964*. Canberra, 1966.

Mines, New South Wales Department of. *Mining and the Environment*. New South Wales Parliamentary Paper No. 135. Sydney: Government Printer, 1972.

National Development, Department of. *The Atlas of Australian Resources*. Canberra: Government Printer, 1970.

Norman, N.R. *Mining and the Economy: An Appraisal of the Gregory Thesis*. Report Prepared for the Australian Mining Industry Council, Canberra, 1977.

Ranger Uranium Environmental Inquiry. *First Report,* 1976, *Second Report,* 1977. Canberra: Australian Government Publishing Service.

Trade and Resources, Department of. *Australia's Mineral Resources*. (A series of occasional publications on individual minerals.) Canberra: Australian Government Publishing Service.

Uranium: Australia's Decision. (Collection of speeches etc. prepared by the commonwealth government.) Canberra: Commonwealth Government Printer, 1977.

Ward, J. and McLeod, I. *Mineral Resources of Australia*. Canberra: Bureau of Mineral Resources, 1980.

BLACK SANDS

A History of the Mineral Sands Mining Industry in Eastern Australia

I.W. MORLEY

The prize to be won from the black sands was rutile and zircon; for over forty years Australia has supplied ninety percent of the world's rutile and still holds more than seventy percent of the world's reserves. The total production of rutile, zircon and other minerals from the sands of eastern Australia in these four decades was worth over a billion dollars in export earnings to Australia.

Full geographical, historical, geological, corporate and technical descriptions of the industry are provided by the illustrations, figures, maps, diagram of intercompany relationships, data and accounts of companies, syndicates and processes in the text and the appendices. The author has a great capacity for research and writes an interesting and readable story with authority and insight from an intimate knowledge of the industry.

COAL IN QUEENSLAND

The First Fifty Years

R.L. WHITMORE

Coal is one of the major energy resources of the future. It is already making an important impact on the economy of Queensland as the state's vast coalfields are rapidly being developed to serve the nations of the world. But the world was not always interested in Queensland's coal, even though its wide occurrence was realized in the 1840s. For almost 150 years it lay practically forgotten except by the small group of individuals who made a living, and occasionally a fortune, out of mining it.

The book has an excellent collection of illustrations, with original documents and plans, and detailed maps of the mined areas. Industrial archaeological techniques were employed to record the remains of the historical sites which were still accessible in 1979, and there are photographs and paintings of the people and places involved in the early mining ventures.